The Tibetan Book of the Undivided Universe

David Bohm's Quantum Philosophy of Wholeness In the Light of Buddhist Metaphysics

Graham Smetham

shunyatapress.com
Brighton, Sussex, England

© 2021 Graham Smetham
All rights reserved.

ISBN: 978-1-716-45177-5

Chapters

Introduction: Bohm's Quantum Worldview - a 'Radical' Buddhist Meditation? **21**

First Turning of the Holomovement - Setting the Scene for Wholeness **33**

The Unfolding of Bohm's Pilot-Wave Theory - NonLocal Connections Implicates Ontological Wholeness **91**

Bohmian Quantum Emptiness **125**

A Matter of Unfolding Mind? The Dzogchen / Yogacara Ground Consciousness Implicates Bohm's Implicate Order **159**

Quantum & Buddhist Views of Free-Will: Bohm, Henry Stapp, Michael Mensky & The Buddhist Spiritual Path **209**

The Holomoving Alterverse & the Consciousness-Only Three Natures of Reality **245**

Quantum Samsara
A Bohmian Unity of Fragmentary Quantum Mirrors? **283**

Dharmakaya, Dharmadhatu & Buddhist Implicate Levels of Consciousness **313**

Visions of Totality & Unbounded Wholeness: West & East **331**

Preface &
Acknowledgements

About twenty years ago the 'unfolding' of the process of reality, which David Bohm called the 'Holomovement', made a dramatic intervention into my life. One evening I settled into meditation posture but, because I was tired, instead of practising my usual focused breathing meditation, I decided to meditate on an internally generated image of sparkling water, in order to generate a spacious calm mind. What happened next astonished me. The vague image of sparkling river water I had managed to produce was replaced by a full colour panoramic vista of a Tibetan plain surrounded by mountains. From the mountain peaks an assembly of Tibetan monk-deities wearing yellow hats flew towards me and surrounded me. At the time I did not know the identity of these deities but I later found out they were 'wisdom-beings', emanations of the wisdom bodhisattva Manjushri. This vision was like nothing I had ever experienced before in my life, in fact it was the first waking vision I had ever had. The life-like precision and vibrancy of the vision was extraordinary, not at all vague or unclear, the experience was as if I had a personal cinema inside my head.

Once I was surrounded by these beings I saw, in great colourful precision and clarity, a 'heart-sphere' of deep red-orange coloured liquid form at my heart chakra. Then tubes of this liquid funneled out from my heart to the hearts of the Tibetan monk-deities surrounding me. This 3-D Technicolor vision lasted maybe minute or two, and then when I opened my eyes I saw that the channels of orange 'nectar' were connecting me to the hearts of the other meditators surrounding me. The vision then faded.

About a week later I was looking at the Buddhist section of a local bookshop. I saw a particular orange slim book on the shelf and an 'inner voice' indicated that this book was important for me. The book was called *Heart Jewel* by the Tibetan Buddhist leader of the New Kadampa

Tradition, Geshe Kelsang Gyatso. I therefore bought it, but did not start looking at it for a few days. When I got around to reading through it I was even further surprised, to say the least, because very similar meditation visions to the one that spontaneously appeared to me were described in the book:

> We then imagine white rays of light, which are hollow like straws, coming from the hearts of Je Tsongkhapa and his two Sons, and merging into one ... We imagine the great wisdom of the Venerable Father and Sons, in the aspect of orange coloured nectar, all the atoms of which are in the aspect of tiny Manjushris. From all these tiny bodies of Manjushri, which are the commitment beings, infinite rays of light radiate throughout the ten directions and draw back the great wisdom of all the Buddhas in the aspect of Manjushri's form body. These wisdom beings dissolve into the tiny commitment beings within our body and become inseparably one with them. All these countless tiny Manjushris then dissolve into our root mind at our heart.[1]

Tsongkhapa (1357-1419) was a Buddhist scholar-practitioner who lived in Tibet during a time when Buddhist teachings are said to have been be in decline. Through his study, practice and attainments, he led a renaissance of Buddhist teachings, and his teachings became the basis for the establishment of the Gelug monastery, and subsequently the Gelug school of Tibetan Buddhism arose on the basis of his teachings.

Manjushri is the Buddhist Bodhisattva of Wisdom. Generally, a 'bodhisattva' is a practitioner on the edge of full enlightenment who deliberately postpones enlightenment in order to remain in samsara, or cyclic existence, in order to work for the enlightenment of all beings. There are also some iconic bodhisattvas, however, who are fully enlightened beings who work for the enlightenment of all beings. Manjushri is such a bodhisattva. In particular, Manjushri is the fully enlightened Bodhisattva of Wisdom, in this context 'Wisdom' is *prajna*, the direct non-conceptual insight into the ultimate nature of reality. During a very intensive Manjushri retreat Tsongkhapa gained a very clear vision wherein Manjushri appeared to him within a mandala.

Earlier, in the year before I was surprised by the Manjushri commitment-beings vision, I had been thinking about perhaps writing a book about the mathematical concept of 'zero'. I have no idea why I came up with the idea of writing a book about zero. I have a Bachelor of Arts degree in Mathematics (Essex University only awarded B.A.s when I was there, fifty years ago) and then went on to do an M.A. / PhD. course in the Philosophy of Religion, specialising in Buddhism, at the University of Sussex. I completed the taught course at Sussex, and even taught a minor course on Science and Religion, but failed to complete the PhD. because of various obstructing life vicissitudes. In one sense, this was a personal disaster as I had a famous publisher very interested in publishing my work.

Anyway, returning to my meditation vision, which took place twenty years after my studies at Sussex University. One of the meditation visualisations described in *Heart Jewel* was one that was performed by Buddhist practitioners who were about to write books about the *Dharma*, the Buddhist doctrines. Thus in the section 'Receiving the Attainment of the Wisdom of Composing Dharma Books', a similar meditation to that already described had an extra feature:

> ...we visualize the wisdom of composing Dharma books of Je Tsongkhapa and his sons flowing from their hearts in the form of orange-coloured nectar, all the atoms of which are in the aspect of tiny Dharma books on the subject of which we are going to write ...[2]

I did not realise at the time, of course, that my meditation vision might actually be an intimation that I was about to write a book about Buddhist philosophy. My intention to write a book about 'zero', however, later developed into the task of researching and writing a book about the Buddhist concept of *sunyata*, the Buddhist term for 'emptiness', and quantum physics. The concepts of zero and sunyata are related. The root of the term *sunyata*, which is translated as 'emptiness', is *sunya*, the zero point, the cosmic seed of emptiness which is 'swollen' with potentiality. One meaning of *sunya,* which is the Indian origin of the concept of zero, is 'the swollen', in the sense of an egg of potentiality which is about to burst into manifestation.

I was so inspired by this vision that I decided to join the New Kadampa Tradition Buddhist Community at the Brighton Bodhisattva Centre, which was located just up a hill from the bookshop where I had bought the copy of *Heart Jewel.* Whilst there, especially at the beginning of my stay, I had dreams of meeting with Tibetan deities and also experiencing 'emptiness' by walking through walls in my dreams. I also had very profound meditation experiences, generated by a regular, sustained and committed meditation schedule. However, I became dissatisfied with the style of life and practice within the NKT and left after a year and a half. I no longer have any affiliation with any particular tradition.

These events, especially the intrusion from the 'implicate order' of karmic traces of a previous life as a Tibetan monk, were life-transforming. And one of the results was my return to serious committed research and writing. Whilst at the Buddhist Center I reread Bohm's *Wholeness and the Implicate Order* alongside Buddhist texts, as well as many other books on physics and Buddhist philosophy, also works on Science and Buddhism, the excellent works of B. Alan Wallace for example. My view to write a book about zero now transformed into a decision to write a definitive book on the subject of the Buddhist concept of 'emptiness', and other Buddhist doctrines, and their relationship to discoveries in quantum physics. This became my first book *Quantum Buddhism: Dancing in Emptiness* [QB], which I self-published in 2010.

One of the insights which contributed to this endeavour was my identification of Bohm's notion of the 'implicate order' and the Buddhist *Yogacara-vijnanavada* concept of the *alayavijnana*, the 'ground' or 'store' consciousness. I knew at the time that this insight was novel and important, and it became a central aspect of QB. This insight is also covered to various degrees in my other books, and it is again treated in great depth and detail, with new perspectives, in this current book which focuses upon the ideas of David Bohm in relationship to Buddhist metaphysics. One of the crucial aspects of Bohm's concept of the 'implicate order' is that events which occur in the manifested 'explicate order' of the everyday world are conceived of as being 'enfolded' as 'active information' within the 'implicate order' which subsequently condition future 'unfolding' into the future 'explicate order'. Thus there is a causal relationship, through a mechanism of internal memory, between past and future. This mechanism is clearly connected to the Buddhist concept of *karma*, which is the view that a cause and effect mechanism operates at all levels of the process of reality, both physical and moral.

This type of mechanism is central within the *Yogacara* consciousness-only Buddhist psycho-metaphysics, it constitutes the karmic mechanism which operates within the ground consciousness. This is an issue which I have explored in my previous books, but do so in this book in much greater detail within the context of Bohm's ideas. Interconnections between Bohm's later 'mystical' quantum perspective and Buddhism have been outlined in my other works, but here these insights take center place and are explored in a depth and detail not found anywhere else.

On the issue of degree of depth and detail, I think this may be an appropriate place to make some remarks concerning the reception of my work, and the state of research in the arena of science and spirituality in general. When I was researching and writing my first book *Quantum Buddhism* [QB], there were not many works of this nature available. And there were no works available which dealt with the issues in the depth, detail and precision that my work achieves, I feel safe in making this claim as it has been reinforced by others. This remains true today, there are no other works with the scope, detail and precision I have deliberately set out to achieve. The reason for this is that I deliberately set out to make sure that my work went beyond what was available at the time, in depth, detail and precision, precisely because there is no point adding a work to those already available which added nothing new, nothing of greater import. Furthermore, it was clear to me that a work of severe detail and precision was necessary to achieve greater credibility than works already extant were accorded within the science and philosophical mainstream.

In the introduction to *Quantum Buddhism* I wrote:

> When I began this project there were very few works dealing with the subject of quantum physics and Buddhism, although there was significant interest in quantum theory and various 'mystical' worldviews, an interest which generated a great deal of animosity in hardened scientific-materialist enclaves. This interest and research in the 'mystical' dimensions of quantum theory, a viewpoint which has vociferous opponents, has been dubbed 'quantum mysticism.' ... When I registered the domain name 'quantumbuddhism' ... a google search on the term 'quantum Buddhism' would not have produced a great many significant hits. Shortly after registration of this domain and the setting up of my site, however, the number of sites purporting to deal in some way with this topic seemed to increase dramatically. Very few of these sites, however, seem to offer any new deep insights which would really justify the appellation *quantum Buddhism*. It was always my intention to firstly offer new, detailed and profound insights and elucidations in the research into parallels, interconnections and mutually reinforcing

> perspectives of quantum physics and Buddhist philosophy ... [QB:11]

Furthermore, a vitally important point is that it is precisely *because* works expounding relationships and interconnections between science and religion and spirituality tend not to be profoundly detailed and precise that detractors are able to employ easy debunking strategies. As I wrote in QB:

> The physicist Peter Woit, author of the book *Not Even Wrong*, a critique of string theory, is clearly outraged that *The Tao of Physics*, along with a very similar book that was published shortly after – Gary Zukav's *The Dancing Wu Li Masters* and 'other books of the same genre' still grace the shelves of major bookstores and are selling very well. Such titles, according to Woit, are part of 'an embarrassing new age cult.' Surveying the literature dealing with this area it is impossible not be struck by the severe polarisation into pro-quantum-mysticism and anti-quantum-mysticism positions. Unfortunately most passionate proponents of the quantum-mysticism worldview tend to be rather loose in the standards of evidence and philosophical rigor that they employ or require. Books like *The Tao of Physics* or *Dancing Wu Li Masters* do not need rigorous argument and detailed exposition to appeal in this quarter. This is perhaps one reason why entrenched detractors of this perspective tend to be quite exasperated, not to say contemptuous, when attempting to keep the unruly worldview under control. ... It is indeed true that the opponents of attempts to question the validity of the unquestioning materialism which has marked the general Western academic attitude up until the present, and is still prevalent, tend to display a marked aggressive attitude, often resorting to sarcastic contempt. Peter Woit, for instance, is so contemptuous of Buddhist philosopher B. Alan Wallace's book *Hidden Dimensions – The Unification of Physics and Consciousness* (2007), ... that he cannot be bothered to marshal any actual reasoning to support his contempt:

After enraging lots of philosophers, I fear that now I'll enrage lots of Buddhists, in particular by having no interest in wasting time discussing Wallace's ideas.[3]

However, the books and essays penned by Wallace, focusing on the significance of the interrelationship between the Buddhist philosophy and modern science, are actually fairly meticulously argued and cogent. It might be said that *Hidden Dimensions* does not present all the detailed argumentation to thoroughly make his case that:

> ...the measurement problem in quantum mechanics, the time problem in quantum cosmology, and the hard problem in brain science are all profoundly related.[4]

> However, if the reader explores the directions indicated in his book, Wallace's contention is well supported, ... Wallace's book *Choosing Reality* (2003) still remains one of the most thoughtful and intelligent investigations into the epistemological and metaphysical interconnections between the Buddhist Madhyamaka and modern Western science. Wallace also edited the collection of essays *Buddhism and Science: Breaking New Ground* which was inspired by the interdisciplinary dialogues, organised by the Mind and Life institute, between Buddhist practitioners (including the Dalai Lama) and philosophers, physicists and cognitive scientists. The depth and rigour of the analysis found in these thoughtful essays generally goes far beyond that found within the tirades targeted by critics at what we can call the 'Quantum Buddhism' perspective. It appears that those who wish to undermine the significant appraisal of the interconnections between the areas of science and Buddhism rarely take on the task of rigorously demonstrating their objections to any serious philosophical examination of the field. [QB:42-43]

So it was precisely because I was aware that the field was at that time causing such a polarisation of this kind, and that works presenting connections between quantum theory and spirituality were lacking in detail and precision, that I set out to produce a more detailed, precise and technical work.

I believe that *Quantum Buddhism: Dancing in Emptiness*, is such a work, and some people have contacted me to congratulate me and say that they consider I have achieved my aim. One of these was a highly qualified physicist who is also a long-term Buddhist/Non-Dual practitioner, very knowledgeable about spiritual philosophies, and who is acquainted, not closely, but acquainted, with Basil Hiley, a physicist who collaborated with Bohm in the later years of Bohm's life. He therefore is very knowledgeable of the work of Bohm and Hiley. Mike Roper contacted me to say that he considered that my work is definitive and unsurpassed in the arena of the interconnections between quantum physics, Buddhism, and spiritual philosophy. He says that he is 'mystified' as to how my work has not been recognised more widely. Indeed, it was in large part Mike's encouragement and insistence which led to my starting on this current work.

Some other people have told me that without my work they do not think they would have really understood the connections involved between quantum discoveries and Buddhist metaphysics. This is because of the detail presented in my work. I know my work has been recommended by a couple of meditation teachers as being definitive in this area, because their students have contacted me to tell me so, a few have travelled to visit me to discuss how I came to be able to do this work. A Google search for the phrase 'Buddhism and quantum physics' brings up an image of my book on the first page of results, and yet the silence from the academic community has been deafening!

Since Mike Roper contacted me in order to come to meet me to discuss my work we have become good friends, and we stay in at least weekly contact to discuss physics, philosophy and spirituality and my work. This evening during our discussion I asked whether I should actually make an effort to discover why my work is ignored, he thought I should do so. To be quite honest I am reluctant to actually hassle people, or make strenuous efforts to promote my work over the internet for example. I was offered the chance a while back to meet the Dalai Lama and it was suggested that I should promote my book to him, but I turned the opportunity down, I just do not like having to hawk it around in this way. One of the people who had come to meet me to discuss my work had told me she had given a copy to the Dalai Lama, so he should have had a copy anyway.

When I had first published *Quantum Buddhism* I did for a while try and promote my work, but met with disinterest. I was in regular contact with a very famous quantum physicist for a while, and he was for a while very supportive and encouraging, but he stopped the contact after reading a draft of my second book, I do not know why, although an essay he published a short time later bore a very close similarity to an article I had written and and sent him! I met B. Alan Wallace in 2011 at a conference and gave him a copy of QB, but he seemed completely disinterested, despite the fact that he regularly gives long lectures, very good lectures, on this subject. And, as already indicated, at least one significant Western Buddhist teacher has suggested to his students that they read my book, a group of his students came to visit me to discuss my work. When I sent an outline of my work to the notable Tibetan Buddhist teacher Ringu Tulku, he praised it and included it on his website,[5] indicating that it was exactly the kind of research needed.

The almost complete lack of interest from academia and other less 'mainstream' channels over the eleven years since my publishing QB does seem very odd, and Mike has indicated that, as I said, he is mystified, and would like to discover a reason for this overwhelming lack of interest for work which both he, and some others, consider goes beyond anything else available. Indeed, one person contacted me after reading QB to say I must be enlightened! I emailed back saying this was most definitely not true. My own meditation level is about first jhana, which is not at all exulted. However, it is sufficient to be able to turn off gross thought-processes in everyday life in order to experience the clarity and luminosity of the energy of pure consciousness, which is a capacity which enhances the significance of books such as this book. This, and my other books can, of course, be understood and appreciated on a purely intellectual level, but they become far more deeply significant when there is some experience of the luminous energetic presence of primordial awareness within one's own mind-stream. In fact, it is certainly the case that materialists are materialists precisely because they have no experience of what Bohm calls 'implicate' levels of consciousness, which are the deeper levels of primordial mind. I was able to activate this potentiality in my mindstream by using the excellent book *Mindfulness, Bliss and Beyond* by the Theravadin monk Ajahn Brahm, and also B. Alan Wallace's wonderful book *Stilling the Mind: Shamatha Teachings from Dudjom Lingpa's Vajra Essence*, putting the instructions into practice.

During the roughly 8 years of research which went into the final phase of the writing of QB I was for a lot of the time in a great deal of pain due to a misdiagnosed medical condition which I have suffered from for most of my life, and has been misdiagnosed by incompetent doctors for most of my life. In order to cope with the pain and discomfort I developed the ability to use alcohol to numb pain whilst maintaining a completely clear and focused mind. People in the past were astounded to see the amount I was able to consume and still be able to conduct complex intellectual discussions and expositions. I became mildly famous at Sussex University for my ability to meditate and then go down the bar and consume large amount of beer whilst expounding complexities of Buddhist metaphysics, mythology and physics. As I have just mentioned I am very far from enlightenment and my life has certainly involved many episodes very, very far from enlightenment, especially in the more distant days. Furthermore, much of my latter life has been lived in poverty, by Western standards, precisely because of long term misdiagnosed illness, and a need for alcoholic self-medication, a need which led to a serious confrontation with possible death a few years ago. And yet, whilst writing QB, books fell off bookshop shelves opened at the exact place I needed to resolve particular quandaries, and some sections of my book chapters were written in my dreams. All very odd!

But, on the other hand, if Bohm's view of the holomovement, and the Buddhist views of the dharmakaya and samsara, the cycle of birth and death which takes place within in the eternal pool of primordial mind-energy, are correct, which seems to be the case, then such events are not so odd, after all. It is quite possible for a sentient being at any point in time to partake of mixed karmic energies from different, although related in some way, karmic sources. According to Buddhism, there are no fixed 'selves', but there are interrelated continuities of mind energies coursing through the universal mind-energy of the dharmakaya, Bohm's 'holomovement'.

Since 2010, when I published QB, there has been an explosion of book writing and internet forums devoted to the mysteries of quantum physics, quantum metaphysics, and also the connections between quantum metaphysics, Buddhism and spirituality. When I was writing QB this was not the case, there were a few books on these subjects but very, very little happening on the internet. Today it is impossible to keep up with all the internet channels devoted to such subjects. I have occasionally left comments on such channels pointing out issues and elucidations which can be found in my work, but again, no response. One begins to wonder about the true motivations behind the explosion of people seeking to get on the internet quantum 'spirituality' discussion and exposition circuit! For a short period I ran a quantum Buddhism meditation group and I posted a few, not very adequate, YouTube videos. I took them down but was asked to leave them, so despite my reservations, have left them; but, in general have been reluctant to join the self-promotion. However, it has been suggested to me I should change my mind on this issue.

In 2018 another less than welcome karmic twist disturbed the serene, and unnoticed, path of my quantum mystical endeavors. One of the people who had been very enthusiastic about my work was the New-Age propheteer Paul Levy. He wrote an Amazon review of QB, part of which read:

> Out of all the books I've received, not to mention all of the ones in my library from years past, "Quantum Buddhism" stands out. It so blew me away that I've already ordered Smetham's next book ... In his writings, Smetham reveals a deep understanding of how the deepest wisdom of the teachings of the Buddha and quantum physics show a precise correlation, continually pointing out these correlations in creative ways that could not be further away from the fuzzy, new age thinking that is characteristic of many such books.

He contacted me to ask which of my books contained the most significant insights. He then set about writing his own version of QB, clearly based on my work. In significant sections he clearly plagiarised my books, virtually reproducing significant sections, sometimes lifting the exact sentences in a few places. In 2018 he published his book *The Quantum Revelation: A Radical Synthesis of Science and Spirituality*, and when I read this book I had a very weird feeling I was reading my own work. And then in the most technical section of Levy's book I realised I *was* reading my own work. Levy does give one reference to me, but this is a smoke-screen or cover-up of the extensive pilfering of my ideas, some of which I can show to be unique to me. Levy's book was showered with praise by many New-Age worthies. The New-Age Priestess Jean Houston suggests that Levy's book is worthy of being compared as a twentieth century Dante's *Divine Comedy*, an evaluation which is itself comical! I contacted the New-Age pundits who hailed Levy as a cosmic genius to point out the real source of his heavenly inspiration, but, again, the response was Cosmic Silence!

Levy then popped up on several New-Agey YouTube channels, and when I commented about his mystical misdemeanors on such channels, with a link to a document clearly proving my case, the comments were turned off! Full details of this New-Age Comedy can be found in the last chapter, 'A Quantum of Plagiarism & some glitches in the Quantum Revelation!' of the book I am publishing just after this book. This companion book is titled: *Quantum Revelations of the Real and Unreal:*

Quantum Buddhist Metaphysics Rectifies New Age Propheteering & Subtle Quantum Materialist Madness.

After publishing my last book, *Quantum Path to Enlightenment*, I decided to take a break from constant research and writing. But then sometime in 2020 Mike Roper got in touch with me to praise my work and suggesting we meet up. He visited Brighton and we spent a couple of hours in Waterstones bookshop discussing the state of research in 'quantum mysticism' and 'quantum Buddhism', and such topics. Since then we have had weekly discussions concerning such issues, as well as the state of the world in general. Mike is a 'chartered physicist' - a 'CPhys'. Until meeting Mike I had no idea such people existed. According to the Institute of Physics:

> CPhys represents the highest standards of professionalism, up to date expertise, quality and safety, and holders have demonstrated the capacity to undertake independent practice and exercise leadership. Chartered physicists have attained an integrated masters degree in physics (or demonstrated equivalent levels of underpinning knowledge), acquired a breadth of physics related competence and have exercised significant levels of responsibility for a sustained period of time. The title also denotes commitment to keep pace with advancing knowledge and with the increasing expectations and requirements for which any profession must take responsibility.[6]

I point this out because my own education in academic physics only goes as far as first year university. After that I concentrated exclusively on Mathematics. Whilst this is perfectly adequate to understand the equations of physics sufficiently for understanding the debates in the interpretations of quantum theory, sometimes those who wish to undermine any kind of quantum-spiritual perspective will claim that only those rare individuals who understand the most complex and subtle equations describing the physical world ever concocted could possibly understand quantum theory. This claim is false, although one should have a reasonable degree of mathematical competence and understanding of physics. But it helps to have one's views endorsed by someone whose qualifications surpass one's own.

When the 'Infinite Potential' film,[7] and the subsequent discussions, about the life and work of Bohm were being broadcast on the internet, our weekly discussions were devoted in large part to Bohm's works and ideas. We were both dissatisfied with certain aspects of the treatment of this subject by the 'Infinite Potential' presentation. In particular, there was a tendency to present Bohm's ideas from an over-emphasised, to my mind inappropriate, New-Age perspective. And this was not the only shortcoming, important technical issues were ignored. Such technical issues are addressed in this book. Although this book does centrally deal with interconnections with Bohm's ideas and Buddhism, the reader will find that along the way very detailed and precise issues of quantum metaphysics are discussed and solved. It was because of my desire to see a more detailed and extensive analysis carried out that I threw myself into this task of researching and writing this current book, with Mike's enthusiastic encouragement and support.

Mike has provided his following endorsement of this book:

> This book, like Graham Smetham's other work is a 'tour de force'. It manages to bring together in one volume three different, but connected strands, and show they form a coherent and consistent 'whole'. His treatment of the subtleties of David Bohm's 'metaphysical insights', the intricacies of Quantum Mechanics interpretation and the equally subtle and intricate underlying 'philosophy' of a very important 'school' of Tibetan Buddhist philosophy, is well above and beyond anything readily available. All this is done in a single volume - weaving these three threads together - to show they indeed form a genuine 'whole'. A wonderful achievement. Where others merely 'suggest' connections - in typical 'New Age' fashion - he actually spells out the details. A unique and valuable contribution to the ongoing efforts to formulate a comprehensive 'metaphysical' picture of 'Reality'. Provided the reader has some background in Physics and Buddhism - there is no better work available, that tackles this monumental task. A consummate achievement!

My profound thanks also go to Erik Scothron, the main editor for Shunyata Press, who has been a long-term collaborator in my work, being hugely significant from the beginning. Although Erik now resides in the Philippines and has less time, and obviously we cannot talk face to face as we did daily during the writing of my first books, he nevertheless does an invaluable job of reading for content and offering feedback, and also proof reading for errors, a task of great tedium, but essential! As this is a self-published book, I am sorry to say that a few errors always slip through. I apologize for any irritation caused.

I hope this work contributes to the task of elucidating the true spiritual nature of reality.

The websites and forum for my work are:

Websites:
quantumbuddhism.org
shunyatapress.com

Facebook forum:
https://www.facebook.com/groups/185195343194

Introduction:
Bohm's Quantum Worldview - a 'Radical' Buddhist Meditation?

After watching the film about the life and work of David Bohm, 'Infinite Potential: The Life and Ideas of David Bohm', I joined the discussion group that was associated with the film. Because of this film, and the subsequent surrounding interest, together with encouragement from a Buddhist retired physicist friend, I had decided to interrupt work on my seventh book in a series on the subject of quantum physics and Buddhist philosophy. The book I was working on at the time was an exploration of some of the excesses and absurdities of quantum New-Age fantasies and also an exposure of some subtle materialist deceptions regarding quantum phenomena which are advanced in opposition to New-Age excesses. That book will be published at a future date. The film-inspired interruption was in order to write this book, which is about the important ideas of quantum physicist David Bohm, explored in the light of Buddhist and pre-Buddhist *Bon-Dzogchen* metaphysics.

I was able to write a couple of chapters quickly, due to a large amount of research I had previously done to be used in current and future projects, and I posted the draft chapters to the 'Infinite Potential Dialogues Inspired by David Bohm' Facebook discussion group to see what kind of feedback I might get. I was surprised when one of the members commented that he thought that it is worthless engaging in such research and one should just 'be', so to speak. It seemed that he thought that this was what David Bohm would want. It immediately seemed odd to me that someone would think that Bohm would not want anyone to seriously research his work, and I also wondered why someone with such a view would bother to join a discussion group devoted to discussing Bohm's ideas!

But, upon further reflection, it was also a reasonable question to ask, assuming that the right motive was involved. Luckily, I had recently read a book that offered a very cogent answer. The following is taken from the introduction to *Adorning Maitreya's Intent: Arriving at the View of Nonduality*:

> In the *Cula-Malunkyovada* Sutta, for instance, in the parable of the poisoned arrow, the Buddha tells a disciple that metaphysical speculation can be a dangerous waste of time by likening it to a soldier wounded by a poisoned arrow who refuses to be treated until he knows everything about the arrow and the soldier who shot

it. ... Why, then, one might reasonably ask, has the tradition produced this gigantic collection of writings of a philosophical nature? ... in the context of the Buddhist tradition, philosophical argumentation has a soteriological function. It is an aid to liberation designed to remove confusion about the path and reality itself. ...[8]

According to this point of view, then, it is necessary to have some degree of metaphysical insight in order to have a direct insight into the nature of reality.

I have spent many years in research and writing about the interconnections between quantum physics and Buddhist metaphysics. For me, the clarification of what in Buddhism is called the 'view' - our intellectual understanding of the nature and functioning of reality - has helped my practice, meager as it may be! And the development of Bohm's ideas is particularly relevant here because of the way that his attempt in 1952 to create a more 'realistic' and deterministic 'classical' quantum 'interpretation' had the seeds of its own dissolution within it, which led automatically in the direction of a vision of wholeness. This is a fascinating story that I have focused on in the third chapter of this book. I do not think it is a story that has been told in quite such a detailed fashion elsewhere.

For me, the very writing of this book has been a kind of 'analytic meditation' in the sense that it has made me far more aware, at a deeper level, of the truth of the interconnected nature of the process of reality. According to Acharya Lama Tenpa Gyaltsen:

> ...when we do analytical meditation, which involves effort and reasoning, it is very important to be mindful that one's awareness remains in the center of one's body. Otherwise, we will just engage in a superficial conceptual investigation rather than in genuine analytical meditation. When your mind is resting and clear, proceed to the analysis. How do we conduct the analysis? We select an appropriate example of the object of analysis and examine it using our reasoning.[9]

Applying such a procedure to the ideas of David Bohm can have great benefit!

In the practice of the Buddhist *Lamrim*, the path to enlightenment style of meditations, there are two phases in the meditation process. Firstly, the meditator must generate, through 'analytical meditation', a deep mental conviction or feeling regarding the meditation topic: the precious nature of human birth, or the certainty of death for examples, and once a mental image of the appropriate conviction-feeling is generated it is then held in focus single-pointedly with fixed meditation.[10] I have found the research and subsequent exposition of David's Bohm's thinking in the context of Buddhism to be like a very extended and deep 'analytical meditation' which has given me what seems to be, at least to me, a deep insight into some important aspects of the nature of the process of reality. It is my hope that the reader may experience something similar through the reading of this book.

The Dalai Lama has made some very clear endorsements of the view that there are close links between Buddhist metaphysical insights and the discoveries of quantum physics. For example:

> Broadly speaking, although there are some differences, I think Buddhist philosophy and Quantum Mechanics can shake hands on their view of the world.[11]

And:
> ... there is an unmistakable resonance between the notion of emptiness and the new physics. If on the quantum level, matter is revealed to be less solid and definable than it appears, then it seems to me that science is coming closer to the Buddhist contemplative insights of emptiness and interdependence.[12]

And a recent report tells us that according to the Dalai Lama:

> Spirituality Without Quantum Physics Is An Incomplete Picture Of Reality.[13]

And, as the Dalai Lama considered Bohm to be "one of his scientific gurus", we can only conclude that the ideas of Bohm must have significant relevance for the Buddhist worldview.

As I began to explore the terrain it quickly became apparent that Bohm's insights, embodied in works such as *Wholeness and the Implicate Order*, contain great significance for Buddhist metaphysics, and, also, this connection works in the other direction, Buddhist and pre-Buddhist *Bon-Dzogchen* metaphysics contains much to illuminate modern quantum metaphysical perspectives. We shall see that this is particularly true for the Buddhist (and pre-Buddhist *Bon*) *Dzogchen* ("Great Perfection" or "Great Completion") and *Yogacara-Vijnanavada* (Consciousness-Only) worldviews.

Note that where the term *Dzogchen* is used in this current work it generally refers to both Buddhist and the earlier pre-Buddhist *Bon* versions. Pre-Buddhist and Buddhist perspectives will be conflated, unless differences are significant, for the purpose of exposition. Although the primary focus of the discussion in the context of Buddhism/Bon is Buddhist *Yogacara-Vijnanvada* (consciousness-only yogic vehicle) and *Dzogchen* (Buddhist and pre-Buddhist), there are other associated perspectives that are implicitly relevant. Thus, for example, Buddhist *Mahamudra*, the 'Great Seal' tradition, is closely connected with *Dzogchen*. As Khenchen Thrangu points out in his book *Pointing out the Dharmakaya*:

> Two different lineages of the meditation of looking directly at mind arose in Tibet. One was mahamudra and the other was the dzogchen lineage. Different teachers have made somewhat different statements about the relationship between these two styles of practice and teaching. ... The instructions in both of these traditions is simply called "guidance on the mind" because in both systems everything hinges on the student's recognition of the nature of mind.[14]

Thus, much of the discussion of *Dzogchen* in this book will also apply to *Mahamudra*, *Dzogchen* has been privileged here simply because the preponderance of the significant texts used are within the *Dzogchen* tradition.

As previously indicated, the significance of conceptual analysis, within the quest for knowledge and experience of the ultimate nature of reality, is questioned by some. The Radical Dzogchen practitioner and apparently 'anti-philosopher' Keith Dowman appears to be uncompromising in his emphasis on the very limited role that intellectual analysis and discourse, and even structured practice, can play in describing and pointing towards ultimate experience. He writes:

> ...elaborate meditation - and along with it all 'spiritual practice' - is superfluous. If spiritual practice is ultimately useless, so also is dogma and cant, whether it be rational and humanistic or religious and apocalyptic. Buddhist belief systems that define a specific starting point, an elaborate path and goal, for instance, provide dogma tailored to the requirements of adherents ... on a graduated path .... spiritual materialism infects the minds of well-intentioned

people susceptible to attachment to the intent of coherent oral or written soteriological teaching. Radical Dzogchen, free of belief, cannot be dogmatic, indeed it is entirely pragmatic in that the view arises spontaneously in response to the requirements of every unique moment.[15]

The term 'view' here refers to the core understanding of the nature of the process of reality. Dowman is indicating that, within the more establishment-focused gradualist modes of Buddhist practice, the 'view' actually becomes a distraction, and perhaps obstructive, because of excessive reliance on conceptual modes of analysis. Perhaps it might be expected, then, that Dowman would not have much time for 'quantum mystical' insights, but this is not the case:

> 'Quantum mysticism' has provided useful metaphors that move the rational mind towards acceptance of the anomalous phenomenology of nonduality. ... When we are told, for example, that the electron, which revolves around the nucleus of the atom, moves in and out of different orbits without apparent cause..., leaving no trace of its previous revolution, our rational intellects may space out and a moment arise adventitiously when the nature of mind can shine through and an existential understanding of 'nonabiding' may arise.[16]

Here, Dowman indicates how an appreciation of a quantum fact can open up an opportunity for a person's mind to make its own non-conceptual 'quantum leap' to instantaneously perceive a deep existential truth about the process of reality.

It is in this sense that I consider the explorations of this book may be thought to be a kind of 'radical' Buddhist meditation. For example, Dowman indicates that the contemplation of one particular quantum phenomenon: the quantum electron orbit instantaneously 'jumping' from one level to another, without a continuous path in between, may lead to an opening of the mind to understanding its own nature. The same situation is, to my mind, operative to some degree, with a greater scope of view, within the development of Bohm's thought.

We have already intimated that Bohm's 1952 article attempted to try and produce a more 'realistic' and 'deterministic' account of quantum reality, an attempt to expunge some of the quantum 'weirdness' from quantum reality, but it produced an internally unstable system of thought which contained the seeds of its own necessary further development. It is the details of the inner movement and endpoint of the move from the 1952 'pilot-wave' theory to the later spacious quantum view of 'wholeness', as explicated in Bohm's book *Wholeness and the Implicate Order*, which provides us with an analytic 'meditation' which can also open one's mind to the spaciousness of the interconnected universe and the experiential spaciousness of primordial Mindnature which forms its inner nature.

Mindnature, the primordial field of the process of reality, is described in the excellent book on the pre-Buddhist Bon-Dzogchen worldview: *Unbounded Wholeness: Bon and the Logic of the Nonconceptual*. This is a book, echoing down from Tibet through hundreds of years, which, as its title indicates, is fully resonant with Bohm's later perspective. The following is a short passage from *Unbounded Wholeness*:

> Being wholly uncontrived, mindnature neither improves upon enlightenment nor becomes flawed in samsara. Always present in all beings, it is the abiding condition itself, otherwise described as unbounded wholeness.[17]

The story of the necessity of internal development of Bohm's thought towards its final 'undivided universe' universe-view is just one part of the 'radical meditation' that is involved when we explore the details of Bohm's thought and its interconnections with Buddhism, especially Dzogchen, and also pre-Buddhist Bon-Dzogchen.

The interconnections between Bohm's perspective and that of Dzogchen in particular is spectacular, and moving between the two metaphysical arenas to savour the mutual resonance is certainly a spur to a meditative frame of mind, an opening of mental spaciousness as described by Dowman above. For example, the resonance between the overall worldview and themes, although expressed in different idioms, from *Wholeness and the Implicate Order* and the wonderful Dzogchen text translated and presented with the title *Unbounded Wholeness: Dzogchen, Bon and the Logic of the Nonconceptual* is remarkable. The resonance between these two books is an aspect of the final chapter of this book, the preceding chapters explore many other significant connections between Bohm's ideas and Buddhism and Bon Dzogchen.

In particular, we may highlight the notion that the dualistic world of the appearance of matter and the functioning of embodied mind derives from a deeper, mind-like, potential mind-energy source. Here is a passage from Bohm's *Wholeness and the Implicate Order*:

> The new form of insight can perhaps best be called Undivided Wholeness in Flowing Movement. This view implies that flow is, in some sense, prior to that of the 'things' that can be seen to form and dissolve in this flow. One can perhaps illustrate what is meant here by considering the 'stream of consciousness'. This flux of awareness is not precisely definable, and yet it is evidently prior to the definable forms of thoughts and ideas which can be seen to form and dissolve in the flux, like ripples, waves, and vortices in a flowing stream. As happens with such patterns of movement in a stream some thoughts recur and persist in a more or less stable way, while others are evanescent. The proposal for a new general form of insight is that all matter is of this nature: That is, there is a universal flux that cannot be defined explicitly but which can be known only implicitly, as indicated by the explicitly definable forms and shapes, some stable and some unstable, that can be abstracted from the universal flux. In this flow, mind and matter are not separate substances. Rather, they are different aspects of one whole and unbroken movement.[18]

Here we find some central significant Bohmian themes of the essential wholeness of the process of reality, mind and matter as arising from a deeper common source, the arising from this common source of relatively autonomous forms, including sentient beings, which take part in the flux taking place within the totality of interconnected wholeness.

Similar themes can be found to be significant within Buddhist metaphysics and Bon Dzogchen. Thus, in *Unbounded Wholeness* we read:

> ...this wholeness, which cannot be totalized or bounded, is thoroughly compatible with diversity. Indeed, diversity, though ordinarily considered the antithesis of unity, is here offered as proof that unbounded wholeness exists. Being so diverse, and constantly changing besides, means that unbounded wholeness admits of no defining characteristic or stable identity, in this sense it is indefinable and unspeakable. ...[19]

This perspective finds a clear resonance within the much later views expressed by Bohm, who describes his notion of the 'holomovement', which is the flow of wholeness:

Thus, in its totality, the holomovement is not limited in any specifiable way at all. It is not required to conform to any particular measure. Thus, the *holomovement is undefinable and immeasurable.*[20]

Returning to the description of Mindnature from the Dzogchen text *Unbounded Wholeness*, we read:

Mindnature, clear light which is this wholeness,
Primordially not a substantial thing,
Primordially clear conscious open awareness
For me to say "This" is unfitting.

...

The inability to describe unbounded wholeness in any one way ... dramatically testifies to wholeness's decentered multiplicity and thus to its incommensurability with conceptual limitation. Unbounded wholeness can, and must, be called both definite and indefinite; this is the principle of wholeness.

...

Definite and indefinite ... here turn out not to be a mutually exclusive binary. Likewise, other dyads such as Buddhas and sentient beings, conventional and ultimate, or conditioned and unconditioned are all ... "facets of wholeness", not mutually exclusive ... indefiniteness ... as an evocation of multifaceted reality, continuously brimming with shapes and colors even as it remains an unmitigated whole.[21]

The following passage is taken from another Dzogchen work *Everything is Light*, translated and introduced by Keith Dowman:

The zero-dimension awakens in zero-appearance. The sole holistic sphere of dharmakaya blends with the undivided field of appearances into an elixir that shines like sunlight in the sky. ... The field that is cognized by the zero-essence is the spaciousness of reality, the dharmadhatu ... without center or periphery, without top or bottom, without any spatial bias. ...unchanging matrix of unelaborated spaciousness. ... The basis of an inconceivable field is itself inconceivable, and so no fixed concept can arise in pure presence ... The nondual subjective and objective aspects comprise the spontaneity of the self-envisioned field, which is invested by the clear light of present awareness. ... the elixir of spun essence is the vision of present awareness that is like magical illusion.... In that way [is] related the nature of ultimate spaciousness to awareness of the interconnected universe.[22]

As we shall see, by moving between such deeply related perspectives, one from a Western scientific-philosophical context and the other from the 'mystical' Buddhist Dzogchen metaphysical spirituality, sometimes with specific detailed interconnections, each throws light upon the other and thus deepens and widens the mutually illuminating metaphysical viewpoints. In this way, the following chapters do constitute a kind of extended analytic meditation. Whereas many attempts to connect quantum phenomena with spiritual traditions do so in a piecemeal and limited fashion, taking scraps of quantum evidence to weave a quantum-mystical 'new-age' type worldview, Bohm's later vision interconnects with and embraces the Buddhist Dzogchen mystical view in particular in a much more thoroughgoing and comprehensive manner.

In this context, Hee-Jin Kim's description of the Zen master Dogen's attitude to philosophy and enlightenment seems apposite:

> Dogen viewed the philosophical enterprise as an integral part of the practice of the Way. ... if and when intellect was purified and reinforced by the samadhi [meditation] of self-fulfilling activity. Our philosophic and hermeneutical activities were no longer a means to enlightenment, but identical to enlightenment itself, for to be was to understand - one was what one understood. Thus the activity of philosophizing, like any other expressive activity, was restated in the context of our total participation in the self-creative process of Buddha-nature.[23]

And Kim gives a direct quote from Dogen, a quote also appropriate to the thought of David Bohm:

> The monastics of future generations will be able to understand a non-discriminative Zen based on words and letters, if they devote efforts to spiritual practice through words and letters and words and letters through the universe.[24]

The significance of this insight will, I hope, become clear during the course of the following explorations.

First Turning of the Holomovement
Setting the Scene for Wholeness

The work of the maverick and important physicist David Bohm has been accorded fresh impetus recently with the release of the excellent film dealing with his life and work: "Infinite Potential: The Life and Ideas of David Bohm". This film is the first of a series of films and discussions promoted through the Infinite Potential internet-based platform (infinitepotential.com) and organised by the Fetzer Franklin Fund with the filmmaker Paul Howard. The description of the film portrays Bohm as:

> ... the man Einstein called his "spiritual son" and the Dalai Lama his "science guru." A brilliant physicist and explorer of Consciousness, Bohm's incredible insights into the underlying nature of reality and the profound interconnectedness of the Universe and our place within it are truly transformational.

This, well-deserved, accolade is made by a group of physicists and philosophers who are enthusiastic about the 'later' quantum metaphysical perspective presented by Bohm in his works beginning with his ground-breaking book *Wholeness and the Implicate Order* and culminating with *The Undivided Universe,* which was written in the final years of his life in collaboration with the physicist Basil Hiley.

In this later and final phase of Bohm's life he was significantly impressed and influenced by what some might consider to be the 'mystical' perspective of the popular and influential 'mystic' and philosopher Jiddu Krishnamurti, whose teachings derive from the doctrines of Buddhism, woven together with insights Krishnamurti gleaned from his own investigations of psychology and philosophy. So, it is clear that Bohm's later engagement with his work in physics took place within a context of a spiritual worldview informed by Buddhist metaphysics. We shall find that significant aspects of Bohm's ideas conform closely with Buddhist metaphysical doctrines. In this book we shall survey some significant deep resonances.

Bohm did not always carry out his scientific explorations in the midst of a penumbra of spiritual interests, and neither does everyone appreciate his later mystical engagement with physics. The science writer for the *Scientific American* John Horgan wrote in his 2018 article 'David Bohm, Quantum Mechanics and Enlightenment':

> Some scientists seek to clarify reality, others to mystify it. David Bohm seemed driven by both impulses. He is renowned for promoting a sensible (according to Einstein and other experts) interpretation of quantum mechanics. But Bohm also asserted that science can never fully explain the world, and his 1980 book Wholeness and the Implicate Order delved into spirituality. Bohm's interpretation of quantum mechanics has attracted increasing attention lately.[25]

The recent "increasing attention" primarily focuses on Bohm's later work. But, it is a sometimes overlooked fact of Bohm's intellectual development that in an earlier phase of his scientific explorations, during the period that his interest in the welfare of humanity had led him to misguidedly embrace Marxism, he advanced what amounted to a much less 'spiritual' deterministic quantum-materialist version of quantum theory. This was his 'pilot wave' interpretation, which he proposed in his 1952 article '*A Suggested Interpretation of the Quantum Theory in Terms of 'Hidden' Variables*'. This attempted pristine world of 'real' quantum waves pushing around tiny particles of matter was proposed in order to counter and deflate what some physicists at the time considered to be a too vague, imprecise, and perhaps 'mystical' viewpoint that had been enshrined in the Copenhagen 'interpretation' by the 'founding fathers' of quantum mechanics Neils Bohr and Werner Heisenberg.

In a lecture given at the 1928 International Physics Conference at Lake Como Bohr indicated a central aspect of the Copenhagen interpretation of quantum theory as follows:

> Now the quantum postulate implies that any observation of atomic phenomena will involve an interaction with the agency of observation not to be neglected. Accordingly, an independent reality in the ordinary physical sense can neither be ascribed to the phenomena nor to the agencies of observation.[26]

A significant feature of this formulation is that it asserts that there is no separate "independent reality in the ordinary physical sense" to either side of the interaction between observer and observed. This view seems to suggest that observing minds and the observed 'matter' at the quantum level in some way interpenetrate and are inseparable.

Another significant aspect of the Copenhagen viewpoint is the 'Complementarity Principle', which asserts that atomic phenomena have both wave and particle properties, and that these properties manifest in a mutually exclusive manner. They are therefore said to be 'complementary'. The behaviour of such phenomena as light and electrons depends on the experimental arrangement, it is sometimes wavelike and sometimes particle-like; but it is impossible to observe both the wave and particle aspects simultaneously. Taken together, however, they present a fuller description than either of the two taken alone. In this context Bohr referred to:

> The existence of different aspects of the description of a physical system, seemingly incompatible but both needed for a complete description of the system. In particular, the wave-particle duality.[27]

And, according to Bohr, this is:

> The phenomenon by which, in the atomic domain, objects exhibit the properties of both particle and waves, which in classical, macroscopic physics are mutually exclusive categories.[28]

And he referred to:

> The apparently incompatible sorts of information about the behavior of the object under examination which we get by different experimental arrangements can clearly not be brought into connection with each other in the usual way, but may, as equally essential for an exhaustive account of all experience, be regarded as 'complementary' to each other.[29]

In observations such as these Bohr seemed to be inclined to import a mystical flavour of the East into quantum physics. As Bohr himself wrote:

> For a parallel to the lesson of atomic theory regarding the limited applicability of such customary idealisations we must in fact turn ... even to that kind of epistemological problems with which already thinkers like Buddha and Lao Tzu have been confronted, when trying to harmonize our position as spectators and actors in the great drama of existence.[30]

Thus it seems that Bohr was indeed aware of a 'mystical' perspective within his quantum worldview.

Not all physicists were, or are, happy with this state of affairs, a situation which Bohr, a Danish physicist, had a primary role in perpetrating. For this reason, Adam Becker, in his excellent book on the history of quantum theory: *What is Real,* has titled his chapter on the Copenhagen hegemony: 'Something Rotten in the Eigenstate of Denmark'! This is an allusion to a line from Shakespeare's play *Hamlet*: "Something is rotten in the state of Denmark". However, Becker has, for philosophical-literary effect, used the word "Eigenstate", which refers to a possible state of a quantum system, to replace the original word "state". This allusion by Becker highlights the controversial Copenhagen view that prior to 'measurement' a quantum entity hovers in a ghostly 'super-position' of many possible/potential 'eigenstates', of which none of them are really real!

In the early 1950s Bohm was one of the physicists who disliked this move towards a more 'mystical' physics, this may have been influenced by a leaning towards a Marxist influenced subtle materialism. In the conclusion to his 1952 article he wrote (in the following quote 'Ψ-field' is the mathematically described quantum wave-field):

> The usual interpretation of the quantum theory implies that we must renounce the possibility of describing an individual system in terms of a single precisely defined conceptual model. We have, however, proposed an alternative interpretation which does not imply such a renunciation, but which instead leads us to regard a quantum-mechanical system as a synthesis of a precisely definable particle and a precisely defined Ψ-field which exerts a force on the particle.[31]

Bohm's use of the term "renounce" here is perhaps a response to Bohr's assertion concerning the fact that he thought that a "renunciation of the visualization of atomic phenomena is imposed upon us" by the nature of reality, a view which, clearly, Bohm did not share. As can be seen in Bohm's conclusion, whereas Bohr thought that the quantum world mystically-morphed between appearances of waves or appearances of particles, but never both, Bohm suggested that both waves and particles were there at all times, the waves pushing particles around. However, as we shall see, this sparse quantum mechanistic worldview later transmuted into a glorious spiritual vision of an Undivided Universe which is enlivened by the presence of a kind of primordial consciousness projecting itself into embodied limitation.

In the book *The Undivided Universe* co-authored by Bohm and Hiley, which was in the final stages of completion when Bohm died, they describe Bohm's final metaphysical vision of an interconnected holistic universe as follows:

> We may suppose that the universe, which includes the whole of existence, contains not only all the fields that are now known, but also an indefinitely large set of further fields that are unknown and indeed may never be known in their totality. Recalling that the essential qualities of these fields exist only in their movement we propose to call this ground the *holomovement*. It follows that ultimately everything in the explicate order of common experience arises from the holomovement. Whatever persists with a constant form is sustained as the unfoldment of a recurrent and stable pattern which is constantly being renewed by enfoldment and dissolved by unfoldment. When the renewal ceases the form vanishes.[32]

Note that, if this description is taken in isolation, without elucidation from other descriptions from Bohm's work and some other sources, such as discussions with colleagues, it may be misleading. This is because the assertion that the holomovement is the '*ground*', and that "everything in the explicate order of common experience *arises from* the holomovement", gives the impression that the explicate order, which includes the everyday world of our experience, *arises* out of the *ground* of the holomovement. This would seem to indicate that the explicate order is separate from, being projected out of, so to speak, the holomovement.

However, this is not the correct picture. When all of the various metaphors and descriptions in Bohm's writings and talks are taken into account, it is clear that the holomovement is the movement of the totality, and therefore the explicate order is an internal aspect of the holomovement. The holomovement contains both implicate orders and the explicate order. Because the contents of the explicate order, which unfold from the implicate order(s), have a greater stability, the explicate order has a degree of separation from the implicate orders. But all the orders are within the holomovement. Bohm's friend and collaborator F. David Peat describes this:

> Bohm believes that the Implicate Order has to be extended into a multidimensional reality; in other words, the holomovement endlessly enfolds and unfolds into infinite dimensionality. Within this milieu there are independent sub-totalities (such as physical elements and human entities) with relative autonomy.

The layers of the Implicate Order can go deeper and deeper to the ultimately unknown. It is this "unknown and undescribable totality" that Bohm calls the holomovement.[33]

Thus, we see that "independent sub-totalities (such as physical elements and human entities) with relative autonomy", which are the contents of the explicate order, unfold out of the implicate orders, but the entire process remains within the holomovement.

In his book *Wholeness and the Implicate Order*, Bohm gives the following description:

> It is being suggested here, then, that what we perceive through the senses as empty space is actually the plenum, which is the ground for the existence of everything, including ourselves. The things that appear to our senses are derivative forms and their true meaning can be seen only when we consider the plenum, in which they are generated and sustained, and into which they must ultimately vanish. This plenum is, however, no longer to be conceived through the idea of a simple material medium, such as an ether, which would be regarded as existing and moving only in a three dimensional space. Rather, one is to begin with the holomovement, in which there is the immense 'sea' of energy ... This sea is to be understood in terms of a multidimensional implicate order, ... while the entire universe of matter as we generally observe it is to be treated as a comparatively small pattern of excitation. This excitation pattern is relatively autonomous and gives rise to approximately recurrent, stable and separable projections into a three-dimensional explicate order of manifestation, which is more or less equivalent to that of space as we commonly experience it.[34]

This entire process, including the projection of a 3-D world, takes place within the holomovement.

To illustrate his idea of the relationship between the implicate order and the manifested 'explicate' order Bohm used the example of glycerine machine, whose primary function is to illustrate laminar fluid flow:

> ...two concentric glass cylinders, with a highly viscous fluid such as glycerine between them, which is arranged in such a way that the outer cylinder can be turned very slowly ... A droplet of insoluble ink is placed in the fluid and the outer

cylinder is turned, with the result that the droplet is drawn out into a fine thread-like form that eventually becomes invisible. When the cylinder is turned in the opposite direction the thread-like form draws back and suddenly becomes visible ...[35]

The state of the apparatus when the droplet is drawn into an invisible thread is representative of the 'enfolded' implicate order. Turning the cylinder back 'unfolds' the implicate order until at a certain point the manifest 'explicate' order of the drop will become apparent. Individual drops can be enfolded by the process, each being enfolded in a closely aligned sequence so that when the cylinder is turned to unfold the enfolded drops they will manifest as if there was a single moving drop. The drop appears to be a single moving entity but this illusion is mistaken. In actuality the appearance of a single moving drop is a succession of enfolded drops which manifest sequentially from the implicate order. Note that before the drop is 'unfolded' it is in an 'enfolded' state wherein bits of it exist spread over the volume of the glycerine. This corresponds to Bohm's view that what appears to be independent 'particles' are appearances, which are enfolded in a smeared out fashion within the implicate order, in a similar way to a hologram.

A hologram is a special type of photograph that creates a complete three-dimensional image when it is illuminated in the right manner by a beam of light. All the information which produces the 3-D scene is encoded into the pattern of light and dark areas on a two-dimensional piece of film. This example of a hologram is another example used by Bohm to illustrate an aspect of his ideas:

> We proposed that a new notion of order is involved here, which we called the implicate order (from a Latin root meaning 'to enfold' or 'to fold inward'). In terms of the implicate order one may say that everything is enfolded into everything. This contrasts with the explicate order now dominant in physics in which things are unfolded in the sense that each thing lies only in its own particular region of space (and time) and outside the regions belonging to other things. The value of the hologram in this context is that it may help to bring this new notion of order to our attention in a sensibly perceptible way; but of course, the hologram is only an instrument whose function is to make a static record (or 'snapshot') of this order. The actual order itself which has thus been recorded is in the complex movement of

electromagnetic fields, in the form of light waves. Such movement of light waves is present everywhere and in principle enfolds the entire universe of space (and time) in each region (as can be demonstrated in any such region by placing one's eye or a telescope there, which will 'unfold' this content). ... this enfoldment and unfoldment takes place not only in the movement of the electromagnetic field but also in that of other fields, such as the electronic, protonic, sound waves, etc. There is already a whole host of such fields that are known, and any number of additional ones, as yet unknown, that may be discovered later.[36]

And, in his book, co-authored with F. David Peat, *Science, Order, and Creativity,* Bohm added a further 'superimplicate' order to his conception to emphasize a hierarchical nature of implicate orders that he conceived of as the overall basic structure of the process of reality:

...a generative order, in the form of the superimplicate order, lies at the foundation of physics ... In the first implicate order this is basically a movement of a field, and yet, through the information in the second implicate order, the movement is organised into a particlelike behavior. ... all of the so-called elementary particles can be treated in this way, as quantum mechanical fields that are organised by information in their superimplicate orders which make possible the creation, sustenance, and annihilation of particlelike manifestations. They are thus relatively constant and autonomous particlelike features of the holomovement that emerge through the generative order.[37]

The diagram on the next page shows a schematic top-down layout of Bohm's view of the descent (it could be around the other way, the direction is irrelevant) from the fundamental source of the manifested world, which Bohm called the superimplicate order, through movements in the first implicate order which are subsequently organized to become particles, or 'particlelike manifestations', within the second implicate order. The 'particles' then descend through further implicate orders wherein they are further organized until they finally manifest as the 'explicate' order of the everyday world. This diagram and description is very basic, Bohm actually indicates that each type of particle would have its own field. But we are concerned with basic principles here.

The philosopher Paavo Pylkkanen, in his book *Mind, Matter, and the Implicate Order*, describes Bohm's vision, which derives from quantum field theory:

> Underlying each particle is a movement in a field. This movement enfolds information about the whole universe into a small region where the field manifests itself as a particle-like entity. Because the field is also spread, in principle, throughout the universe, information about the particle-like entity can be found in every region of the universe. In this sense, the whole universe is enfolded in everything, and everything is enfolded everywhere in the whole universe. The implicate order thus prevails as the most fundamental order of the universe currently known to us.[38]

------------ **BOHM'S HOLOMOVEMENT** --->>>>>>

SUPERIMPLICATE ORDER

{HOLOGRAPHIC INFORMATION FIELD}

↓

FIRST IMPLICATE ORDER

{movements}

↓

SECOND IMPLICATE ORDER

{organizing}

↓

PARTICLES or PARTICLE-LIKE MANIFESTATIONS

↓

EXPLICATE ORDER

{EVERYDAY WORLD}

→→→→→→→→→

According to Bohm, the process of reality is an intimately interconnected multitudinous 'flux' of potentiality which underlies the manifestation of apparent manifested 'fragments' or 'sub-totalities' which unfold from and within this interconnected 'wholeness':

> The new form of insight can perhaps best be called Undivided Wholeness in Flowing Movement. This view implies that flow is, in some sense, prior to that of the 'things' that can be seen to form and dissolve in this flow. One can perhaps illustrate what is meant here by considering the 'stream of consciousness'. This flux of awareness is not precisely definable, and yet it is evidently prior to the definable forms of thoughts and ideas which can be seen to form and dissolve in the flux, like ripples, waves and vortices in a flowing stream. As happens with such patterns of movement in a stream some thoughts recur and persist in a more or less stable way, while others are evanescent. The proposal for a new general form of insight is that all matter is of this nature: That is, there is a universal flux that cannot be defined explicitly but which can be known only implicitly, as indicated by the explicitly definable forms and shapes, some stable and some unstable, that can be abstracted from the universal flux. In this flow, mind and matter are not separate substances. Rather, they are different aspects of one whole and unbroken movement.[39]

This is the 'holomovement', a vast, interconnected universal mind-like organism moving and 'unfolding' in time. The 'holomovement' is a central concept in Bohm's overall quantum metaphysical worldview. It brings together the holistic principle of "undivided wholeness" with the view that this undivided whole is also in a state of development, or dynamic flux, or becoming. The term 'holomovement' itself enfolds all the various levels or layers that Bohm speaks of.

As previously indicated, this Bohmian vision of the undivided wholeness of the universal holomovement was a development of a more 'realistic' and sparser quantum worldview proposed by Bohm in 1952, in which he suggested that quantum particles were guided around the universe by 'pilot-waves'. Bohm's earlier suggestion was in response to the types of explanations being used at the time, and previously, to explain strange quantum behavior. The prevailing 'interpretation' concerning the strange functioning that appears at the quantum level, which had been revealed by investigations such as the double-slit-experiment, was the Copenhagen interpretation. The Copenhagen interpretation was proposed by the physicist Niels Bohr around 1920. In essence, it suggests that, before a 'measurement' is performed, a quantum particle doesn't actually exist in one state or another, it is in a state of 'semi-existence', or indeterminate-existence, in all possible states at the same time. But when a 'measurement' or an 'observation' takes place the semi-existent or indeterminate set of possible-particles is forced to adopt just one of its possible states, the others disappear, Subsequently there is just one 'real' particle.

Several significant physicists at the time found this point of view, or 'interpretation', disconcerting. The physicist Roger Penrose has said of such a proposal in his book *Shadows of the Mind*:

> Taken at its face value, the theory seems to lead to a philosophical standpoint that many (including myself) find deeply unsatisfying. At best, and taking its descriptions at their most literal, it provides us with a very strange view of the world indeed. At worst, and taking literally the proclamations of some of its most famous protagonists, it provides us with no view of the world at all.[40]

Bohm was one of the dissenting physicists, and he suggested that the strange behaviour of subatomic particles might be a result of the operation of quantum field forces acting on particles. What appeared to be strange non-classical quantum weirdness might be caused by 'hidden' features that did not conflict with ordinary ideas of causality and reality; the quantum realm may operate more in line with mechanisms that operate above the quantum level, i.e. mechanisms within 'classical' reality, which is the 'reality' of the everyday world. So Bohm proposed a quantum ontology (i.e. what kind of 'stuff' really exists and what it does) which involved continuously 'real' particles being guided on the crests of quantum waves. It was this more 'realistic' quantum 'ontological interpretation' which later developed spectacularly into the grand holomovement of implicate and explicate orders within an undivided universe.

Given the fact that in his later metaphysical explorations Bohm saw connections between his perspective and 'Eastern' and 'mystical' notions, it should be no surprise that parallels and connections may be found between his final metaphysical ideas and certain schools of Buddhist psycho-metaphysics, as well as Chinese Taoism and Hindu

Yoga philosophy. Indeed, in his later vision Bohm seems to be more in accord with Bohr's supposed 'mystical' perspective to some degree. Thus, in *Science, Order, and Creativity* Bohm and Peat point out that:

> Thus, in Buddhism, each person is directed through reflection and meditation, to be aware, moment to moment, of the whole train of his or her thoughts. It is stated that in this process the fundamental "groundlessness" of the self can be seen. In this way a key piece of "misinformation" can be cleared up, i.e the almost universal assumption that the self is the very ground of being. This leads to Nirvana, in which there is a blissful unification with the totality.... Approaches of this kind move in the direction of the transcendent .. of union of the individual with the ultimate totality.[41]

And they write with regard to Krishnamurti:

> His writings go extensively and deeply into the question of how, through awareness and attention to the overall movement of thought, the mind comes to a state of silence and emptiness, without any sense of division between observer and the observed.[42]

In *Wholeness and Implicate Order* Bohm wrote:

> ... the easily accessible explicit content of consciousness is included within a much greater implicit (or implicate) background. This in turn evidently has to be contained in a yet greater background which may include not only neurophysiological processes at levels of which we are not generally conscious but also a yet greater background of unknown (and indeed ultimately unknowable) depths of inwardness that may be analogous to the 'sea' of energy that fills the sensibly perceived 'empty' space. Whatever may be the nature of these inward depths of consciousness, they are the very ground, both of the explicit content and of that content which is usually called implicit. Although this ground may not appear in ordinary consciousness, it may nevertheless be present in a certain way. Just as the vast 'sea' of energy in space is present to our perception as a sense of emptiness or nothingness, so the vast 'unconscious' background of explicit consciousness with all its implications is present in a similar way. That is to say, it may be sensed as an emptiness, a nothingness, within which the usual content of consciousness is only a vanishingly small set of

facets.[43]

Here we find a direct link into Buddhist psycho-metaphysics.

In Buddhism the term 'emptiness' is a central technical term (*sunyata*) designating the ultimate nature of the process of reality, an immaterial 'ground' of potentiality from which both mind and matter emerge. The following description of 'emptiness' is from the *Yogacara,* consciousness-only, perspective, a viewpoint which presents 'emptiness' as a positive phenomenon, the creative underlying field of the process of reality, and not as a mere 'absence' of dualistic features of the process of reality (as in the *Madhyamaka* Buddhist school which embraces a more 'negative' viewpoint of the ultimate nature):

> That is a unique feature of the Yogacarin presentation of emptiness, because emptiness is normally understood as a complete negation or a completely negative term rather than something positive. Here, once subject and object are negated, emptiness, which is reality, is affirmed in its place. A short passage from the *Madhyantavibhanga* says, "Truly, the characteristic of emptiness is nonexistence of the duality of subject and object, and the existence of that nonexistence." "The existence of that nonexistence" is reality. Duality is removed, but emptiness itself is another kind of existence.[44]

Here a resonance with Bohm's characterisation of the implicate order is apparent. Bohm highlights the lack of "division between observer and the observed" and this is clearly stated as the central feature of the Yogacarin view of emptiness. According to Buddhist *Yogacara* the ultimate nature of the process of reality is stated to be a "positive" phenomenon of non-dual awareness which manifests when "duality of subject and object" is negated. It is the sphere of nondual mind which manifests when all dualistic (i.e. 'observer-observed' division) movements within mind are pacified. In Bohmian terms we can think of the explicate order as having been experientially removed in order to directly perceive the implicate order(s).

The following passage, taken from a commentary to the *Diamond Sutra*, which has the title *Describing the Indescribable*, by the Chinese Buddhist teacher Hsing Yun, would also seem to be appropriate in this context:

> Dust clouds the metaphorical pool of enlightened awareness. ... Lakshana rush into the mind and appear before it like clouds of dust-like lakshana; impure intentions are based on deluded visions of dust. Dust clouds the mind on all levels; matter is dust, illusion is dust, and thoughts and perception also are dust. Only the Tathagata sees the 'vast realm of emptiness' in which all of this floats in the clarity of perfect awareness.[45]

Here, the "metaphorical pool of enlightened awareness" represents the mind in its state of pristine undivided consciousness which has qualities of crystal-like clarity and empty-luminosity. Such assertions of the experiential qualities of the undivided field of consciousness are not speculative. They can be experienced as a result of correct and committed meditation practice. The term 'lakshana' means 'marks', or 'signs', but here is best rendered as 'disturbances'. Perceptions of the material world, thoughts, and intentions for various actions within the material world all "rush into the mind" and cloud it as if with dust. Only an enlightened being, a 'Tathagata', one who has 'gone-beyond', has a mind which has "the clarity of perfect awareness" that is able to directly perceive the "vast realm of emptiness" within which the dualistic world floats. Here the "vast realm of emptiness" would correspond to Bohm's most subtle implicate order.

One of the Buddhist psycho-metaphysical perspectives that will be employed in this investigation is *Dzogchen* (rDzogs-chen), this is a view especially significant to the work of Bohm. The term *Dzogchen* is translated as the 'Great Perfection' or the 'Great Completion'. This term indicates an appreciation of the importance of a view of Totality, as in Bohm's notion of the holomovement. The great Buddhist scholar Herbert Guenther, who specialised in *Dzogchen*, was an expert explorer in this interface of science and Buddhist spirituality. In his book *The Teachings of Padmasambhava* Guenther refers to Bohm's work eleven times, and in the introduction to his book *Ecstatic Spontaneity: Saraha's Three Cycles of Doha* Guenther writes the following:

> The fact is, modern physics has become ever more "mystical," not least of all because of the exposure of some of its most outstanding representatives to Eastern thought. David Bohm's association with J. Krishnamurti is well known; Erwin Schrödinger was deeply impressed by Indian philosophy; Neils Bohr chose as his emblem the Taoist ying-yang symbol; C.G. Jung's collaboration with Wolfgang Pauli led to the idea of

synchronicity; and so on. Does it not seem fitting that these pioneers be called by the name that history has given to many of its great visionaries - mystics?[46]

Perhaps in some respects this is a relevant observation. However, it is also necessary to bear in mind that the number of such 'mystical' physicists who activated direct experience of the deepest implicate levels of the process of reality within their own mindstreams is negligible, assuming there are any at all.

In his book *From Reductionism to Creativity rDzogchen and the New Sciences of Mind* Guenther referred to Bohm's work in the context of the Dzogchen worldview. In the following we can identify Guenther's notion of the "inner dynamics of Being" with Bohm's holomovement:

> It is this ... inner dynamics of Being that eventually pushes it, figuratively speaking, over the instability threshold into its actuality so the virtually operative actuality in Being now assumes a true actuality that may be called Being's "eigenstate". This process is termed *gzhi-snang* which, borrowing a term coined by David Bohm, I render as "holomovement," which in the rDzogs-chen context means that Being in its totality (*gzhi*) lights up (*snang*), and in this lighting-up makes its presence felt. The implication is that, as paradoxical as it may sound, Being is nowhere else than in the what-is, ... this means that we are the whole and yet only part of it.[47]

In this description the term 'eigenstate' is a technical term within quantum theory applying to the measurement process. Before a measurement is carried out, for example, an electron's momentum and position are both uncertain (Heisenberg's Uncertainty Principle), and it is only possible to make a precise measurement of one of these characteristics. This means that it is only possible to know the precise value of the eigenstate of position, or the eigenstate of momentum. Both of these eigenstates may take, when measured, a range of discrete values called eigenvalues, but only one of the eigenstates can be known at one time. The term "eigenstate" is derived from the German word "eigen" meaning "inherent" or "characteristic". When a measurement is performed on position, for example, this eigenstate will manifest as being in just one of its possible eigenvalues. Thus we can say that eigenstates are possible modes of manifestation from the multiple uncertainty of quantum potentiality.

In the above description by Guenther, Absolute Being is conceived of as originally 'existing' in a virtual state of non-manifestation. This state of non-manifestation contains a multitude of virtual possible manifestations, just as a state of quantum potentiality also contains a set of potential eigenstates. Thus, these virtual states of possible manifestations within Absolute Being can be compared to potential quantum eigenstates of manifestation within the quantum field of Unmanifest Being. In addition to this characterisation, Guenther also indicates that Absolute Being contains an internal tendency or necessity towards the instability of manifestation, and therefore a consequent manifestation into actuality of one of the virtual potentialities into a manifested 'eigenstate' of actuality takes place. Physicists Stephen Hawking and Leonard Mlodinow, in their book *The Grand Design: New Answers to the Ultimate Questions of Life*, implicitly indicate the operation of this internal quantum unfoldment tendency when they point out that:

> We are the product of quantum fluctuations in the very early universe.[48]

The 'big bang' was the first moment of a cascade of a tumultuous sequence of 'unfoldments' within the preexisting quantum field of potentiality, which eventually gave rise to the current universe. This manifestation of an 'eigenstate' internal to Being, is a manifestation within the wholeness of Being, and is clearly described, paradoxically, as both containing the essential nature of the wholeness of Being at the same time as being a movement and manifestation within Being as a whole.

This situation is compared by Guenther to Bohm's notion of the 'holomovement' of implicate and explicate orders. This corresponds to Guenther's metaphor of unmanifested (implicate) and manifested (explicate) 'eigenstates' within the wholeness of the overall holomovement of the process of reality. Bohm, in *Wholeness and the Implicate Order*, uses his notion of the "holomovement" as his central metaphor for the way in which the Absolute Whole of the process of reality consists of an 'implicate order' lying behind (metaphorically) the 'explicate orders' of manifestation. This perspective can be compared with Guenther's Dzogchen account of Absolute Being in its modes of 'virtual actuality' and subsequent apparent full actuality of manifestation, thereby being in a movement of unfolding manifestation. Thus we have the Absolute Wholeness, or 'Absolute Being' (the word 'Being' here does not denote a static 'Being'), in motion, denoted by the 'holomovement'.

The following section is taken from the introduction to Guenther's book *The Teachings of Padmasambhava*. It describes a direct experiential activation of a holistic dimension of being, a dimension that can be identified with Bohm's more scientific-metaphysical approach:

> [Padmasambhava] formulated a holistic vision that transcends the traditional division between the physical and the mental, the emotional-instinctual and the spiritual, in two related disciplines that remain experiential through and through. One he referred to by the name.of ... *spyi-ti yoga* . Let us begin with the *spyi-ti* experience. This is how Padmasambhava lets the teacher "Utterly free from the limitations set by the categories of rational thought" explicate to the "Little Man (who is the whole's) self-manifesting Light" ... No less revealing is the following statement ... that links this *spyi-ti* experience and teaching with the (whole's) intensity/energy of which Padmasambhava has repeatedly spoken:

> *spy*i means the totality of that which exists without exception, *ti* means (the whole's) intensity/energy becoming a vortex; *spyi* means (the experiencer's) spiritual-excitation-excitability, ego-logical mentation, overall psychic background, and (its) divisive concepts, *ti* means (the whole's) intensity/energy becoming a vortex, *yo* means the totality (of all that is) being indivisible and unpremeditated, *ga* means (the whole's) intensity/energy from whose vortex the giving birth to thoughts/meanings arises. Therefore (this experience) outshines all other spiritual pursuits by its brilliance; Therefore the *spyi-ti* experience is completeness with respect to (the experiencer's) psychic-spiritual constitutedness.[49]

The 'teacher utterly free from limitations' in this description refers to the influences of subtle implicate orders which leave traces of wholeness within the 'fragmentation', as Bohm would say, within the explicate order. The explicate order is the abode of "Little Man" who has been projected into the movement of the explicate order by the projective energies - the "self-manifesting Light" - within the "vortex", which corresponds to Bohm's holomovement. The entire energetic-vortex-holomovement gives rise to dualistic thoughts and meanings which create division within the explicate order. However, it is through the spiritual practice which manifests the "*spyi-ti* experience" that completeness can be restored.

As we have seen, in his book *Wholeness and the Implicate Order* Bohm calls the totality of all that is by the term the 'holomovement', which encompasses what he calls the 'implicate order', and the 'explicate order' which unfolds from the hidden implicate order(s) into the experience of everyday life. The holomovement is the process which he conceives of as encompassing both orders. As the holomovement progresses the implicate order explicates, or 'unfolds', the explicate order of experience into the everyday world, and subsequently events within the explicate order can also enfold in some way back into the implicate order. In the following passage Bohm describes the basic overview of his vision:

> Our basic proposal was then *what is* is the holomovement, and everything is to be explained in terms of forms derived from this holomovement. Although the full set of laws governing its totality is unknown ... these laws are assumed to be such that from them can be abstracted relatively autonomous or independent sub-totalities of movement ... having a certain recurrence and stability of their basic patterns of order and measure. ... we have contrasted implicate and explicate orders, treating them as separate and distinct but ... the explicate order can be regarded as a particular or more distinguished case of a more general set of implicate orders from which the latter can be derived. What distinguishes the explicate order is that what is thus derived is a set of recurrent and relatively stable elements that are *outside* of each other.[50]

In other words, the holomovement consists of the overall implicate order, and, as the holomovement progresses, perhaps according to internal laws that we cannot know, the operation of these internal laws causes an unfoldment of a set of implicate levels within the implicate order, and this unfoldment generates, at more manifest levels, explicate levels of everyday type experience. This explicate level of the experiential world is characterized by relatively stable elements which appear to be completely separate from each other, even though at the deeper implicate levels all aspects of the holomovement are fundamentally interconnected. It is interesting to compare this Bohmian vision to the following view from the eighth century Tibetan pre-Buddhist Bon-Dzogchen text which is presented in *Unbounded Wholeness*:

> The inability to describe unbounded wholeness in any one way ... dramatically testifies to wholeness's decentered

> multiplicity and thus to its incommensurability with conceptual limitation. Unbounded wholeness can, and must, be called both definite and indefinite; this is the principle of wholeness. ... Definite and indefinite ... conditioned and unconditioned are all ... "facets of wholeness," not mutually exclusive ... indefiniteness ... as an evocation of multifaceted reality, continuously brimming with shapes and colors even as it remains an unmitigated whole.[51]

Again we find a dramatic resonance between Buddhist/Dzogchen worldviews and Bohm's perspective.

In their book *The Undivided Universe* Bohm and his associate Basil Hiley give the following account, in which they address the issue of consciousness:

> ...the intuition that consciousness and quantum theory are in some sense related seems to be a good one ... Our proposal in this regard is that the basic relationship of quantum theory and consciousness is that they have the implicate order in common. The essential features of the implicate order are ... that the whole universe is in some way enfolded in everything and that each thing is enfolded in the whole. However, under typical conditions of ordinary experience, there is a great deal of *relative* independence of things, so they may be abstracted as separately existent, outside of each other, and only externally related. However, more fundamentally the enfoldment relationship is active and essential to what each thing is, so that it is internally related to the whole and therefore to everything else. Nevertheless, the explicate order, which dominates ordinary 'common sense' experience as well as classical physics, appears to stand by itself. But ... it cannot be properly understood apart from its ground in the primary reality of the implicate order, i.e. the holomovement. All things found in the explicate order emerge from the holomovement and ultimately fall back into it. ... It takes only a little reflection to see that a similar sort of description will apply even more directly and obviously to consciousness...[52]

Thus we see that there is a very significant correspondence between Guenther's description of how in the Buddhist 'Great Completion' metaphysical worldview Absolute Being is conceived of as a repository of enfolded virtually operating 'actualities' which, because of inner dynamics, become unstable as virtually hidden and therefore flow out into non-virtual manifestation, and Bohm and Hiley's account of how the implicate order, operating within the dynamics of the holomovement, produces an explicate descent to manifest as the explicate order of the everyday world. In both these cases the parts are said to in some sense contain the whole at the same time as also appearing to be an independent part of the whole.

The Buddhist *Yogacara* school, a tradition which emphasizes the investigation of cognition, perception, and consciousness by means of meditative and yogic practices, considers that all phenomena derive from a deep level of Mind-Energy. And in this tradition we discover similar formulations to those previously cited:

> The metaphysical doctrine of the ancient Yoga tradition puts forth an understanding of the creation, progression and eventual destruction of the Universe that seems surprisingly modern, to the extent in which it agrees with leading edge advances in science, quantum mechanics and cosmology. Those who go deeply into this subject, will find this doctrine rooted in a profound understanding of a great mystery called PARAMĀRTHA, which in Indian philosophy means 'the Absolute', devoid it is said of all attributes, and essentially distinct from manifested finite Being. The manifestation (*pravrtti*) and re-absorption of the Universe, or domain of finite Being, and how the latter relates to the transcendent infinitude of the Absolute has been central to Yogacara inquiry from the beginning of its history ...
> It is believed that by means of proliferation (*prapanca*, differentiation), the innate essence of being in three forms (*trisvabhava*) manifests or is transformed, as it were, into active mentation in the act of Creation. This is then explained as the coming into being of *alayavijnana*, universal or cosmic consciousness, which is a concept that has also been held in the Western philosophy by many great thinkers, from Plato, Plotinus and others... [53]

Thus we can discern the possibility of a rich, mutually enhancing and invigorating intellectual, and potentially experiential, mutual elucidation and illumination, through the exploration of these deep metaphysical accounts of the structure and process of reality.

As we have seen, according to Bohm the holomovement can be characterised, or perhaps non-characterised:

> Thus, in its totality, the holomovement, is not limited in any specifiable way at all. It is not required to conform to any particular measure. Thus, the *holomovement is undefinable and immeasurable.* To give primary significance to the undefinable and immeasurable holomovement implies that it has no meaning to talk of a *fundamental* theory, on which *all* of physics could find a permanent basis, or to which *all* the phenomena of physics could similarly be reduced.[54]

Here Bohm appears to indicate that as soon as any measurement or determination might occur, then the holomovement would cease to be the holomovement precisely due to an enforced limitation. Paradoxically, this does not mean, however, that nothing can be said about the holomovement! Bohm obviously wrote and said quite a lot on its behalf.

I have been surprised by some interactions I have had on discussion forums where some people have asserted the complete and utter transcendent indefinable nature of the holomovement, apparently not realizing that if this were really the case then Bohm himself would have been unable to say anything about it. What Bohm means with the 'undefinable' claim is that from the point of view of the explicate order the inward workings and interconnections of the implicate order within the holomovement are not fully definable using concepts as they are employed in a manner appropriate to the explicate order, especially by the techniques of the scientific community. But clearly Bohm did take steps to 'define' what he meant by the term. A few paragraphs back in his book from the above quote Bohm actually offered a kind of definition in the form of a metaphorical description:

> It will be useful ... to consider some further examples of enfolded or *implicate* order. Thus, in a television broadcast, the visual image is translated into a time order, which is 'carried' by the radio wave. Points that are near each other in the visual image are not necessarily 'near' in the order of the radio signal. Thus, the radio wave carries the visual image in an implicate order. The function of the receiver is then to explicate this order, i.e. to 'unfold' it in the form of a new visual image.[55]

Thus we see that, according to Bohm, the manifested everyday world of matter and consciousness is encoded, or 'enfolded', at a deeper level of the process of reality, within some kind of 'carrier' medium, this is, in Bohm's terminology an 'implicate order'. And an important aspect of Bohm's example of the radio wave is that the 'order', or the arrangement, of the information as it is encoded is entirely different to the way in which it appears when it is 'unfolded'. The process of the enfoldment into the carrier medium, and unfoldment from the implicate order of the carrier medium into the 'explicate order' of the everyday world of experience makes up the overall 'holomovement'. So, we can conclude that the holomovement encompasses the coordinated movement of both the implicate and the derived explicate orders.

In the early pages of *Wholeness and the Implicate Order* Bohm wrote that:
> ...there is a universal flux that cannot be defined explicitly but which can be known only implicitly, as indicated by the explicitly definable forms and shapes, some stable and some unstable, that can be abstracted from the universal flux. In this flow, mind and matter are not separate substances. Rather, they are different aspects of one whole and unbroken movement. In this way, we are able to look on all aspects of existence as not divided from each other, and thus we can bring to an end the fragmentation implicit in the current attitude toward the atomic point of view, which leads us to divide everything from everything in a thoroughgoing way. Nevertheless, we can comprehend that aspect of atomism which still provides a correct and valid form of insight, i.e. that in spite of the undivided wholeness in flowing movement, the various patterns that can be abstracted from it have a certain relative autonomy and stability, which is indeed provided for by the universal law of the flowing movement.[56]

Thus, we see how Bohm conceives of mainstream physics as describing limited sub-domains, the most dominant being the atomic view, which function with relative degrees of independence, but which also function within the more encompassing context of the much greater realm of the interdependent unity of the holomovement. And it is important to note that, when taken within the context of the limited domains of appropriateness, the abstracted patterns are valid.

According to Bohm, the explicate realms of matter and consciousness derive from a deeper more unified source:

> If matter and consciousness could in this way be understood together, in terms of the same general notion of order, the way would be opened to comprehending their relationship on the basis of some common ground. Thus we could come to the germ of a new notion of unbroken wholeness, in which consciousness is no longer to be fundamentally separated from matter.[57]

For Bohm, the implicate order underlies the duality of matter and individuated consciousness. It is the ground from which they both arise as manifested coordinated 'explicate' aspects of the deeper implicate order that is usually beyond the direct reach of dualistic awareness. The realms of individuated mind and the appearance of matter mutually unfold from the implicate order into what Bohm termed the 'explicate' order. Mind and matter, subject and object, are coordinated manifestations from the deeper enfolded order of the holistic non-local 'implicate order.'

The 'non-local' nature of the implicate order, and the explicate orders which unfold from it, is one of the implicate order's essential characteristics. And it is the issue of quantum non-locality which is a significant issue for Bohm's reworking of quantum theory. The term 'non-locality' refers to the fact that, in pre-Bohmian quantum interpretations, there is a possible instantaneous 'communication' between 'entangled' particles within a quantum field, however vastly far apart their positions may be. However, in non-Bohmian views this 'spooky', as Einstein called it, possible interconnection does not manifest in the 'classical' everyday world. In Bohm's undivided universe quantum metaphysics, on the other hand, such 'spooky' interconnections appear to be part of the everyday world. In fact they are an essential part of Bohm's worldview, Bohm and Hiley call their vision one of an 'Undivided Universe' precisely because such universal instantaneous interconnections between any two points exist to some degree in the everyday world.

Having briefly surveyed some aspects of Bohm's later 'mystical' phase in relation to some resonant passages from Buddhist metaphysics, it is worthwhile tracing Bohm's intellectual journey from his quantum-materialist 1952 article towards his later viewpoint. To begin this task, it is illuminating to explore and clarify how the work of Bohm, and his later associate Basil Hiley, can be viewed through contrasts and elucidations with other quantum perspectives. This will be an implicit theme throughout this book. By understanding the viewpoints that Bohm was motivated to replace, or clarify, we can comprehend how the fully developed Bohmian perspective can actually enfold other perspectives under its holistic unity. We shall see that other quantum perspectives may become partial viewpoints within the overall Bohmian holomovement. This seems completely in line with Bohm's motivation that various fragmentary viewpoints on the process of reality should be amenable to a higher order unity of harmonization.

As a point of departure we can ask the question: what prompted Bohm in 1952 to drastically rework the quantum Copenhagen worldview generally accepted at the time? Bohm and some other physicists, notably Einstein, were worried and disturbed by the details of the transition from quantum state to everyday 'classical' 'particle' state proposed by the Copenhagen interpretation of quantum mechanics, established primarily by Neils Bohr and Werner Heisenberg around 1925-1927. This viewpoint claims that a quantum 'particle' does not actually become a real, 'classical' - everyday - 'particle' until its quantum state is measured. Prior to being measured, on this view, a particle is not really a particle at all, it is a large array of possible particles, each with differing possible positions and momentums. Such a quantum particle does not exist in one definite state or another, but in all of its possible states at the same time. It is only when we measure, or 'observe', its state that a quantum particle is forced to adopt one of its possible states, and that is the state that is observed. All the other possible states that semi-existed before the measurement disappear, and the adopted state becomes the 'real' 'classical' state of the particle. The prior state of existing in all possible states at the same time is called a 'superposition', and the mathematical description of the superposition is called a 'wavefunction'. When a quantum object is measured, or 'observed', the wavefunction 'collapses' and the object is forced into one of the states contained in its wavefunction.

Some physicists and philosophers seem content to leave the details of the mechanism underlying the transition from quantum potentiality to a measured reality as being unknowable:

> It is a fundamental quantum doctrine that a measurement does not, in general, reveal a preexisting value of the measured

property. On the contrary, the outcome of a measurement is brought into being by the act of measurement itself, a joint manifestation of the state of the probed system and the probing apparatus. Precisely how the particular result of an individual measurement is brought into being - Heisenberg's transition from the possible to the actual - is inherently unknowable. Only the statistical distribution of many such encounters is a proper matter for scientific inquiry.[58]

Others have suggested that in some sense the human beings performing the measurement have a primary role in creating the nature of the transition. Thus, the German physicist Pascual Jordan declared that:

> Observations not only disturb what has to be measured, they produce it... We compel [the electron] to assume a definite position... We ourselves produce the results of measurements.[59]

Thus the Copenhagen view does come in different flavors. But in general this approach to the situation can create the impression that the consciousnesses of observers can have direct effects on quantum possibility states in order to 'create' a fully 'real' material reality. Thus, it almost seemed, to some disgruntled physicists, that reality was being magically conjured up from quantum unreality. Even worse, for some concerned participants in the quantum debate, was the implication that consciousness had a creative role at the quantum level. Such ideas have led towards some very implausible suggestions by recent 'New Age' quantum prophets. And, even the important and highly influential physicist John Wheeler suggested that this magical solidification of quantum unreality by perception can happen backwards in time.

Wheeler came to this conclusion after pondering a cosmic thought experiment. In this thought experiment, simply described, a beam of light which originates in a distant quasar is bent around the two sides of an intervening galaxy on its way to earth, When it reaches Earth the split beam is used in a split-beam quantum experiment. In the famous 'double slit experiment' it can be shown that the way in which a quantum 'wave-particle' manifests, either as a wave or particle, depends upon whether or not the experimenters have knowledge of the path taken between the source and the measuring apparatus by the quantum 'wave-particle'. Knowledge of which slit the 'wave-particle' passes through produces a particle, but lack of path-knowledge enables the quantum wave to remain a wave and therefore travel through both slits, producing an interference pattern of light and dark bands. Wheeler's thought experiment involves a split-beam experiment on a cosmic scale. If particles were created in this experiment with cosmic beams then Wheeler concluded that the experiment had forced the whole cosmic experimental setup to have been using particles from the moment the light left the distant quasar source centuries backwards in time. Thus, in Wheeler's evaluation, the choice of whether waves or particles were travelling across the universe had been delayed until the experiment on Earth had been performed, and therefore the nature of reality seems to have been determined backwards in time, determined by conscious observation on Earth. However, there is a perhaps more plausible, but still remarkable, explanation for this scenario. In this alternative view it is only the perceptions which change, particles are not actually 'created' backwards in time.

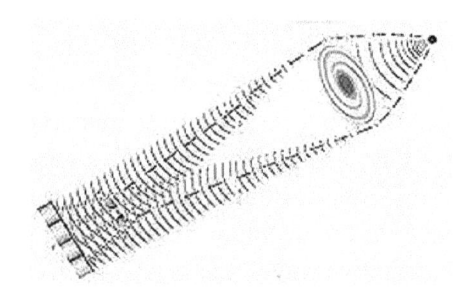

Wheeler's cosmic 'delayed choice' experiment.

A less cosmic version of this experiment has now been carried out by Anton Zeilinger and his team at the University of Vienna,[60] and Wheeler's ideas seem to have been verified. Some physicists have therefore concluded that it is likely that the process of the universe constitutes a self-manifesting loop involving observation:

> Physics gives rise to observer-participancy; observer-participancy gives rise to information; information gives rise to physics.[61]

The physicist Freeman J. Dyson describes this view:

> Wheeler would make all physical law dependent on the participation of observers. He has us creating physical laws by our existence.[62]

Thus, we see that, on the basis of this cosmic thought experiment, Wheeler sometimes indulged in what some consider to be flights of quantum 'mystical' speculation.

The later John Wheeler was a great master of asserting the seemingly incredible, only to retreat to safer shores of credibility when perhaps the academic waves got a bit turbulent. Towards the end of his life Wheeler seemed fond of masterful quantum-mystical intellectual wheeler-dealing. For example, in his book *Geons, Black Holes & Quantum Foam: A Life in Physics*, Wheeler presented his 'Wheeler U' diagram, and explained how he envisioned sentient observation could 'collapse' the wave-function of reality backwards in time as follows:

> My diagram of a big U (for universe) attempts to illustrate this idea. The upper right end of the U represents the Big Bang, when it all started. Moving along down the thin right and up along the thick left leg of the U symbolically traces the evolution of the universe, from small to large - time for life and mind to develop. At the upper left of the U sits, finally, the eye of the observer. By looking back, by observing what happened in the earliest days of the universe, we give reality to those days.[63]

The crucial issue here, of course, is exactly what Wheeler meant by the claim "we give reality to those days". Does he really mean the collective consciousness of the evolved sentient beings 'create' reality, backwards in time, from quantum unreality? However, in the very next paragraph Wheeler retreats from the apparent extremity of his flight of quantum mystical speculation:

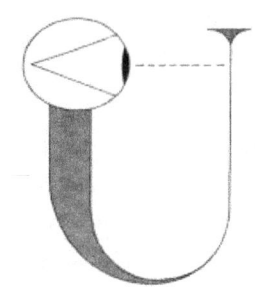

The Wheeler U

> The eye could as well be a piece of mica. It need not be part of an intelligent being. The point is that the universe is a grand synthesis, putting itself together all the time as a whole. Its history is not a history as we usually conceive history. It

is not one thing happening after another after another. It is a totality in which what happens "now" gives reality to what happened "then," perhaps even determines what happens then.[64]

These two perspectives advanced by Wheeler are not equivalent, the second assertion clearly nullifies an important feature of the first. The first depiction is a quantum psycho-metaphysical account in which consciousness is a creative force acting backwards in time in order to select in some way which quantum potentialities become manifest at previous times in the evolution of the universe. This is a view that has been later reiterated, and also subsequently denied, in their book *The Grand Design* by Stephen Hawking & Leonard Mlodinow, in a chapter titled 'Choosing Our Universe'.

In Wheeler's case, his second observation seems to quickly cancel out any suggestion of quantum consciousness mystically acting backwards in time. In this sanitised second version some unspecified influence emanating from chunks of stone performs the job of lifting reality out of quantum indeterminacy. This version is still a bit mystical, but not quite as fully New-Agey, so to speak, as Wheeler's more courageous quantum creative consciousness perspective! Wheeler explained his caution by gently indicating that he was under the influence of a kind of anti-quantum-consciousness scientific-academic thought police at the time:

> In these later years, I have dared to think about and write about and ask about the physical world in terms that some of my colleagues consider outside the scope of science - science as it is now accepted, defined, and practised. Is the universe a self-excited circuit, made real by observation?[65]

So it looks as if Wheeler really might have believed his fully New-Age first version, but thought it better not to antagonise his less mystically inclined colleagues! Although it should also be born in mind that in 1979, Wheeler spoke to the American Association for the Advancement of Science (AAAS), requesting that parapsychology should be expelled, as he considered it to be a pseudoscience.

Perhaps it is difficult to know exactly what to make of such contrasting and seemingly contradictory attitudes, but they should probably be viewed against the prevailing scientific collective consciousness of the time. The physicists Bruce Rosenblum and Fred Kuttner, in their book

Quantum Enigma: Physics encounters consciousness, point out in regard to the attitude of mainstream physics since the 1950's that:

> In physics departments a conforming mind-set increasingly meant that an untenured faculty member might endanger a career by serious interest in the fundamentals of quantum physics. Even today it is best to explore the meaning of quantum mechanics while also working a 'day job' on a mainstream physics topic.[66]

In other words, there was an overwhelming anti-quantum-metaphysical-speculation and definitely anti-quantum-mysticism prejudice within the academic world which determined what physicists felt might be acceptable research and speculation within their discipline.

Rosenblum and Kuttner remain very cautious about just how significant the encounter between physics and consciousness is, apparently for fear of being branded as 'New Age' fantasy merchants. Because of this fear of academic criticism, Rosenblum and Kuttner seem to have been forced to hedge their bets. At one point in their book we read that:

> ...physics' encounter with consciousness, demonstrated for the small, applies to everything. And that 'everything' can include the entire Universe.[67]

An assertion which appears to be quite far-reaching. But then they seem to get worried about going too far into dangerous speculative territory:

> ... we argue that it is a social responsibility of the physics community to openly present physics' mysterious encounter with consciousness, the quantum enigma. Only by so doing can we challenge the purveyors of pseudoscience who use the mysteries of quantum mechanics to promote their quantum nonsense.[68]

The significant issue which clearly needs addressing, of course, is exactly when and why does the exploration of the connection between consciousness and physics become 'pseudoscience'. Surely one would have hoped the answer to this would be: 1) when there is absolutely no evidence for such a connection; or 2) there are no serious convincing arguments that can be made for claims made in this arena.

Bohm, of course, was working at a time when such restrictive views were operative. But he was certainly a physicist who took his "social responsibility ... to openly present physics' mysterious encounter with consciousness" seriously, and he did wonderful work in this field of intellectual exploration. Wheeler's autobiography, in which he profoundly hedged his quantum mystical bets, was published in 1998, but the prevalence of a anti-quantum-mystical subtle-materialist worldview is still operative to a large extent within scientific circles, and it is perhaps remarkable that Bohm was able to swim against this current at the time he was writing his later work.

Seemingly 'mystical' speculations on the part of admired physicists such as Wheeler have had an impact on the growth industry of New-Age fantasy indulgence. A recent New-Age-quantum-intoxicated Wheeler devotee, inspired by some of Wheeler's claims, has been heaped with accolades from the New-Age community by making ludicrous claims which take Wheeler's visions towards a quantum theatre of the absurd. In his book *The Quantum Revelation: A Radical Synthesis of Science and Spirituality*, the New-Age prophet Paul Levy informs his readers that:

> When a physicist observes an elementary particle – which from the quantum point of view "causes" the particle to exist – it is as if the physicist is "dreaming up" the quantum entity in the same way that a dreamer dreams up their own dreamscape. At the same time (if we let our creative imagination run wild) it is as if the elementary particle is reciprocally dreaming, as it dreams up the physicist to observe it and hence, bestow upon it existence. The physicist and the subatomic particle are ... mutually dreaming each other up ...[69]

This is certainly "imagination run wild"! Run, in fact, into a intellectual wilderness of self-indulgence. A less dreamy and more rational account of the vital importance of the nature of electrons for our existence is provided by physicist Sabine Hossenfelder:

> Fermions are extreme individuals. No matter how hard you try, you will not get two of them to do the same thing in the same place - there must always be a difference between them. ... This is why electrons, which are fermions, sit on separate shells around the atomic nuclei. If they were bosons, ... they would instead sit together on the same shell, leaving the universe without chemistry - and without chemists, as our own existence rests on the little fermions refusal to share space.[70]

And this is not the only extraordinary fact about electrons which accounts for the existence of physicists, chemists, people, animals and the entire universe. The physicist and science writer Michio Kaku tells us that:

> The reason why molecules are stable and the universe does not disintegrate is that electrons can be in many places at the same time. ... electrons can exist in parallel states hovering between existence and non-existence.[71]

The ability of electrons to be in two places at the same time, whilst still maintaining an identity as a single entity, is a feature of quantum behavior that is absolutely crucial for the functioning of reality. It is this feature of electron functioning that allows them to hold molecules together. So the remarkable abilities of tiny, barely existent, "hovering between existence and non-existence" electrons are essential to the functioning of the entire universe.

The fact that the functioning of reality depends crucially on wisps of electronic almost-nothingness that refuse to be in the same place as another wisp of electronic almost-nothingness, and can be in many places at the same time, might indeed move one towards universal dream-interpretation of reality, but it does not imply that electrons go to bed at night. As Hossenfelder says: "Helium atoms don't get hungry and are just as well tempered on Monday as on Friday."[72] She might have added, for good measure, the fact that electrons do not dream, and certainly do not dream up physicists who then dream up the electrons who then dream up physicists etc. etc.

This detour into quantum fantasy indicates how some presentations of the nature of quantum functioning can lead to absurd New-Age claims. Levy also uses the work of Bohm in his New-Age fantasy rampage, but he cannot find quotes from Bohm with the same New-Age wow-value as he can with Wheeler:

> To quote David Bohm, the "inseparable quantum interconnectedness of the whole universe is the fundamental reality." An expression of this undivided wholeness - the fundamental reality - is that consciousness is no longer separated from matter but somehow essential to it. Consciousness is not one thing and matter another thing that it interacts with; on the quantum level conciousness and matter are indistinguishable. As Bohm points

out, if we don't see this "it's because we are blinding ourselves to it." [73]

Such a view can certainly derived from the insights of Bohm, as we have seen, according to Bohm the explicate realms of matter and consciousness derive from a deeper more unified source.

The quantum physicist Amit Goswami, a retired professor of theoretical physics at the University of Oregon and a self-styled pioneer of his new paradigm of science called "science within consciousness", describes the phenomenon of the 'collapse of the wavefunction', unashamedly as the "observer effect":

> In quantum physics, objects are depicted as possibilities (a possibility wave); yet when an observer observes, the possibilities collapse into an actuality (the wave collapses into particle, for example). This is the observer effect.[74]

And Fritjov Capra, in his famous book *The Tao of Physics,* published in 1975, had no doubts concerning the role of consciousness at the quantum level:

> At the atomic level 'objects' can only be understood in terms of the interaction between the processes of preparation and measurement. The end of this chain of processes lies always in the consciousness of the human observer.[75]

And the significant physicist Roger Penrose, in his 1994 book *Shadows of the Mind* also seems to suggest something like this, when he states that the quantum measurement scenario implies that:

> At the large end of things, the place where 'the buck stops' is provided by our *conscious perceptions*.[76]

But, in his earlier 1989 book *The Emperor's New Mind* Penrose seemed unconvinced by this view:

> Is the presence of a conscious being necessary for a 'measurement' *actually* to take place? I think that only a small minority of quantum physicists would affirm such a view.[77]

The fact that Penrose was motivated to consider this possibility however shows that when the evidence is looked at it is possible to come to such conclusions. As physicists Rosenblum and Kuttner say:

> Consciousness and the quantum enigma are not just two mysteries; they are *the* two mysteries; ... Quantum mechanics

seems to connect the two.[78]

And the significant physicist Eugene Wigner, co-recipient of the 1963 Nobel Prize for Physics was also impressed by the apparent role of consciousness (although he later changed his mind about this when he came across the 'decoherence' viewpoint):

> When the province of physical theory was extended to encompass microscopic phenomena, through the creation of quantum mechanics, the concept of consciousness came to the fore again; it was not possible to formulate the laws of quantum mechanics without reference to consciousness.[79]

These are a selection of remarks by significant physicists who have suggested at some point in time that Bohr's Copenhagen's viewpoint requires some active role for consciousness in bringing quantum semi-real wavy potentiality into the realm of the hardcore everyday world we all appear to live in.

The essential issue here is the fact that from the point of view of Bohr's Copenhagen interpretation, although Bohr himself did not claim precisely this, it seems that, to put the issue in crude terms, an essentially large set of 'unreal' (from the point of view of the 'classical' everyday world) possible particles, hovering between existence and non-existence, transform when measured or 'observed' to become a really real particle; one of the possible particles becomes the 'real' particle, and all the other possible particles vanish. In this scenario there seems to be an essential element of subjectivity involved, it seems as if the experimenter's decisions during an experiment can determine what becomes 'real' and what does not. Some physicists went so far as to suggest that human and animal minds might be responsible for the 'collapse' of the wavefunction.

However, there were also physicists who were deeply sceptical of such claims. For example the important physicist John Bell, pondering upon Wheeler's claim that consciousness collapses the universal wavefunction backwards in time, made the wonderful sceptical remark:

> Was the wave function of the world waiting to jump for thousands of years until a single-celled living creature appeared? Or did it have to wait a little longer, for some better qualified system ... with a PhD?[80]

But such sharp sarcastic comments do not cut through the views of some who do believe that consciousness does have a significant role at the quantum level. For example, Amit Goswami has asserted, like Wheeler before him, that consciousness can act backwards in time to solidify actuality from out of the quantum realm of potentiality, and this means that consciousness can actually affect the history of the universe backwards in time:

> This is what quantum physics demands. In fact, in quantum physics this is called "delayed choice." And I have added to this concept the concept of "self-reference." Actually the concept of delayed choice is very old. It is due to a very famous physicist named John Wheeler, but Wheeler did not see the entire thing correctly, in my opinion. He left out self-reference. The question always arises, "The universe is supposed to have existed for fifteen billion years, so if it takes consciousness to convert possibility into actuality, then how could the universe be around for so long?" Because there was no consciousness, no sentient being, biological being, carbon based being, in that primordial fireball which is supposed to have created the universe, the big bang. But this *other* way of looking at things says that the universe remained in possibility until there was self-referential quantum measurement—so that is the new concept. An observer's *looking* is essential in order to manifest possibility into actuality, and so only when the observer looks, only then does the entire thing become manifest—including time. So all of past time, in that respect, becomes manifest right at that moment when the first sentient being *looks*.[81]

So here Goswami attempts to give an answer to Bell's sarcastic remark about the universe having to wait around for a sentient being with a PhD to make an appearance and start looking around. According to Goswami time itself, as well as the universe, comes into manifested being when there is an internal self-referential act of self-perception within a timeless quantum possibility-field.

As previously intimated, Stephen Hawking and Leonard Mlodinow's 'theory of everything' proposed in their book *The Grand Design*, suggests a variation of Wheeler's viewpoint. According to Hawking and Mlodinow (H&M):

> Quantum physics tells us that no matter how thorough our observation of the present, the (unobserved) past, like the

> future, is indefinite and exists only as a spectrum of possibilities. The universe, according to quantum physics, has no single past, or history. The fact that the past takes no definite form means that observations you make on a system in the present affect its past. ... the universe doesn't have just a single history, but every possible history, each with its own probability; and our observations of its current state affect its past and determine the different histories of the universe...[82]

In the H&M quantum metaphysical scenario a spontaneous universal creative act projects all possible futures into a universal possibility or potentiality space. At the point of creation everything that possibly can happen becomes potential, so at the point of creation all possible future histories of the universe come into being as potentialities:

> In this view, the universe appeared spontaneously, starting off in every possible way. Most of these correspond to other universes .... Some people make a great mystery of this idea, sometimes called the multiverse concept, but these are just different expressions of the Feynman sum over histories.[83]

And, a crucially significant feature of the H&M presentation is the fact that the "observers are part of the system"[84] and, furthermore, "we create history by our observations, rather than history creating us".[85] So the observers, or what John Wheeler called "observer-participants", are able to weed out possible universes by the impact of their 'observations', and thereby select those which remain in the possibility mix, even backwards in time. Thus, one of the central chapters in *The Grand Design* is titled 'Choosing Our Universe':

> The idea that the universe does not have a unique observer-independent history might seem to conflict with certain facts that we know. There might be one history in which the moon is made of Roquefort cheese. But we have observed that the moon is not made of cheese, ... Hence histories in which the moon is not made of cheese do not contribute to the current state of our universe, though they might contribute to others. This might sound like science fiction but it isn't.[86]

And these quotes really do come from H&M's book.

A lot of people, who think they know what Stephen Hawking's views are, might be quite surprised if they actually bothered to read this book. I was once in a debate on a Richard Dawkins fan forum with a dyed in the wool materialist Dawkins-worshipper, and I used the H&M quotes just cited to illustrate my point about the primacy of consciousness in the process of reality. The guy I was debating refused to believe me, saying he knew all about Stephen Hawking, who was a materialist who declared God to be unnecessary, and therefore I must be lying. I replied I was holding the book in my hands and he could probably locate the quotes on the internet, but he still absolutely disbelieved me!

The prevalence of views which embrace a creative role for consciousness at the quantum level has led to a reaction on the part of physicists and philosophers who see, through their mode of view, no significant evidence or reason to come to such a conclusion. Thus the theoretical physicist Jean Bricmont, who is an advocate of the early 1952 version of Bohm's 'ontological' quantum worldview, writes of what he views as misguided quantum New-Age claims:

> The mysterious character of quantum mechanics has led to numerous abuses, misinterpretations, speculations and extrapolations, perhaps more than any other scientific theory. It would take an encyclopedia to cover all of them … We have seen the two "mysteries" of quantum mechanics concern the role of the observer and actions at a distance. A third alleged "novelty" supposedly introduced by quantum mechanics is the death of determinism. Almost all the abuses or invalid extrapolations of quantum mechanics rely on one or more of these ideas.[87]

And Bricmont is also critical of physicists who he considers to be insufficiently circumspect about indulging in quantum-mystical type views. Thus he admonishes Wheeler for his excessive reverence for the inscrutability of Bohr's quantum lack of clarity:

> A more extreme example of Bohr worship has been exemplified by John Wheeler, who compared Bohr's wisdom to nothing less than "that of Confucius and Buddha, Jesus and Pericles, Erasmus and Lincoln".[88]

And Bricmont tells us that when we are considering:

> …non-scientific extrapolations and exploitations of quantum mechanics, one should not minimise the role played by those scientists who have emphasized the "disappearance of objectivity" or even of "reality" supposedly implied by the quantum discoveries.[89]

Also, Bricmont writes that the majority of physicists:

> ...do not accept the notion of a consciousness totally independent of the brain. Besides, even if one were to accept the idea that mind, independent of the body, intervenes in the collapse process, there is nothing whatsoever in quantum mechanics to suggest that our conscious choices affect the collapse of the wave function one way or another. So there is no reason to take seriously this sort of link between consciousness and quantum mechanics.[90]

And here we find an important feature of Bricmont's anti-quantum consciousness position; it seems that the most unacceptable proposal from his point of view is the notion that "our conscious choices affect the collapse of the wave function one way or another". So it seems that, according to Bricmont, even if some kind of connection between consciousness and the quantum realm were to be shown, it would not be such that people would be wandering around sending beams of consciousness out to 'collapse' quantum wavefunctions in a manner desired by the person doing the beaming! Humans dreaming up electrons which then return the favour by dreaming up humans is not an option!

However, this sober analysis of the quantum situation seems to have been missed by over-enthusiastic quantum prophets such as Levy who, casting aside warnings from practising physicists, without any argument as to why such warnings can be ignored, and with abandonment approaching contempt, rush to tell their readers that:

> One of the greatest sovereign powers that we all wield as human beings, although often unknowingly or without awareness, is the power of where to place our attention. As if we all have an unknown superhero power, the very power of creation lies invisibly enfolded within our field of attention. Quantum physics reveals to us that turning the gaze of our attention towards anything is a powerful creative act that alters, energizes, and potentiates whatever our gaze falls on. Focusing our attention is an act of creation in and of itself. Our beam of attention intersects and interacts with the multidimensional probability waves that hover in a ghostlike state of unrealized potentiality that comprise matter in its unobserved state. Once imbued with our attention, whatever we are looking at instantly materializes into a particular and perceivable appearance.[91]

Here Levy pushes the evidence far beyond anything it can possibly support. As Bricmont says, there is no evidence that human beings can beam beams of attention-consciousness to collapse quantum wavefunctions at will. The evidence for a quantum role for consciousness is far more subtle, and is only found within very delicate and contrived quantum experiments.

Whilst it is possible to set up very contrived and complicated quantum experiments which can give the appearance that something like this may be happening, the notion that each individual can through an act of direct intention affect quantum reality in everyday life is far beyond the available evidence. As the significant physicist Wojciech Zurek points out:

> ...while the ultimate evidence for the choice of one alternative resides in our illusive "consciousness," there is every indication that the choice occurs much before consciousness gets involved and that, once made, the choice is irrevocable.[92]

Simply put, Zurek is pointing out that experiments at the quantum level indicate that consciousness is in some way involved in 'choosing' quantum alternatives, but at the 'classical' everyday level the way that quantum alternatives actually manifest seem quite independent of what particular beams of consciousness people might be projecting around.

It might be thought that some claims by some significant physicists actually support a Levy type extreme New-Agey perspective. For example, according to Rosenblum and Kuttner, the evidence obtained from quantum level experiments shows that:

> The object was not there before you found it there. Your happening to find it there *caused* it to be there.[93]

But such claims only apply in the context of complex, delicately designed quantum experiments. As Zurek and others indicate, there is no reason to extend this evidence to claim that wavefunctions can be 'collapsed', and the material world manifested, as a matter of conscious willpower in everyday life.

However, within the context of quantum experiments there is a fairly impressive number of significant physicists who suggest that consciousness is implicated in some way. And this means that consciousness must also be implicated in the 'creation' of the apparently external 'material' world in some way, even though it is not a construction of individual willpower. For example, Bernard d'Espagnat, tells us that:

> The doctrine that the world is made up of objects whose existence is independent of human consciousness turns out to be in conflict with quantum mechanics and with facts established by experiment. [94]

Here is the conundrum, there is no evidence that individual minds wander around intentionally constructing their own material realities at will. There is no evidence that human beings can actually beam rays of consciousness to collapse wavefunctions in everyday life, which is the claim contained in the Levy description quoted above. But nevertheless, as the previous overview indicates, there is suggestive evidence that consciousness may be involved, operating at the quantum level, in the construction and functioning of the material world in some way.

These kinds of perspectives clearly have a subjective element, either subtle or crude. At some level, they seem to suggest, consciousness intervenes to produce experienced 'reality' from the realm of quantum spread-out wavy potentiality. This division into the pre-measurement situation of a quantum bundle of possible experiences, and the post-measurement situation of having experiential possession of just one element of the possible elements in the quantum bundle, leads to the idea that such a situation is one wherein we are concerned with mere knowledge of one of a possible state of affairs, rather than having a hardcore 'real' reality one can really get to grips with, so to speak. It seems we are in the mist of flimsy quantum epistemology which gives us knowledge of a possibility and can never get our hands on a really satisfying reality which has an ontological credibility! As Bohm and Hiley write in the introduction to *The Undivided Universe*, in the Copenhagen viewpoint it seems that:

> ...quantum theory is concerned only with our *knowledge* of reality ... in more philosophical terms, it may be said that quantum theory is primarily directed towards *epistemology* which is the study of how we obtain our knowledge... It follows from this that quantum mechanics can say little or nothing about reality itself. In philosophical terminology, it does not give what can be called an *ontology* for a quantum system. Ontology is concerned with what is and only secondarily with how we obtain knowledge ...[95]

In other words, this style of quantum theory does not supply as with a really real reality! It only seems to provide a questionable 'knowledge' about something essentially unknowable.

One way of looking at this is to consider that if we accept that we only get an experienced 'reality' after a 'measurement' takes place then we really cannot say that the pre-measurement state is 'real' in the same way as the post-measurement state is 'real', so it seems that reality is measured into reality from a state of unreality. And this does not give any kind of coherent 'ontology', because it does not tell us exactly what kind of 'stuff' the world is made of. As Roger Penrose has declared:

> Undoubtedly the world is strange and unfamiliar at the quantum level, but it is not unreal. How, indeed, can real objects be constructed from unreal constituents?[96]

It can easily seem, to those accustomed to believe in a really 'real' reality, that in the quantum measurement situation we come to have mere knowledge of we know not what. Werner Heisenberg, in this vein of thinking, asserted that quantum physicists: "no longer deal with the elementary particles themselves but with our knowledge of them".[97]

Physicist Henry Stapp, who was around at a time so that he was able to discuss such issues with Heisenberg, has later concluded from his discussions and his own further explorations that:

> We live in an *idealike* world, not a matterlike world.' The material aspects are exhausted in certain mathematical properties, and these mathematical features can be understood just as well (and in fact better) as characteristics of an evolving idealike structure. There is, in fact, in the quantum universe no natural place for matter. This conclusion, curiously, is the exact reverse of the circumstances that in the classical physical universe there was no natural place for mind.[98]

Stapp considers that quantum discoveries have completely overturned materialist accounts of the process of reality, and also believes that quantum mechanics offers insight into the nature of free will:

> Philosophers of mind appear to have arrived, today, at less-than-satisfactory solutions to the mind-brain and free will problems, and the difficulties seem, at least prima facie, very closely connected with their acceptance of a known-to-be-false understanding of the nature of the physical world, and of the causal

role of our conscious thoughts within it.[99]

A crucial phrase here is, of course, *'known-to-be-false'*. For Stapp it is incomprehensible that anyone in the scientific or academic community in general could embrace any kind of materialist outlook in the age of quantum physics:

> ...the re-bonding [between mind and matter] achieved by physicists during the first half of the twentieth century must be seen as a momentous development: a lifting of the veil. Ignoring this huge and enormously pertinent development in basic science, and proclaiming the validity of materialism on the basis of an inapplicable-in-this-context nineteenth century science is an irrational act.[100]

However, there appears to be a spiritually destructive materialist perspective which still pervades much of modern science!

In order to explain the nature of the 're-bonding' between mind and matter Stapp employs a formulation by John von Neumann (1903-1957), who was considered to be one of the world's foremost mathematicians at the time he proposed it (1925-1930), of the quantum 'measurement' process, which is the quantum process through which an actual experienced reality appears to emerge from the potentialities that are contained within the quantum wavefunction of possibilities:

Process 1: The 'free choice' of the experimental setup, Heisenberg called this phase "a choice on the part of the 'observer' constructing the measuring instruments and reading their recording". This choice is "not controlled by any known physical process, statistical or otherwise, but appears to be influenced by understandings and conscious intentions."[101] Whilst this process was originally delineated as a phase within the experimental setting, Stapp also indicates that such a 'free choice' of 'probing actions' is a part of the general human condition:

> Probing actions of this kind are performed not only by scientists. Every healthy and alert infant is engaged in making wilful efforts that produce experiential feedbacks ... Thus both empirical science and normal human life are based on paired realities of this action-response kind...[102]

The hugely significant point in this "process 1" "free choice" is that it poses a question to which 'reality' can feedback a 'yes' or a 'no', and the fact that the choice of the question is free means that the "free choice" actually determines the nature of the possible feedbacks. Thus the "free choices" of how experiments are set up determine the nature of experienced reality for each experiment:

> ...the process is active: it injects into the physical state of the system being acted upon properties that depend upon the intentional chosen action of the observing agent.[103]

Stapp calls this "process 1" 'a dynamical psychophysical bridge'.[104] It is important to note here that such decisions regarding experimental arrangements do not entail beaming wavefunction-collapsing rays of conscious intentionality!

Process 2: The deterministic quantum evolution of the potentialities within the quantum wavefunction.

Process 3: is what Paul Dirac called a "choice on the part of nature." It is the 'yes' or 'no' feedback from the experimental set-up – yes reality is this way, or no reality is not this way; Stapp indicates that complex questions can be reduced to yes-no choices.

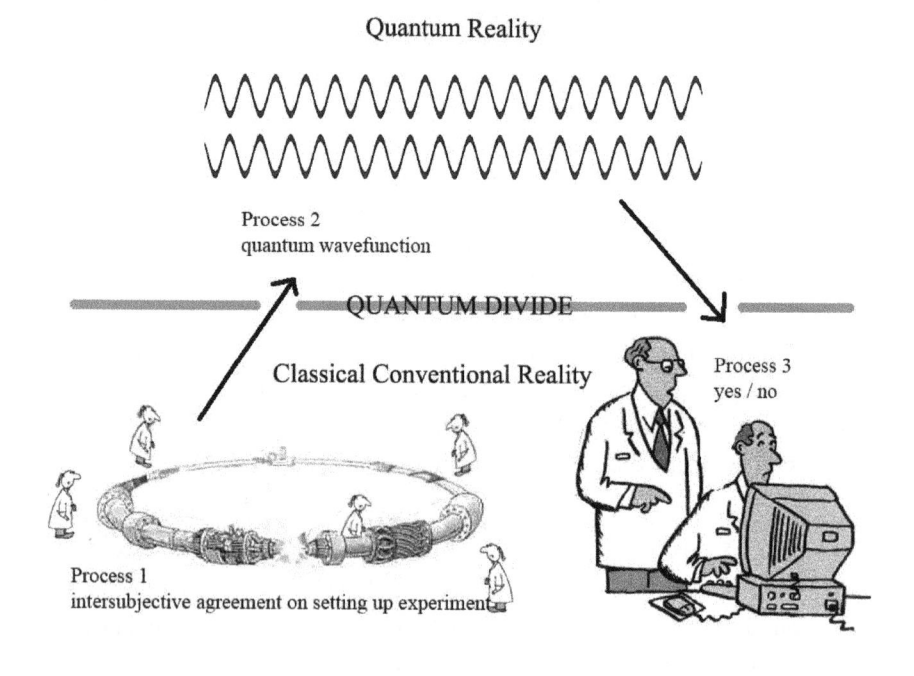

Quantum Reality

Process 2
quantum wavefunction

QUANTUM DIVIDE

Classical Conventional Reality

Process 3
yes / no

Process 1
intersubjective agreement on setting up experiments

John Wheeler also advanced a quantum metaphor involving questions and yes-no answers. He suggested the quantum metaphysical situation could be compared to someone playing a quantum-variant version of twenty-questions. The usual version is that one person leaves the room while the rest of the group are supposed to agree on some person, place or thing that the room-leaver must discover through questioning the people in the room when the leaver returns. The person who left the room then re-enters and tries to guess what the agreed upon entity is by employing a series of questions that can only be answered with a yes or a no. But in Wheeler's quantum version the group decides to play a trick on the questioner. The room-members do not agree an entity before the questioning game begins. When the leaver returns to begin questioning the first person to be queried thinks of the target for the guessing game *after* the questioner asks his question. This first person then gives an answer to the first question. Each subsequent person does the same, making sure that the response given is consistent and coherent with the immediate question and also with all previous questions and answers. The word that the questioner eventually comes up with does not denote an object agreed upon by the gathering at the outset, it emerges out of a space of linguistic possibility as a result of the coherent, yet creative, set of answers given.

Wheeler says of this, putting himself in the place of the questioner: "The word wasn't in the room when I came in even though I thought it was". Science writer John Horgan writes of this in his *Scientific American* article 'Do Our Questions Create the World?':

> In the same way, the electron, before the physicist chooses how to observe it, is neither a wave nor a particle. It is in some sense unreal; it exists in an indeterminate limbo. "Not until you start asking a question, do you get something," Wheeler said. "The situation cannot declare itself until you've asked your question. But the asking of one question prevents and excludes the asking of another." Wheeler has condensed these ideas into a phrase that resembles a Zen koan: "the it from bit." In one of his free-form essays, Wheeler unpacked the phrase as follows: "...every it--every particle, every field of force, even the spacetime continuum itself--derives its function, its meaning, its very existence entirely--even if in some contexts indirectly--from the apparatus-elicited answers to yes-or-no questions, binary choices, bits."[105]

It was this situation within quantum metaphysics, in which reality seemed to hang in a haze of being an "indeterminate limbo", or mind-like mathematical possibilities, before a 'measurement', so that quantum experiments did seem just to produce knowledge out of quantum thin air, so to speak, that provoked metaphysical dissent among some physicists.

Erwin Schrödinger, the originator of the fundamental quantum wave-equation, remarked concerning this state of affairs that:

> ...the reigning doctrine rescues itself or us by having recourse to epistemology. We are told that no distinction is to made between the state of the natural object and what I know about it, or perhaps better, what I can know about it if I go to some trouble. Actually - so they say - there is intrinsically only awareness, observation, measurement.[106]

This is an interesting insight from the point of view of *Yogacara* consciousness-only Buddhism and Buddhist *Dzogchen* because of the notion, common to both of these, that the process of reality fundamentally derives from a deep layer of energy which has the nature of awareness. Furthermore, according to such Buddhist worldviews the experiences which emerge from the deep level of mind-energy depend upon the actions, which are essentially yes-no decisions, made by sentient beings. This mechanism is, of course, an central aspect of *karma*, a central Buddhist doctrine.

Thus, the Dzogchen psycho-metaphysical vision of the Dzogchen philosopher-practitioner Dudjom Lingpa tells us that:

> ...all sensory experiences of samsara and nirvana manifest as specific forms that come and go within the expanse of the space of supreme emptiness. The ground aspect of the dharmakaya, buddha nature, becomes evident as the supreme principle that pervades all of samsara and nirvana. This is the ground aspect of awareness as supreme freedom from limitations ... samsara and nirvana are the phantasmagoria of a single awareness...[107]

The term 'samsara' denotes what Bohm would call the explicate order, wherein, according to the Buddhist worldview, the great portion of humanity live lives of various degrees of dissatisfaction, interspersed with momentary pleasures. It also denotes the cycle of birth and death, Buddhism asserts the rebirth of streams of consciousness, although Buddhism also denies the existence of any fixed 'self' that continues from life to life. As we shall see later 'nirvana' corresponds to mental states beyond the explicate, corresponding to deep implicate realms of consciousness.

It is also noteworthy that this description indicates that both samsara and nirvana "manifest as specific forms that come and go within the expanse of the space of supreme emptiness". Compare this with the summary of David Bohm's article 'The Implicate Order: A New Order for Physics' which reads:

> The author suggests that emptiness is really the essence. It contains implicitly all the forms of matter. The implicate order really refers to something immensely beyond matter as we know it — beyond space and time. However, somehow the order of time and space are built in this vacuum. At present there is no law that determines the vacuum state.[108]

Bohm, of course, is not the only physicist who, at some point in their career, saw some kind of connection between quantum discoveries and Buddhism. Here is an observation made by Robert Oppenheimer:

> ... discoveries in atomic physics are not in the nature of things wholly unfamiliar, wholly unheard of or new. Even in our own culture they have a history, in Buddhist and Hindu thought a more considerable and central place. What we shall find [in modern physics] is an exemplification, an encouragement, and a refinement of old wisdom.[109]

And Oppenheimer made the following observation when discussing the Heisenberg Uncertainty Principle:

> If we ask, for instance, whether the position of the electron remains the same, we must say 'no;' if we ask whether the electron's position changes with time, we must say 'no;' if we ask whether the electron is at rest, we must say 'no;' if we ask whether it is in motion, we must say 'no.' The Buddha has given such answers when interrogated as to the conditions of man's self after his death; but they are not familiar answers for the tradition of seventeenth and eighteenth-century science.[110]

As we shall see in more detail in a later chapter, according to a core Tibetan Buddhist metaphysical perspective the ultimate nature of all physical phenomena hovers between existence and non-existence:

> From certain single perspectives
> [The Buddha] taught them as either 'nonexistent' or 'existent.'
> From both perspectives,
> He expressed them as 'neither existent nor nonexistent.'
> Since they do not exist as they appear,

> He talked about their 'nonexistence.'
> Since they appear in such ways,
> He spoke about their 'existence.'[111]

This state of hovering between extremes of existence – neither existent, nor non-existent, nor both existent and non-existent, nor neither – but not fully occupying any of these possibilities is both the hallmark of the unobserved quantum realm and of Buddhist 'emptiness'. And such a strange existential configuration really is the inner nature of the quantum world. As Jeffrey Alan Barrett tells us in his book *The Quantum Mechanics of Minds and Worlds*:

> …a neutral K meson is typically not a K^0
> meson, not a $-K^0$ meson,
> not both and not neither.[112]

The unobserved quantum world hovers in an indeterminate state hovering between existence and nonexistence.

Erwin Schrödinger had a long-term interest in the Hindu Vedas.[113] However, despite his interest in Vedic mysticism, Schrödinger, as the previous quote above from him indicates, was dissatisfied with an awareness-based 'epistemological' view of the quantum situation. In fact, Schrödinger said that if he were not convinced that Bohr really did believe his Copenhagen viewpoint: "I would call it intellectually wicked."[114] And Schrödinger wrote to Max Born, who was a supporter of Bohr: "Have you no anxiety about the verdict of history. Are you so convinced the human race will succumb before long to your folly?" And Einstein shared a view in alignment with Schrödinger, declaring that the Copenhagen account of quantum functioning "operates with an incomplete description of physical systems."[115]

Schrödinger and Einstein were soon joined by the important quantum physicist John Bell, who expressed his "anxiety" concerning quantum "verdicts" concerning the nature of reality in angst-driven observations such as:

> One wants to be able to take a realistic view of the world, to talk about the world as if it were really there, even when it is not being observed. I certainly believe in a world that was here before me, and will be there after me, and I believe you are part of it! And I believe that most physicists take this point of view when they are being pushed into corners by philosophers.[116]

Jean Bricmont, in his book *Making Sense of Quantum Mechanics*, in which he extols the virtues of Bohmian Mechanics, writes:

> What we need is a theory which tells a story about what is going on in the world, even when we do not "observe" it, and which makes the same predictions as ordinary quantum mechanics, whenever we do make "observations" or experiments. If such a theory existed, then all the confusing talk about the centrality of observations would disappear and we could analyse that theory in order to see how it helps us to understand the quantum world. Amazingly, such a theory actually does exist, and has existed ... since the the beginning of quantum mechanics ...proposed by Louis de Broglie ... and developed by David Bohm in 1952.[117]

In other words, Bohm's 1952 approach to quantum mechanics, which unknowingly to Bohm at the time mirrored ideas earlier proposed by Louis de Broglie, appeared to banish the mystical creative power of the 'observer' within the functioning of the quantum realm.

The abstract to David Bohm's 1952 paper, '*A Suggested Interpretation of the Quantum Theory in Terms of "Hidden" Variables*', reads:

> The usual interpretation of the quantum theory is self-consistent, but it involves an assumption that cannot be tested experimentally, viz., that the most complete possible specification of an individual system is in terms of a wave function that determines only probable results of actual measurement processes. The only way of investigating the truth of this assumption is by trying to find some other interpretation of the quantum theory in terms of at present "hidden" variables, which in principle determine the precise behavior of an individual system, but which are in practice averaged over in measurements of the types that can now be carried out. In this paper and in a subsequent paper, an interpretation of the quantum theory in terms of just such "hidden" variables is suggested. It is shown that as long as the mathematical theory retains its present general form, this suggested interpretation leads to precisely the same results for all physical processes as does the usual interpretation. Nevertheless, the suggested interpretation provides a broader conceptual framework than the usual interpretation, because it makes possible a precise and continuous description of all processes, even at the quantum level. This broader

conceptual framework allows more general mathematical formulations of the theory than those allowed by the usual interpretation. Now, the usual mathematical formulation seems to lead to insoluble difficulties into the domain of distances of the order of 10^{-13} cm or less. It is therefore entirely possible that the interpretation suggested here may be needed for the resolution of these difficulties. In any case, the mere possibility of such an interpretation proves that it is not necessary for us to give up a precise, rational, and objective description of individual systems at a quantum level of accuracy.[118]

As we can see, Bohm refers to his new approach as a type of "hidden variables" account of quantum functioning. The phrase "hidden variables" refers to the possibility that, as Einstein had claimed, the current quantum theory of the time was "incomplete", and therefore there were hidden elements, that, when found, would remove notions of observers 'creating' reality, or that reality was only subjective knowledge that was produced when minds somehow interacted with an unknowable quantum mist. The issue of the feasibility of "hidden variables" had a history of debate and controversy, these issues were encapsulated in "no hidden variables" theories, but the details need not concern us here. Bohm made the claim that his "hidden variables" approach avoided any issues that might have been thought to be problematic.

In the case of Bohm's new theory the "hidden" part is a 'pilot wave' which guides 'real' particles, which are self-existent particles, completely independent of observers. The way in which the pilot wave behaves is such that the particles behave exactly as they appear to behave in the Copenhagen scenario, but they are continuously existent particles that do not seem to magically appear on the scene when a 'observer' gets observing! In his pro-Bohmian Mechanics book *Quantum Sense and Nonsense*, Bricmont explains concerning Bohm's 1952 theory:

➢ It is a "hidden variables" theory.

➢ Its "hidden variables" are not hidden at all (hence the expression "hidden variables" is quite a misnomer in this case).

➢ There is no fundamental role whatsoever for the "observer" in that theory.

➢ The theory is not contradicted by the no hidden variables theorems.

It is a sort of statistical interpretation of quantum mechanics, but a consistent one.

- ➤ The de Broglie-Bohm theory is entirely deterministic.
- ➤ It accounts for all the observations used to justify the validity of ordinary quantum mechanics.
- ➤ It allows us to understand the "active role" of the measuring devices, meaning that a measurement in general does not record some pre-existing value of the system being "measured", as the "no hidden variables" theorems imply. But it does so without making it a philosophical a priori.
- ➤ It explains to some extent where the nonlocality of the world comes from.[119]

The last two points: 1) results of measurements depend on interactions, not prior independent properties, and 2) the feature of "nonlocality", Einstein's "spooky action at a distance", require further elucidation which will follow shortly.

Here are a two more useful explanatory quotes from Bricmont's excellent book, taken from the chapter 'The de Broglie-Bohm Theory in a Nutshell':

> The de Broglie-Bohm theory is simply a theory of matter in motion, just like Newton's theory. Of course, the way particles move is different ... but there is nothing philosophically new.[120]

And:

> In the de Broglie-Bohm theory, the *complete physical state* of a particle or a system of particles is given both by its wave function, which is the same as in ordinary quantum mechanics, *and* the positions of the particles. They both change in time, in the following way:
>
> 1. The wave function evolves according to the usual rules, *but nothing special happens to it during measurements.*
>
> 2. The motion of particles is guided by their wave function.[121]

Thus, it is thought by some that this perspective is, in some sense, more 'realistic'. When John Bell read Bohm's 1952 article he was moved to proclaim that:

But in 1952 I saw the impossible done. It was in papers by David Bohm. Bohm showed explicitly how parameters could indeed be introduced, into non-relativistic wave mechanics, with the help of which the indeterministic description could be transformed into a deterministic one. More importantly, in my opinion, the subjectivity of the orthodox version, the necessary reference to the 'observer,' could be eliminated. But why then had Born not told me of this 'pilot wave'? If only to point out what was wrong with it? Why did von Neumann not consider it? More extraordinarily, why did people go on producing 'impossibility' proofs, after 1952, and as recently as 1978? Why is the pilot wave picture ignored in text books? Should it not be taught, not as the only way, but as an antidote to the prevailing complacency? To show us that vagueness, subjectivity, and indeterminism, are not forced on us by experimental facts, but by deliberate theoretical choice?[122]

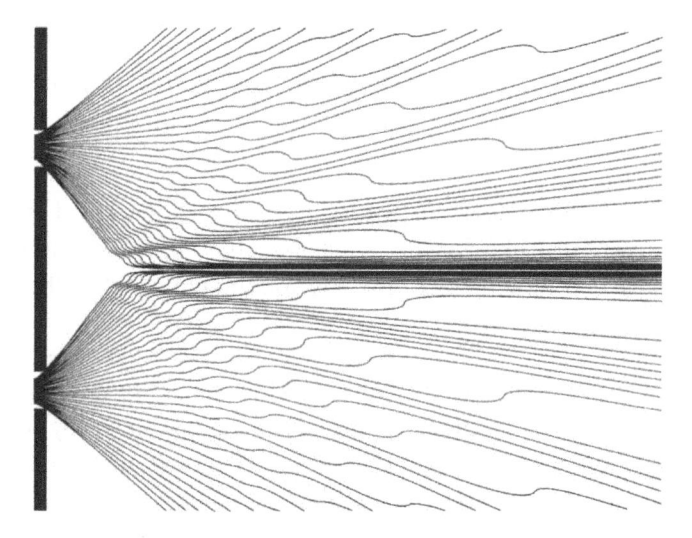

Bohmian pilot wave guided particle paths in the two slit experiment.

One can detect here that Bell was inclined to suspect that there was some kind of conspiracy on the part of some, perhaps mystically inspired or inclined, cabal of physicists to promote the Copenhagen view. Andrew Whitaker, in his book *The New Quantum Age,* comments that:

> What is exciting about the Bohm interpretation is that, totally contrary to the claims of Bohr and von Neumann, the fairly simple set of ideas produces the standard results of quantum theory ... but remains entirely realist. In fact, Bohm's interpretation is also deterministic, though for Bell in particular that was rather unimportant - it was the realism that was important.[123]

John Bell's determination to live in a really 'real' world is illustrated by a telephone call with BBC interviewers, prior to an interview. Bell warned the interviewer, when told that he might be asked if quantum theory showed that 'reality does not exist', that he was: "an impatient, irascible sort who tolerates no nonsense".[124] This seems to suggest that Bell, like others then and now, had an ingrained, a priori preference for a 'realistic' interpretation of quantum phenomenon. But it is interesting to note that Einstein, who had famously also expressed a preference for a more deterministic perspective, with remarks such as "God does not play dice!", wrote to Max Born about Bohm's proposal expressing his opinion that:

> That way seems too cheap to me. This path seems to me too easy ... a physical fairy-tale for children, which has rather misled Bohm and de Broglie.[125]

It seems Einstein was a hard man to please!

We saw above that Bricmont refers to the "interactive" nature of measurement outcomes in the Bohmian worldview. In their book *The Undivided Universe* Bohm and Hiley explain:

> The probability of a particular result of the interaction between the instrument and the observed object is shown to be exactly the same as that assumed in the conventional interpretation. But the key new feature here is that of the *undivided wholeness* of the measuring instrument and the observed object, which is a special case of the wholeness ... with quantum processes in general. ...it is no longer appropriate, in measurements to a quantum level of accuracy, to say we are simply 'measuring' an intrinsic property of of the observed system. Rather what happens is that the measuring apparatus and that which is observed participate irreducibly in each other, so that the ordinary and common sense idea of measurement is no longer relevant.[126]

So, although the Bohmian approach has perhaps rescued 'reality' in some sense, it is not our good old "ordinary and common sense idea" of 'reality' we were so fond of! In fact, it seems to be the case that the claimed 'realism' of Bohm's 1952 viewpoint is often overplayed.

In Bohm's new quantum reality the result of measurements depend upon the way in which "the measuring apparatus and that which is observed participate irreducibly in each other". In other words, quantum entities do not have their own independent properties all to themselves all the time, so to speak. The claimed 'real' properties that show up in measurements depend upon which other bits and pieces of the universe they are interacting with:

> The pilot wave theory is contextual. That is, it recognizes that the outcome of measurements cannot be accounted for simply by hidden variables in the observed system alone. The variables of the measuring equipment (which expression can include human observers) and the interaction between the equipment and the system all play a part in bringing about the result.[127]

So, it seems that Bohm did not manage to banish all traces of quantum unconventionality from reality! In fact it seems that in some aspects 'observers' may get a quantum piece of the action, so to speak.

According to Bohm, the 'pilot wave' theory "provides a broader conceptual framework than the usual interpretation, because it makes possible a precise and continuous description of all processes, even at the quantum level". In other words, it provides a continuous mathematical description of the behaviour of the pilot waves which are guiding the particles, and there are no sudden changes, or jumps, in ontological status, all aspects remain 'real' all the time. Physicist Henry Stapp describes Bohm's pilot wave theory:

> Bohm's model is simple and instructive. It shows that we need not cling to the idea, advanced by the founders of quantum theory, that nature cannot be described in a thoroughly comprehensible way in terms of properties that are always well defined and that evolve in accordance with well-defined deterministic ways. Bohm's model does violate one of the basic precepts of classical physics: that the force on a particle located at a point generally depends strongly upon the precise positions, *at that very instant,* of many other particles all over the universe. This instantaneous connection contradicts the idea of classical relativ-

istic physics that no influence can act ... faster than light.[128]

So there is a price to pay for adopting Bohm's method of reworking quantum mechanics into a supposedly more 'realistic' demeanour. We now have, it is claimed, 'real' particles moving around, and they are not dramatically dependent on observers, or at least not on the minds of observers for their very existence, and the way they move around is guided by real pilot waves. But, we also have to accept that in some circumstances, particles may instantaneously influence each other over vast cosmic distances, which is not a classical phenomenon previously considered as operating in the process of reality. Indeed, as Stapp points out, it seems to contradict relativity theory, which says that points in space cannot have interconnections that operate faster than the speed of light.

This non-local instantaneous interconnection between distant aspects of the Bohmian undivided universe derives from the way in which Bohmian mechanics necessarily deals with the phenomenon of quantum entanglement. Within conventional quantum mechanics, entangled particles are particles that have interacted so that they share a quantum wavefunction. They then remain interconnected so that actions performed on one particle will instantaneous affect the other, even when separated by great distances. This is the phenomenon that Einstein referred to as "spooky action at a distance". In the non-Bohmian type of physics this non-local interconnection only applies at the quantum level and disappears at the classical level, because of the way in which quantum phenomena transition to 'classical' behaviour. But in Bohm's undivided universe, these "spooky" instantaneous interconnections remain operative even at the classical level. This topic will be elucidated in the next chapter.

And there is, as Stapp also points out, one other unsettling feature of Bohm's proposal:

> Bohm's model does however retain one feature of classical physics that can be regarded as objectionable ... This is the need for an arbitrary-looking choice of initial conditions. In particular, some definite initial position for each of the particles in the universe must be chosen.[129]

This difficulty can be surmounted in Bohm's later 'implicate order' perspective, as we shall see at a later point.

It seems that Bohm quickly realized that further development of his ideas was essential. As we have noted, one important issue it highlighted was the very counter-intuitive notion of instantaneous connections between 'particles' across vast cosmic distances, and in pre-Bohmian quantum theory such 'spooky' interconnections only had significance in the quantum world, they did not operate in the everyday classical 'world'. But the mathematical formulations that Bohm had to employ in his new theory meant that the 'spooky' non-local cosmic interconnections were projected into 'classical' reality. And such universal instantaneous interconnections between cosmically distant particles are not an obvious feature of the everyday world, as Basil Hiley and F. David Peat tell us in their introduction to *Quantum Implications: Essays in Honour of David Bohm*:

> ... this relationship may depend on the quantum states of even larger systems, ultimately going on to the universe as a whole. Within this view separation becomes a contingent rather than a necessary feature of nature. This is very different from the way we perceive the macroscopic world around us, where separation seems basic.[130]

One does not expect electrons on the other side of the universe to be disturbed when one raises an afternoon cup of tea!

So we see that in Bohm's 1952 pilot-wave quantum metaphysics a universally holistic network of interconnections is a natural and necessary result of the theoretical development. And in 1957 Bohm began to update his views in his book *Causality and Chance in Modern Physics*, abandoning his Marxist sympathies and moving in a more spiritual direction. At this time Bohm declared that communists "began to minimise the importance of spirit", and Bohm later told an interviewer that at the time, in line with his new intellectual direction, "I also read Buddhism or oriental philosophy, Indian philosophy, yoga and probably some of the Christian philosophers".[131] He also came across the works of Krishnamurti around this time.

In the preface to the 1984 edition of *Causality and Chance in Modern Physics* Bohm explained how his attempt at formulating a fully causal, deterministic and realistic account of quantum functioning led him towards wider vistas of quantum metaphysics. He starts by telling us that:

> ... the development of classical physics ... led to to the notion that the universe may be compared to a gigantic mechanism. ...

> more recent developments in physics, notably relativity and quantum theory, do not fit in with such a mechanistic philosophy. Rather they very strongly suggest the need for a radically new over-all approach, going beyond mechanism. The usual interpretation of quantum theory does not give a clear idea of how far-reaching is this change, because it functions solely as a mathematically algorithm ... an alternative interpretation is discussed, in which an electron (for example) is assumed to be a particle that is always accompanied by a new kind of wave field. ...[132]

Here Bohm is referring to his 1952 pilot-wave proposal. But he now indicates that the initial pilot-wave viewpoint is of a provisional nature, awaiting important further insights:

> This interpretation in terms of particle plus field was regarded, however, as furnishing only a provisional mode of understanding the quantum theory ... In these studies ... it became clear that even the one-body system has a basically non-mechanical feature, in the sense that it and its environment have to understood as an *undivided whole*.[133]

And, Bohm also tells us that:

> The law of the whole can be shown to imply that at the ordinary level of experience ... the whole falls *approximately* into a structure of relatively independent sub-wholes, interacting more or less externally and mechanically. Nevertheless, in a more accurate and more fundamental description, quantum wholeness and non-locality are seen to be the major factors.[134]

In the next chapter we shall investigate how non-local connections necessarily imply a quantum metaphysics of wholeness.

The Unfolding of Bohm's Pilot-Wave Theory

NonLocal Connections Implicates Ontological Wholeness

Erwin Schrödinger posed his famous dead and alive cat paradox in order to highlight the apparent absurdity of the quantum state of superposition, wherein many possibilities for manifestation exist at the same time, although none of them are really real, so to speak. A 'superposition' is a mathematical description of all the possibilities for manifestation which 'semi-exist' prior to a 'measurement' taking place. Once a measurement occurs only one of the possibilities in the superposition actually makes it into full reality, and all the others disappear. Schrödinger outlined his thought experiment in a letter to Einstein, in which he wrote (ψ denotes the quantum wavefunction):

> Confined in a steel chamber is a Geiger counter prepared with a tiny amount of uranium, so small that in the next hour it is just as probable to expect one atomic decay as none. An amplified relay provides that the first atomic decay shatters a small bottle of prussic acid. This and – cruelly – a cat is also trapped in the steel chamber. According to the ψ-function for the total system, after an hour ... the living and dead cat are smeared out in equal measures.[135]

In this thought scenario the cat's state of 'existence' depends upon the state of 'existence' of the atomic constituents in the uranium. Some of these atoms will be in a superposition of decay and non-decay. As the cat lives or dies depending upon non-decay or decay, the cat itself must, according to the mathematics of quantum mechanics, *if the mathematics applies at the non-quantum 'classical' level*, hover between the state of life and the after-life at the same time.

However, it is important to understand that Schrödinger did not actually believe that cats could be placed into states of being half-dead and half-alive. The description he gave was only a thought experiment that he concocted in order to illustrate what he considered to an absurdity within the Copenhagen account of quantum mechanics. In other words, Schrödinger considered the disparity between the way the world seemed to 'exist' at the quantum level and the way it appeared to be in the everyday world was so great that he could not believe the quantum story as presented by Bohr and Heisenberg could be correct. His cat example, then, was meant, not to suggest that cats could be half-alive and half-dead at the same time, but that the Copenhagen quantum account could not be correct.

In the early days of quantum exploration, expressions of surprise and sometimes disbelief on the part of physicists when confronted with quantum discoveries were not in short supply. The same is sometimes true today. For example, the celebrated mathematical physicist Roger Penrose tells the following story:

> I cannot resist quoting a remark that was made to me by Professor Bob Wald, of the University of Chicago, at a dinner party some years ago: If you really believe in quantum physics, then you can't take it seriously.[136]

Heisenberg, one of the founding fathers and the inventor of quantum matrix mechanics, lamented after dinner:

> Can nature possibly be as absurd as it seems to us in these atomic experiments?[137]

In a their book *Quantum Paradoxes* quantum physicists Yakir Aharanov and Daniel Rohrlich refer to a remark made by the charismatic physicist Richard Feynman, he said that once Einstein's theory of relativity was made public it was not long until many people understood it; in contrast Feynman suggested that nobody understood quantum physics:

> ...you get down a blind alley from which nobody escapes. Nobody knows how it can be like that.[138]

Feynman made this observation in 1965; Aharanov and Rohrlich are writing 40 years later and, according to them, very little has changed in the intervening years. They refer to a Woody Allen joke in which someone goes to a psychiatrist to complain that his brother thought himself to be a chicken. The psychiatrist suggests that the man should commit his brother to an insane asylum; the man replies 'Are you crazy we need the eggs!' Aharanov and Rohrlich observe that:

> Quantum Mechanics is crazy – but we need the eggs!'[139]

Such expressions of disbelief are based on a presumption that the world should function in more or less the same way at all levels, large or very, very small. As Jonathan Allday, in his book *Quantum Reality: Theory and Philosophy*:

> The problem is that the small-scale laws describe a way of behaving that, judged by the standards of everyday experience, is utterly bizarre. It is very difficult to see how all the funny business going on at the atomic scale can lead to the regular, reliable world we spend our lives in.[140]

However, for some people, those accustomed to Eastern philosophy for example, why such a view of astonishment should be so pervasive among quantum physicists is difficult to comprehend. Among Buddhist philosophers it was, since the time of the Buddha, and still is of course, generally accepted that an inherently-existent internally-solid material world, as the classical world seems to be, could not actually function:

> If things were not empty of inherent existence, nothing could function…It is their emptiness of inherent existence that allows everything to operate satisfactorily.[141]

According to Buddhist metaphysics it is precisely because the inner nature of the process of reality has a central aspect of indeterminacy that reality can function as a process! As we shall see, Bohm was aware of this necessity.

The Buddha indicated in his teachings that all phenomena are 'empty' of true substantial-nature, having no inner core of 'real' material, or any other persisting, reality. The following is from the *Phena Sutta:*

> Material-form is like a glob of foam; feeling, a bubble; perception, a mirage; mental-formations, a banana tree [banana trees are hollow]; consciousness, a magic trick …. However you observe them, appropriately examine them, they're empty, void to whoever sees them appropriately. … That's the way it goes: it's a magic trick, an idiot's babbling. … No substance here is found.

These categories, called *skandhas* in Pali, of material form, feelings, perceptions, mental-formations and consciousness make up all the types of phenomena within Buddhist psycho-metaphysics. So we see that the Buddha was aware that the material world was not internally solid, in fact it is akin to a "glob of foam", which "would appear empty, void, without substance: for what substance would there be in a glob of foam?" This is intriguing when one considers that Max Planck, a primary founder of the quantum revolution, thought around 1882 that "in spite of the great successes of the atomistic theory in the past, we will finally have to give it up and to decide in favour of the assumption of continuous matter".[142] He later, of course, discovered he was wrong about this! Matter turned out to be, as the Buddha taught, 'empty'.

One aspect of this quantum emptiness is the phenomenon of superposition. Einstein, like Schrödinger, had also discussed the quantum phenomenon of superposition as it would appear if it manifested on the everyday macroscopic level. One of Einstein's thought experiments is called Einstein's Boxes, which is a bit more involved than Schrödinger's cat example. Physicist Travis Norsen has written about this:

> It is well known that several of quantum theory's founders were dissatisfied with the theory as interpreted by Niels Bohr and other members of the Copenhagen school. Before about 1928, for example, Louis de Broglie advocated what is now called a hidden variable theory: a pilot-wave version of quantum mechanics in which particles follow continuous trajectories, guided by a quantum wave. David Bohm's rediscovery and completion of the pilot-wave theory in 1952 led de Broglie back to these ideas; ... Erwin Schrödinger was likewise doubtful that the quantum wave function could alone constitute a complete description of physical reality. His famous "cat" thought experiment was intended to demonstrate quantum theory's incompleteness by magnifying the allegedly real quantum indefiniteness up to the macroscopic level where it would directly conflict with experience. By far the most important critic of quantum theory, however, was Albert Einstein. [Einstein suggested] another thought experiment which, like ... Schrödinger's cat, is intended to argue against the orthodox doctrine of quantum completeness. This thought experiment – "Einstein Boxes"– is due originally to Einstein, although it has also been discussed and reformulated by de Broglie, Schrödinger, Heisenberg, and others. Given its unique simplicity, clarity, and elegance, the relative obscurity of this thought experiment is unjustified. Einstein's Boxes establishes this conclusion with a more straightforward logical argument.[143]

Einstein's Boxes not only brings quantum 'weirdness' into the everyday world within the context of a thought experiment, it also provides a simple, but dramatic, insight into the nature of the spectacular phenomenon of quantum non-locality. This is especially true of Einstein's thought experiment as presented by Louis de Broglie. As Norsen indicates:

> ...de Broglie's re-formulation of the thought experiment in terms of literal boxes simply exaggerates the spatial separation of two parts of the total wave function, and thus brings out more clearly the implications of the locality assumption.

In his 1964 book *The Current Interpretation of Wave Mechanics: A Critical Study*[144], Louis de Broglie described Einstein's Boxes thought experiment. In the following we shall go through his description with some elucidations.

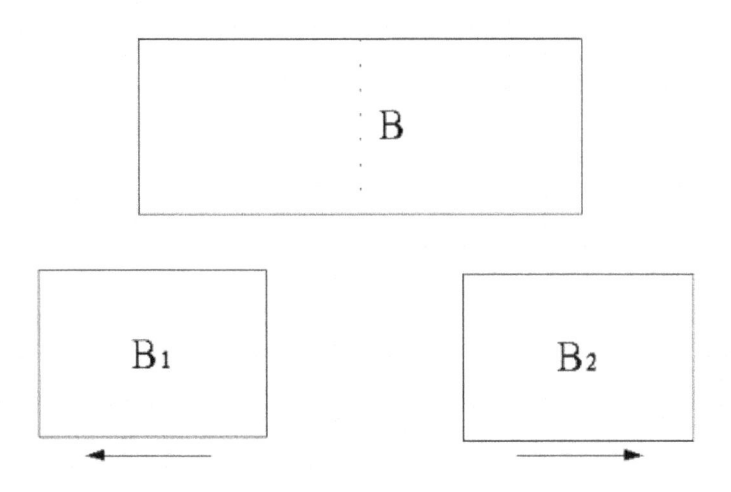

To begin the description de Broglie wrote:

> Suppose a particle is enclosed in a box B with impermeable walls. The associated wave Ψ is confined to the box and cannot leave it. The usual interpretation asserts that the particle is "potentially" present in the whole of the box B, with a probability $|\Psi|^2$ at each point. Let us suppose that by some process or other, for example, by inserting a partition into the box, the box B is divided into two separate parts B_1 and B_2 and that B_1 and B_2 are then transported to two very distant places, for example to Paris and Tokyo. The particle, which has not yet appeared, thus remains potentially present in the assembly of the two boxes and its wave function Ψ consists of two parts, one of which, Ψ_1, is located in B_1 and the other, Ψ_2 in B_2. ...

Here de Broglie refers to an initial situation where a quantum 'particle' is constrained within a box. A quantum particle, according to the Copenhagen view, is not a classical type particle because it does not fully exist as a particle somewhere in the box. Instead 'it' is a quantum haze of potentiality spread out inside the box, and 'it' has a probability (given by the square of the wavefunction $|\Psi|^2$) of being measured into existence for each point in the box. If the box is cut in two to produce two new self-contained boxes which are separated by a large distance, each of the two boxes will now have their own wavefunctions Ψ_1 and Ψ_2. But, although each box has its own wavefunction, there is still only one 'particle' involved, so now there are two interconnected wavefunctions, with one particle smeared out as potentiality across both wavefunctions inside the boxes, *but not in-between the boxes.*

In the next section de Broglie continues:

> The probability laws of wave mechanics now tell us that if an experiment is carried out in box B_1 in Paris, which will enable the presence of the particle to be revealed in this box... According to the usual interpretation, this would have the following significance: because the particle is present in the assembly of the two boxes prior to the observable localization, it would be immediately localized in box B_1 in the case of a positive result in Paris. This does not seem to me to be acceptable. The only reasonable interpretation appears to me to be that prior to the observable localization in B1, we know that the particle was in one of the two boxes B_1 and B_2, but we do not know in which one, ...

Here de Broglie indicates that if an experiment is carried out in Paris, which results in the particle being found to be in the Paris box B_1 then, instantaneously, the wavefunction of potentiality will disappear from the Tokyo box B_2, there will be no possibility of a particle being discovered in B_2. The combined wavefunction of particle-potentiality that has been hovering within the two boxes has now 'collapsed' so that there is just one particle in box B_1. Furthermore, de Broglie clearly states that he finds this situation "unacceptable". For de Broglie the particle must have been fully in one of the boxes before the measurement. He could not accept the idea of a ghost-like hovering in 'semi-existence' within the two boxes, but not in-between the boxes, before the measurement.

Next, de Broglie continues his discussion:

> We might note here how the usual interpretation leads to a paradox in the case of experiments with a negative result. Suppose that the particle is charged, and that in the box B_2 in Tokyo a device has been installed which enables the whole of the charged particle located in the box to be drained off and in

so doing to establish an observable localization. Now, if nothing is observed, this negative result will signify that the particle is not in box B_2 and it is thus in box B_1 in Paris. But this can reasonably signify only one thing: the particle was already in Paris in box B_1 prior to the drainage experiment made in Tokyo in box B_2. Every other interpretation is absurd. How can we imagine that the simple fact of having observed nothing in Tokyo has been able to promote the localization of the particle at a distance of many thousands of miles away?

Here de Broglie describes the truly mind-boggling notion, which is required by the Copenhagen view, that a combined superposed quantum wavefunction for one particle spread between these two boxes, one in Paris and the other in Tokyo, may be 'measured' *negatively* at one end, in this case the Tokyo box B_2, finding no particle there, and that this negative 'measurement' within the box in Tokyo will cause the 'particle' to immediately fully appear in box B_1 in Paris. Again, de Broglie indicates his incredulity that anyone would think such a possibility as anything other than "absurd." How could observing "nothing" in Tokyo create the manifestation of something in Paris?

This "absurd" instantaneous quantum interconnection, which is clearly predicted by the Copenhagen quantum interpretation, is called quantum non-locality. An important aspect of this quantum phenomenon is the fact that it seems that 'entangled' particles demonstrate Einstein's non-local 'spooky' instantaneous interconnections across vast distances. Particles are said to be in a state of quantum 'entanglement' if they have been produced, or have interacted, in a manner which produces a kind of inner connection between the two particles. For example, a photon which encounters a polarizing filter will 'randomly' collapse into a definite state of polarisation and either pass through or be reflected, depending upon the alignment of polarisations. Until it encounters the filter the photon does not have a definite polarisation, instead it is in a superposition of possible polarisations. Its polarisation is randomly determined the moment it reaches the filter. Physicist Brian Greene explains this kind of entanglement:

> The astounding thing is that such a photon can have a partner photon that has sped miles away in the opposite direction and yet, when confronted with the same 50-50 probability of passing through another polarised ... lens, will somehow do

whatever the initial photon does. Even though each outcome is determined randomly and even though the photons are far apart in space, if one photon passes through so will the other.[145]

It is important to comprehend the unexpected nature of this phenomenon. Two 'entangled' photons move a vast distance apart and then, when one is measured into a definite polarisation, the other photon immediately responds as if it 'knows' what has happened to its partner photon. And this immediate response is not affected by any scale of distance between the two photons.

So, it seems that at the quantum level there is some kind of instantaneous communication between entangled particles. This astonishing aspect of the quantum world is quantum 'nonlocality'. Greene has remarked that:

> Numerous assaults on our conception of reality are emerging from modern physics ... But of those that have been experimentally verified, I find none more mind-boggling than the recent realisation that our universe is not local.[146]

In order to reinforce the 'mind-boggling' nature of the quantum phenomenon of nonlocality David Lindley, in his book *Where Does the Weirdness Go*, presents the situation in terms of a pair of gloves, a analogy first employed by John Bell. Suppose someone buys a pair of gloves and then sends the left glove to someone in Hong Kong and the right to someone else in New York. This represents the 'common-sense' situation under the 'locality' view that we are familiar with from our everyday macroscopic world; the 'leftness' of the Hong Kong glove is 'local' to Hong Kong and the 'rightness' of the other glove is 'local' to New York. There can be no instantaneous interconnection between the two gloves which affects their handedness at the moment that one of the packages containing either glove is opened. But if these gloves were quantumly entangled then both gloves would be in a state of hovering between 'leftness' and 'rightness' until one of the packages containing the individual gloves was opened, at which point the gloves would magically adopt a definite 'leftness' and corresponding 'rightness'. Thus there is a non-local instantaneous interconnection between the gloves, an interconnection which defies 'common sense'. It was this quantum defiance of what one expects a really real, straight-down-the-line, no-nonsense, common-sense reality to conform to that led physicists like Einstein, de Broglie, Bell and Bohm in his earlier career, to be suspicious!

In 1935 Einstein, together with Boris Podolsky and Nathan Rosen wrote a paper in which they argued that, contrary to the predictions of quantum physics, particles must have definite local attributes at all times. This paper became famous as the EPR paper. EPR argued that, because the measurement outcome at one end of the entangled pair seems to determine what happens at the other side, even if the 'sides' are cosmically distant, there must be some *local* 'elements of reality', at each side, which pre-determine the measurement outcomes. Einstein, Podolsky, and Rosen formulated this view in their famous Reality Criterion:

> If, without in any way disturbing a system, we can predict with certainty (i.e., with probability equal to unity) the value of a physical quantity, then there exists an element of reality corresponding to that quantity.[147]

Thus Einstein and friends, like de Broglie, Bell, Bohm and some others, could not accept ideas of semi-existent smeared-out 'particles' in instantaneous communication across cosmic distances. There must be "predetermined" "elements of reality" at either end of the quantum conundrum. But, unfortunately for Einstein and his colleagues, the EPR paper actually formed the basis for proving the opposite. In 1964, nine years after Einstein's death and nearly 30 years after the EPR paper, the physicist John Bell produced a crucial mathematical inequality which was used to experimentally test the veracity of the non-local quantum effect, and it was conclusively shown that the quantum world is non-local.

Experiments conducted by Alain Aspect and his team examined the status of Bell's inequality as applied to two entangled particles in the late 1970's and early 1980's and found that the results conclusively proved quantum nonlocality. These experiments, and even more precise experiments subsequently performed, showed that there is an intimate connection between entangled particles which operates instantaneously no matter how distant they may be. Subsequently, astounding experiments have been conducted on three entangled particles. The experiment involves measuring vertical, horizontal, and circular polarisation of the three particles. Quantum theory predicts a 'dance' of these attributes which can in no way be possible if there are locally predetermined programs, or 'local' hidden determining features, for the particle states. Physicist Michael Horne observed concerning this:

> Einstein's 'elements of reality' do not exist. No explanation of the beautiful dance among the three particles can be given in terms of an objectively real world. The particles simply do not do what they do because of how they are; they do what they do because of quantum magic.[148]

Here Horne highlights the fact that the appearance of the "objectively real world" of independent completely separate objects is, from a quantum point of view, misleading. Quantum experiments now show that at a deeper quantum level there are extraordinary 'hidden' interconnections underlying the everyday world: "quantum magic!"

But not everyone was enamoured with 'quantum magic'. John Bell, like Einstein, was someone who was desperate for a 'realist' account of the process of reality. When confronted with evidence that the nonlocal EPR correlations were real and therefore 'spooky' quantum interconnections exist, he declared:

> Most physicists were (and are) unimpressed by [EPR correlations]. That is because most physicists do not really accept, deep down, that the wavefunction is the whole story. They tend to think that the analogy of the glove left at home is a good one. If I find that I have brought only one glove, and that it is right-handed, then I predict confidently that the one still at home will be seen to be left handed. But suppose we had been told, on good authority, that gloves are neither right- or left-handed when not looked at. Then that, by looking at one, we could predetermine the result of looking at the other, at some remote place, would be remarkable. Finding that this is so in practice, we would very soon invent the idea that gloves are already one thing or the other even when not looked at. And we would begin to doubt the authorities that had assured us otherwise. That common-sense position was that taken by Einstein, Podolsky and Rosen, in respect of correlations in quantum mechanics. They decided that the wavefunction, making no distinction whatever between one possibility and another, could not be the whole story. And they conjectured that a more complete story would be locally causal. However it has turned out that quantum mechanics can not be 'completed' into a locally causal theory, at least as

long as one allows, as Einstein, Podolsky and Rosen did, freely operating experimenters. The analogy of the gloves is not a good one. Common sense does not work here.[149]

So, Einstein's belief in, and hope for, independent 'elements of reality', each with their own separate independent on-board bits of information, is false. The universe seems, at least at the quantum level, more like an interconnected pattern, a pattern that contradicts the possibility of absolutely separate, independently existent elements with definite completely independent attributes.

This seems to imply that if no measurements were to take place then the entire universe must become a complex web of quantum entanglement within which every particle would be non-locally connected to all other particles. As Roger Penrose tells us:

> So long as these entanglements persist, one cannot, strictly speaking, consider any object in the universe as something on its own ... Why is it not necessary to consider that the universe is not just one incredibly complicated quantum entangled mess that bears no relationship to the classical-like world that we actually observe?. In practice, it is the continual use of the procedure **R** that cuts the entanglements free ...[150]

Penrose uses the symbol '**U**' to denote the underlying quantum wavefunction that develops through time according to the mathematics of the wavefunction, and the symbol '**R**' to refer to the outcome when a wavefunction collapses into actuality, seemingly through the agency of measurement, denoted by '**M**' in the diagrams on the following pages. I have called these 'Penrose Diagrams', but they are my diagrammatic presentations of Penrose's exposition.

A wavefunction precisely predicts the evolution over time of the state of a quantum system, a 'state' being the large collection of possibilities contained within the wavefunction. But then, as Penrose describes:

> From time to time – whenever we consider that a 'measurement' has occurred – we must discard the quantum state that we have been laboriously evolving, and use it only to compute various probabilities that the state will 'jump' to one or another of a set of *new* possible states.[151]

And, significantly, the new states appear as classically 'physical' states whereas prior to the measurement event the wavefunction is a purely abstract mathematical construction to which a 'classical' physical reality cannot be ascribed.

       M   **M**   **M**
       ⇓   ⇓   ⇓

U> U> U> U> R > U > U> U> U> R > U> U> U> U> R > U> U> U

U = Unitary continuous development of wavefunction
M = Measurement
R = Real momentary discontinuous transition to 'physically real' event.

                   Penrose diagram no. 1

As we have just seen, Penrose tells us that the process **R** also "cuts the entanglements free". In other words, the reason that, according to the Copenhagen view, 'spooky' entanglements do not show up in the everyday 'classical' world is that continuous measurements eliminate entanglements from the classical everyday world. This is illustrated by the next diagram which replaces the symbol **U** with **E**, which stands in for 'universal Entanglement'. Here we see that measurements play the role of 'cutting free' the continuous state of universal interdependent entanglement which would be in operation if the measurements did not take place.

       M   **M**   **M**
       ⇓   ⇓   ⇓

E> E> E> E> R > E > E> E> E> R > E> E> E> E> R > E> E> E>

E = Entangled state of quantum interconnected universe
M = Measurement
R = Real momentary discontinuous transition to 'physically real' event.

                   Penrose diagram no. 2

But, as we know, the primary motivation for Bohm's 1952 pilot-wave proposal was precisely to show that the 'measurement' process as conceived of within the Copenhagen worldview, wherein a stack of 'semi-existent' possibilities get reduced to one 'real' element through a 'measurement' interaction, could be replaced by a new, more 'realistic' proposal in which particles exist all the time, and are guided around by pilot waves. Therefore, there is no Copenhagen-style 'measurement' intervention and no subsequent Copenhagen-style 'reduction'. In this situation Penrose diagram no. 1 becomes Penrose diagram no. 3 and Penrose diagram no. 2 becomes Penrose diagram no. 4. We are left with an undivided, continuously entangled universe!

U> U> U> U> U > U > U> U> U> U > U> U> U> U> U > U> U> U>

U = Unitary continuous development of wavefunction. An undivided universe.

Penrose diagram no. 3

E> E> E > E> E> E> E> E> E> E> E> E> E> E> E> E> E> E>

E = Entangled state of quantum interconnected universe. A continuously interconnected entangled universe.

Penrose diagram no. 4

So, what would a continuously entangled and entangling universe be like. Penrose gives the following account in his massive tome *Roads to Reality*:

> How are we to come to terms with quantum entanglement and to make sense of it in terms of ideas we can comprehend, so that we can manage to accept it as something that forms an important part of the workings of our actual universe? ... A puzzle that must be faced is that entanglements tend to spread. It would seem that eventually every particle in the universe must be become entangled with every other. Or are they already all entangled with every other. Why do we not just experience an entangled mess with no resemblance whatsoever to the (almost) classical world we actually perceive? The Schrödinger equation evolution of system does not help with this. It tends to make things worse, with more and more parts of the universe becoming entangled with any system we start with, as time goes on. ... It is my view that Nature herself is continuously enacting **R**-process effects, without any ... intervention by a 'conscious observer'.[152]

The Schrödinger equation is the fundamental quantum equation which mathematically describes the way in which the potentialities, which 'semi-exist' at the quantum level, develop and change over time. The Copenhagen view indicates that this equation is 'collapsed' into full reality, which Penrose denotes by the term the "**R**-process", perhaps through some kind of mechanism involving consciousness. Penrose, however, suggests that there must be a "**R**-process" internal to the functioning of the universe which is not dependent on consciousness.

The mechanism that Penrose advances as his preferred view is that of gravitational 'objective-reduction', as opposed to a 'subjective-reduction' by consciousness. The basic idea is that there are gravitational aspects within quantum superpositions which cause them to 'collapse' and there-by there is an objective-reduction to a classical state. In this way Pen-rose is able to avoid the agency of consciousness getting a hold on him. However, this seems like a desperate quantum leap in the dark, resorted to merely due to a desire to avoid the 'mystical' intrusion of the agency of mind. For some observers, at one point in his career Penrose was in danger of becoming a quantum mystic! This derives from his view that consciousness has a non-computational aspect of creativity at its core, a view he has explored in books such as *The Emperor's New Mind* and *Shadows of the Mind*. But, it seems that Penrose was able to steer away from such a dangerous 'quantum-mystic' fate for a mainstream physicist by clutching onto to gravity!

Bohm's decision, made forty years prior to Penrose's speculations, to explore a new 'realistic' approach to quantum functioning was, at least in part, a result of being called by Einstein to have a discussion. Bohm had just written a textbook *Quantum Theory* which had been given good reviews. Einstein had read it and was not happy with its Copenhagen perspective. After the discussion Bohm had significantly changed his viewpoint. As Adam Becker tells us in his recent book *What is Real*, Bohm recalled: "I began to wonder, does the [wave function] give a complete description of reality?" And Becker continues:

> Einstein was sure that it didn't. Bohm took that idea and ran with it. In a matter of weeks, he discovered that there was a simple way to rewrite the fundamental equations of quantum theory. The predictions and results remained the same - the new version was mathematically equivalent to the old - but the picture of the world suggested by the math, the story that

it told, was radically different from the Copenhagen interpretation.[153]

It is important to note here that Becker clearly tells us that Bohm was able to transform the equations so that they gave exactly the same results, but they could be 'read' in a different way from the Copenhagen viewpoint. Becker describes Bohm's new revelation:

> In Bohm's interpretation of quantum physics, much of the mystery of the quantum world simply falls away. Objects have definite positions at all times, whether or not anyone is looking at them. ... particles are just particles, and their motions are guided by pilot waves. Particles surf along these waves ...[154]

In Becker's description, it seems for all the world as if Bohm had just seen through all the Bohr-inspired Eastern-mystification and obfuscation that had misled physicists previously, possibly due to hidden hypnotic powers on the part of Bohr, and had miraculously seen the true light of how quantum theory was really just like the everyday world of pots and pans and washing-machines. But this story is misleading It is a story concocted by only focusing on one particular area of Bohm's new 'picture of the world', and disregarding other aspects which are not so conducive to a new, more 'realist' picture.

In this context it is extremely instructive to consider Heisenberg's response to Bohm's 1952 proposal, made in an essay 'Criticisms and Counterproposals to the Copenhagen Interpretation of Quantum Theory':

> When one analyses the papers of ... [those who do not "want to change the Copenhagen interpretation so far as predictions of experimental results are concerned", but try "to change the language of this interpretation in order to get a closer resemblance to classical physics"] it is important to realize from the beginning that their interpretations cannot be refuted by experiment, since they only repeat the Copenhagen interpretation in a different language. .... Along this line, for instance, Bohm has made a counter-proposal to the Copenhagen interpretation. ... Bohm considers the particles as 'objectively real' structures, like the point masses in Newtonian mechanics. The waves in configuration space are in this interpretation 'objectively real' too, like electric fields. Configuration space is a space of many dimensions ... Here

> we meet a first difficulty: what does it mean to call waves in configuration space 'real'? This space is a very abstract space ... but things are in the ordinary three-dimensional space, not in abstract configuration space. One may call the waves in configuration space 'objective' when one wants to say that these waves do not depend upon any observer; but one can scarcely call them 'real' unless one is willing to change the meaning of the word. ... This objective description ... reveals itself as an 'ideological superstructure', which has little to do with immediate physical reality ...[155]

It has to be said that Heisenberg has a powerfully valid point. According to Bohm's pilot wave proposal:

> An N-particle quantum system (N > 1) is a generalisation from the one-particle case. There is only a single guiding wave field represented by the wavefunction ψ, but now ψ is defined on a 3N-dimensional configuration space.[156]

This means that for a 10 particle system, for example, any one of the particles in the system is 'guided' by a quantum pilot wave which has a configuration space of 30 dimensions. As Heisenberg points out, things in the 'real' everyday world move around in three dimensional space, and the forces upon them operate in three dimensional space, but particles in Bohm's worldview are pushed around by forces determined by a 3N-dimensional configuration space, which is 30 dimensions for only 10 particles. According to one calculation a human hair is about 300,000 atoms in width! So can one equally use the term 'real' in these two contexts, as if both situations have exactly the same kind and degree of reality, so to speak, justified?

And yet according to Becker's presentation of Bohm's pilot wave view: "the mystery of the quantum world simply falls away". In another recent book *Through Two Doors at Once: the Elegant Experiment that Captures the Enigma of Our Quantum Reality* the science writer Anil Ananthaswamy provides an equally no-nonsense, down-to-earth and matter-of-fact, or perhaps fact-of-matter, account of Bohm's 1952 quantum worldview:

> Here was an alternative to the Copenhagen view of things: a deterministic theory of particles that move around because of interactions with the wavefunction, which in turn is a real thing and evolves according to the Schrödinger equation.

> Bohm's theory has a definite ontology: the world is made up of particles and wavefunctions, even if wavefunctions are not "physical" in the sense that particles are physical, but nonetheless are real, objective aspects of reality. A particle has a definite position at all times, which means it has a trajectory - in direct contravention of the Copenhagen view of reality. The particle is "guided" by the wavefunction, and thus not just be the usual forces (such as electromagnetism), but by a "quantum potential," a new force felt by the particle because of its interactions with the wavefunction... Moreover, the theory is deterministic: given a particle's position and its wavefunction, you can predict the particle's position at some later time. And even more emphatically, the particle's trajectory is objective reality - it exists independent of an observer.[157]

This description clearly emphasises the 'real', 'deterministic', 'objective' demeanour, the fact that it is "independent of an observer" and the fact that "Bohm's theory has a definite ontology", but it does not mention the fact that the dimensionality of the so-called 'objective', 'real', 'deterministic' guiding wave is so extravagant that, at least to some frugal minds, it pushes the meaning of the word 'real', used with the implication of more 'classical', towards the seriously implausible, certainly from a 'classical' point of view. For in Bohm's perspective the guiding wave for the universe would be, for all intents and purposes, infinitely dimensioned, as the nature of the guiding wave for the entire universe may depend threefold upon the position of every particle in the universe!

Also, consider the issue of "empty ghost waves". Lee Smolin, in his recent book *Einstein's Unfinished Revolution: The search for what lies beyond the quantum*, points out:

> ... there are several reasons that pilot wave theory is not entirely convincing as a true theory of nature. One is the empty ghost branches, which are parts of the wave function which have flowed far (in the configuration space) from where the particle is and so likely will never again play a role in guiding the particle. These proliferate ... but play no role in explaining anything we have actually observed in nature. Because the wave function never collapses, we are stuck with a world full of ghost branches. There is one distinguished

> branch, which is the one guiding the particle, which we may call the occupied branch. Nonetheless, the unoccupied ghost branches are real.[158]

When one comprehends this extravagance and counter-intuitiveness of the 1952 Bohmian quantum metaphysics, with vast numbers of redundant supposedly 'real' "ghost waves" zooming around all over the universe, apparently most of which have no purpose, although some occasionally are able to "recapture its particle and cease to be empty",[159] for example, Heisenberg's objections to the Bohmian quantum metaphysical worldview can be better appreciated. The idea that Bohm's new quantum perspective is more classically 'realistic' "simply falls away", to employ a phrase used by Baggott to suggest that Bohm had rescued the reality of the kitchen sink from quantum mystery.

The point is that the everyday common-sense, kitchen-sink down-to-earth credentials of the 1952 Bohmian perspective seem to be frequently wildly over-inflated. And we still have not looked closely at the issue of Bohmian nonlocality yet! But just to anticipate what is to come shortly, according to Norsen: "The pilot-wave theory is manifestly non-local in the following sense: the velocity of each particle, at a given instant, depends upon the instantaneous positions of all other particles (at least when there is entanglement)". And this non-local interdependence could involve incomprehensibly vast numbers of particles across the universe!

There are 2 main points of Heisenberg's criticism of Bohm's proposal:

1) It is actually Copenhagen in essence but the equations and language have been rearranged to conform to a 'realist' 'ideological superstructure';

2) The claim that the pilot-wave (or 'quantum potential') is 'real' in a classical everyday sort of sense is dubious.

Consider the first Heisenberg objection, which is that Bohm's proposal is in essence the Copenhagen perspective reworked into different language with massaged equations, reworked according to a 'ideological superstructure'. The following passage from *Foundations of Quantum Mechanics: An Exploration of the Physical Meaning of Quantum Theory* by Travis Norsen is important because it begins to clarify why Heisenberg's first objection has some force:

> …probably the most important virtue of the pilot-wave theory is that we do not need to divide the world up into "quantum

system" ... and "classical environment" ... in order to understand measurements and their outcomes. Instead, we are free (indeed, required!) to enlarge the "quantum system" ...until it includes literally everything - the entire universe. This is of course in contrast to ordinary quantum mechanics which ... seems to require one to introduce what Bell called a "shifty split" (i.e., an artificial division of the world into distinct "quantum" and "classical" realms, with special ad-hoc exceptions to the usual dynamical rules when the two realms interact). The claim, then, is that unlike orthodox quantum mechanics, the pilot-wave theory is not afflicted with a "measurement problem".[160]

In the Copenhagen scenario as usually envisaged, the quantum and classical realms are conceived of as essentially separate, and somehow the quantum realm then transmutes into the classical through the intervention of a measurement, and, because of this, quantum effects such as non-locality are lost in the transition, so they do not turn up uninvited into the classical realm. In the Copenhagen view, therefore, nonlocality effects are restricted to the quantum level, and the everyday classical level remains reassuringly everyday, with no outlandish quantum intrusions, so to speak. Bell's "shifty split" acts as a kind of gatekeeper keeping quantum weirdness corralled within the quantum level. Bohm's pilot-wave quantum worldview, on the other hand, conflates this division so that quantum functioning becomes part of the 'classical' world to a much greater extent.

The Copenhagen division appears to be also a division in ontology, i.e. what kind of 'stuff' reality is made up from. Pre-measurement quantum 'stuff' is not 'real' 'stuff' in the way that post-measurement 'classical' stuff really is 'real' 'stuff'. Pre-measurement quantum 'stuff' is more like potential 'stuff', not fully manifested 'stuff'. And this was and is one of the central issues for some physicists. Norsen presents his discomfort with this division:

> ...all this should make one feel very uncomfortable. To begin with, there is a kind of schizophrenic division of the world into two "realms" (the microscopic quantum part, and the macroscopic classical part) which seem to have completely different ontologies and completely different dynamical laws. And then there are apparently special dynamical rules which come into play when the two realms interact, during a

"measurement". If you say "during measurements, quantum wave functions cease to obey Schrödinger's equation and instead collapse" that is already weird and troubling, but it becomes downright *meaningless* if you can't specify exactly what kinds of physical processes count as "measurements".[161]

So, by bringing together the two "realms" into the one all-encompassing quantum-classical "realm", with the classical functioning now being inside the quantum functioning, and, furthermore, reworking the quantum equations with brilliant mathematical ingenuity so that some aspects of particle functioning, primarily position, can be continuously specified, Bohm produced a version of quantum-classical functioning purely within the classical realm wherein particles appeared to be always particles. But another crucial result of this conflation was that quantum effects such as nonlocality now turned up as a significant part of what was previously a non-quantum 'realm' of the 'classical' world. This had to happen because without Bell's 'shifty-split' there is nowhere for nonlocality to be located except the classical world!

But in Bohm's new 'realistic' reality it did seem that the "measurement" problem disappeared, particles were always particles. This feat was welcomed with joy by some, such as John Bell who exclaimed that he had seem the "impossible done". Einstein, however, was not happy with Bohm's proposal, he called it "too cheap", perhaps because the one quantum phenomenon that irritated him most was nonlocality:

> It is not clear exactly what Einstein meant as "too cheap", but it seems likely that the theory did not strike him as a step in the right direction since it failed to eliminate (but in some ways exacerbated) the one feature that Einstein found most unacceptable in orthodox quantum theory: nonlocality.[162]

As we shall see, Bohm quickly realised that his new formulation indicated that nonlocality was a primary feature of his new quantum worldview, and this led him in a new direction of quantum metaphysical development.

As we have seen, a primary aim of Bolm's pilot-wave endeavour was the elimination of the strange split ontology and the removal of the claimed 'measurement problem'. And it has been claimed that Bohm achieved this goal. But, as we have also seen, Heisenberg considered that Bohm's achievement was in essence Copenhagen dressed in new, ideologically correct, which is to say 'realist', clothing. If Heisenberg were to be correct about this, then it would be reasonable to expect that the measurement conundrum had not been banished, but perhaps tucked away in one of the pockets of the new Bohmian clothing, And it is possible to make a case that this is indeed the case. In fact the measurement conundrum appears to have been hidden in a new feature of Bohmian Mechanical functioning called 'Contextuality'.

Whilst it is the case that particles have definite positions at all times within Bohm's pilot-wave formulation, this does not mean that all possible 'properties' which can be 'measured' have intrinsic values which manifest as the same value in different measurement situations. In his book *Making Sense of Quantum Mechanics* Jean Bricmont explains the situation, in the context of the quantum property 'spin':

> ...the value up or down of the spin that actually results from the measurement is "contextual": that value does not depend only on the quantum state and the original particle position but on the concrete arrangement of the "measuring" device. Here the scare quotes are used because we finally see the truth of something we have emphasized several times: there is no intrinsic property of the particle that is being "measured", in general, in a "measurement". Of course, since the system is deterministic, once we fix the full initial state (the quantum state and the position) of the particle and the experimental device, the result of the experiment is predetermined. But that does not mean the spin value we "observe" is predetermined ... So not only is the word "hidden variable" misleading [because the 'variables' in Bohm's proposal are not actually 'hidden'], since except when one measures positions, one does not observe or measure any intrinsic property of the system ...[163]

And then Bricmont makes an admission which is interesting, given the fact that he is a hardcore and committed supporter of the 'realism' of Bohmian Mechanics. This situation, Bricmont says, "vindicates in some sense Bohr's emphasis on" and he then quotes Bohr:

> ...the impossibility of any sharp distinction between the behaviour of atomic objects and the interaction with the measuring instruments which serve to define the conditions under which the phenomena appear.

This clearly shows, as Heisenberg wisely pointed out, that the devilish spectre of Bohr's Copenhagen view has not been fully exorcised from Bohmian Mechanics.

It seems that, by brilliantly massaging equations, Bohm was able to make particles have definite positions and trajectories, but other aspects of Copenhagen quantum weirdness seem to transform into different forms of weirdness. The Cambridge University philosopher of quantum physics Jeremy Butterfield, for example, described quantum 'contextuality' in very dramatic terms in a internet interview when he gave the analogy of trying to determine whether a car was a BMW or a Porsche. If this everyday large scale issue were to be 'contextual' the situation would, indeed, be very strange, for the determination depends upon:

> ...what else you are trying to determine. ... the result you get might depend on whether you are also looking at colour or age. If I want to also know the colour I might find it is a BMW, ... but if I also want to know its age I could find out its a Porsche...[164]

In other words, in a situation of quantum contextuality, using a classical level analogy, if you try and determine car-type and car-colour at the same time you might end up finding a BMW, but try and find the car-type and car-age, of the same car, it might turn out to be a Porsche! Again, this is a very non-classical mode of existence!

Bricmont further discusses non-contextuality as applying only to the position of particles:

> The statement that the measurement of an observable depends upon concrete experimental arrangement used to "measure" it, is true ... for all quantum mechanical "observables", other than position.[165]

Which means that Bohm only produced a classical-type 'realism' in a very limited domain, the domain of position and trajectory of particles. And then Bricmont quotes John Bell on the problem concerning the term "measurement":

> ...the word comes loaded with meaning from everyday life. Meaning which is entirely inappropriate in the quantum context. When it is said that something is 'measured' it is difficult not to think of the result as referring to some pre-existing property of the object in question. This is to disregard Bohr's insistence that in quantum phenomena the apparatus as well as the system is

essentially involved.

This is a crucial point, the 'properties' in Bohm's pilot-wave worldview, other than position, are not independent elements of reality, they arise as 'contextual' results of interactions between the measured system and the measuring apparatus. Finally Bricmont quotes from Detlef Dürr and other Bohmian philosophers:

> "Properties" that are merely contextual are not properties at all, they do not exist, and their failure to do so is in the strangest sense possible.

In the light of statements like this, concerning Bohmian 'properties' other than position, it is difficult not to agree with Heisenberg that the 'realist' nature of Bohmian mechanics is over-played, perhaps even wildly exaggerated. As the philosopher Bernado Kastrup has pointed out concerning a contextual physical reality, which is a physical reality "without non-contextuality":

> It is nonetheless fair to say that, never before in the history of Western thought, at least since the Enlightenment, has the idea of a definite physical world independent of observation looked so precarious. Non-contextuality, if not dead, is in life support. And here are the key implications, ... without non-contextuality, there are a) no physical objects or events, with definite properties and occupying definite positions in space-time, before observation. The world as it is in itself, independently of observation, exists instead in a so-called 'quantum superposition': a state of overlapping potentialities or tendencies, but no concrete 'physicality' in the sense we ordinarily attribute to the word. There is also (b) no tenable way to carve out separate objects and events in the world-in-itself. ... prior to observation, everything in the world remains quantum-mechanically entangled and, as Jonathan Schaffer observed, "there is ... good reason to treat entangled systems as *irreducible wholes*".[166]

Thus, Bohm's attempt to eliminate quantum 'unreality' and indeterminism actually led him in the direction towards a quantum philosophy of wholeness.

This leads us to Heisenberg's second objection, which is that the claim that the Bohmian pilot-wave scenario can be considered as fully 'real' is over-stated, if not dubious. Florian J. Boge points out that, although the

Bohmian view is presented as a no-nonsense classical-type replacement for the vague and mysterious Copenhagen account, there is much equally vague about it:

> ...the wavefunction seems to play ... a *physical* role, and ... it somehow influences the particles, thus correlating them 'non-locally'. Given that ψ is defined on configuration space though, this does not tell us how it manages to transmit its 'non-local influence' to the particles.[167]

And:

> ...we *still* do not know what the wave function 'really is', and how it 'interacts' with particles located in space(time). In fact, the accounts of what exactly the wavefunction represents differ grossly among 'Bohmians'.[168]

And science writer Jim Baggott, in his excellent book *beyond measure: modern physics, philosophy and the meaning of quantum theory*, points out:

> There are many practical objections to the de Broglie-Bohm theory. For example, the wave in the de Broglie-Bohm theory can exert a strong influence on the particle through the form of the quantum potential, but there is no reciprocal action of the particle on the wave, seemingly at odds with classical mechanics in the form of Newton's third law of motion.[169]

Can we take this as a final nail in the quantum coffin of the view that Bohmian mechanics is a thoroughly Newtonian-type 'realistic' account of the everyday world which completely banishes any trace of quantum weirdness from our world? The fact that Bohm's 1952 proposal has been joyfully hailed as bringing the really real world of Newton back to life actually only indicates the desperation among some sections of the scientific and philosophic community to avoid facing the real implications of the quantum lack of real classical kitchen-sink reality!

The truth is that, as Heisenberg points out, Bohm's proposal provides a way to introduce a small degree of classical type functioning for positions and trajectories, thus providing an 'ideological superstructure' for a limited realist account, at the same time as importing a whole load of quantum 'weirdness' into the classical everyday world. In particular, nonlocality becomes a central aspect of the new worldview. Jim Baggott writes about Bohm's theory:

> ...the theory was originally developed simply to demonstrate that a causal (not necessarily entirely classical) interpretation of quantum phenomena is possible in principle. ... some serious problems arise if we attempt to assign properties other than position to the Bohm 'particle', such as charge, and the theory is not without measurement problems.[170]

But this does not mean that Bohm's efforts were in vain, of course, far from it. Although Bohm did not provide a satisfactory 'realist' classical-type quantum interpretation, the subsequent development of his 1952 perspective naturally developed towards a wonderful quantum metaphysics of diversity within wholeness.

One of the aims of Bohm's 'ontological' theory, a theory which supposedly, according to some advocates, gives an (almost) 'realistic' quantum vision, is the elimination of the 'collapse of the wavefunction'. As we have seen, this supposed 'collapse', the 'measurement problem', was, and is, for many, and a disturbing issue. But, Bohm's account does not fully achieve this, it only achieves a different kind of 'collapse' scenario, tucked away off-stage so to speak. Bricmont, in a section of his *Making Sense* book titled 'What About the Collapse of the Quantum State?', declares that:

> The short answer ... is that there is never any collapse of the quantum state in the de Broglie-Bohm theory, but there is an effective collapse or a collapse "in practice", which coincides with the one in ordinary quantum mechanics ...[171]

So, according to Bricmont, in his *Making Sense of Quantum Mechanics* book, there definitely is *never* any 'collapse', except there is an "effective collapse" which is a "collapse in practice"! Does that make sense? It seems that metaphysical nonsense can stand in for sense, such is the desperation to rescue the appearance of really real material reality from the clutches of quantum mystery!

In the context of the wavefunction 'collapse' the philosopher Florian J. Boge explains the situation, making reference to the Bohmian philosopher Detlef Dürr, as follows:

> A first thing to note is that "in a universe governed by Bohmian mechanics there is a priori only one wave function, namely that of the universe, and there is a priori only one system governed by Bohmian mechanics, namely the universe itself" [Dürr 2012]. To make sense of practice,

however, wherein the wavefunction of the universe never occurs, Dürr et al. introduce the notion of *conditional* and *effective wave functions*, used to describe the behavior of subsystems. ... The use of the conditional wave function also leads to an 'apparent collapse'.[172]

Here we see how Heisenberg was correct in his view that the 1952 Bohm proposal can be used as a way of providing a 'ideologically correct' quantum-metaphysics or 'language game'. The term 'language game' is a term coined by the philosopher Wittgenstein; it denotes a way of speaking adopted as a convenience, often for societal, sociological or political ends. Also a 'language game' may be a way of speaking adopted to further a favoured philosophical position. In the context of quantum theory, for example, it is possible to adopt a mode of language in the quantum context which supports either a subtle 'materialist' position, or a more 'mystical' perspective. According to Heisenberg, Bohm's viewpoint was a quantum 'language game' used to assuage the fears of those who tremble at the possibility that quantum reality might not be quite as real as they would like it to be!

And, here we also find clear signs that the seeds of Bohm's later, much more extensive and spectacular, 'wholeness and the implicate order' quantum metaphysical perspective are contained within the 1952 'ontological' 'pilot-wave' theory. As Detlef Dürr points out: "in a universe governed by Bohmian mechanics there is a priori only one wave function, namely that of the universe". Bohm's attempt to produce a more ontologically unidimensional 'realist' account of quantum functioning actually produced a more fully universal and comprehensive quantum perspective of interconnected wholeness, within which there are 'substructures' operating very much in the mode of Bohr's viewpoint, wherein the 'observed' system and 'observing' apparatus forms a necessary whole. And these holistic substructures, functioning within the interrelated context of the holistic universe, turn out to have 'properties' that are 'contextual', and not fully self-contained and independent, precisely because of their dependence upon a universal context.

In their book *The Undivided Universe* Bohm and Hiley describe this situation:

> The probability of a particular result of the interaction between the instrument and the observed object is shown to

be exactly the same as that assumed in the conventional interpretation. But the key new feature here is that of the *undivided wholeness* of the measuring instrument and the observed object, which is a special case of the wholeness to which we have alluded in connection with quantum processes in general. Because of this, it is no longer appropriate, in measurements to a quantum level of accuracy, to say that we are simply 'measuring' an intrinsic property of the observed system. Rather what actually happens is that the process of interaction reveals a property involving the whole context in an inseparable way.[173]

And the fact of quantum nonlocality is a central feature leading to this necessary conclusion. When the mathematical analysis of many-body systems from the Bohmian perspective is pursued it turns out that:

> ... we find that this leads to further new concepts. The most important of these are *nonlocality* and *objective wholeness*. That is to say, particles may be strongly connected even when they are far apart, and this arises in a way which implies that the whole cannot be reduced to an analysis in terms of its constituent parts.[174]

Thus we see that the central issue of nonlocality, which came to the fore in the pilot-wave perspective, necessarily required the development of Bohm's ideas in the direction of his later perspective.

In a very real and ironic sense Bohm's attempt to break with Bohr's quantum perspective backfired. This is because it actually, through its own internal necessity, showed Bohr's Copenhagen view to be a necessary facet of a much more comprehensive quantum-universal interconnected metaphysical picture of wholeness. In the General Introduction to the collection of essays in honour of David Bohm: *Quantum Implications,* Basil Hiley and F. David Peat (editors) describe the development of Bohm's ideas:

> In Bohm's original perception, this notion of interconnectedness was rather vague and ill-defined but with its continual reappearance in different forms, the notion slowly took shape, ultimately leading to a very radical and novel way of looking at reality. This view eventually crystallised into what he now terms the implicate order. The first formal indication of Bohm's departure from orthodoxy can be traced to his

reformulation of quantum mechanics published in *Physical Review* in 1952. But the ideas that lay behind that formulation seem to many to be totally against the spirit of his later work on the implicate order, so much so that they find it hard to see any connection at all. It is true that those papers were more intent on demonstrating that there was another logically coherent interpretation of the quantum mechanical formalism, other than the usual one. But it is the ideas implicit in this reformulation that have connections with the notion of the implicate order.[175]

The fact that many people overlook the seeds of implicit development in Bohm's 1952 approach is precisely because, as we have seen, there is a group of physicists, philosophers and science writers who wrongly promote, in lectures, technical books, and various popular books on quantum physics and metaphysics, the idea that Bohm achieved an almost Newtonian classical-type version of quantum mechanics with his 'pilot-wave' theory, thus supposedly thwarting the dangerous spread of Bohr's alleged vague dalliance with Eastern mysticism. An excellent example of this mistaken view is Adam Becker's incorrect claim that in the pilot-wave theory "the mystery of the quantum world simply falls away". In this context it is useful to take into account the insight of Alexander Wendt presented in his book *Quantum Mind and Social Science*:

> Bohm himself saw his ontology as non-classical, and nowhere is this more apparent in his view - emphasized more in his later works - that the wave function has a primitive form of mentality. Curiously, this suggestion is often completely ignored in the interpretive debate *within* the Bohm Interpretation, giving it a more materialist spin than perhaps he intended.[176]

We have previously seen aspects of this tendency.

It seems that Bohm clearly quite quickly saw the necessary implications of his early work, and in the conclusion of the first chapter of his 1957 book *Causality and Chance in Modern Physics* Bohm wrote: "However, since the natural laws imply some kind of interconnection of all aspects of the world, as well as their approximate autonomy".[177] And on the last page we find a clear adumbration of the later notion of the 'implicate order' and the 'holomovement' and the primacy of wholeness:

> In conclusion, a consistent conception of what we mean by the absolute side of nature can be obtained if we start by considering the infinite totality of matter in the process of becoming as the basic reality. This totality is absolute in the sense that it does not depend upon anything else for its existence or for a definition of any of its characteristics. On the other hand, just what it is can be defined concretely only through the relationships among the things into which it can be analysed approximately. Each relationship has in it a certain content that is absolute ...[178]

And in Bohm's later perspective Bohr's central insight is reinstated. Hiley and Peat, important developers and promoters of Bohm's later perspective, praise Bohr's original insight mightily in the following terms, which rather than denigrate Bohr, elevates him as a forerunner for Bohm's later ideas:

> Perhaps Bohr's deepest perception was not wave-particle duality, nor complementarity, but *wholeness*. Bohr writes, 'The essential wholeness of proper quantum phenomenon finds indeed logical expression in circumstances that any attempt at its well-defined subdivision would require a change in the experimental arrangement incompatible with the appearance of the phenomenon itself'. Remember of course that for Bohr the word 'phenomenon' refers only to observations obtained under circumstances whose description includes an account of the whole experimental arrangement'.[179]

Here Bohr indicates the fundamental quantum aspect of *wholeness*, the fact that 'measured' and the 'measurer' cannot be taken as being entirely separate entities but are themselves interconnected at a deeper level.

Another important aspect of wholeness is the fact that, by bringing the quantum level into the 'classical' world, so to speak, Bohm highlighted a subtle deep level of interdependence within all aspects of the universe. As Hiley and Peat have pointed out, it is the nonlocal nature of the Bohmian quantum worldview, the 'causal interpretation', that has caused consternation among some physicists:

> A particular objection appears to arise from what scientists call the nonlocal nature of this approach. This can be explained in the following way. When several particles are treated in

the causal interpretation then, in addition to the conventional classical potential that acts between them, there is quantum potential which now depends *on all the particles*. Most important, this potential does not fall off with distance between particles, so that even distant particles can be strongly connected. This feature, in which very distant events can have a strong influence, is what is meant by a *nonlocal* interaction and is strongly at variance with the whole spirit of classical mechanics.[180]

Here we see that because Bohm's 'causal interpretation', or 'pilot-wave theory', brings quantum features into the classical realm, then what was once just a classical realm becomes a classical plus quantum realm. So not only are there classical potentials acting on particles, there is now also quantum potentials which depend upon all other particles, no matter how distant the particles may be.

Hiley and Peat continue:

There is a great reluctance on the part of physicists to consider such nonlocality seriously, even though it does lie at the heart of the formal implications of quantum theory. ... there does not appear to be any intrinsic reason to rule out nonlocal forces. ... The quantum potential cannot be ... used to carry any signal between distant effects and therefore its instantaneous connection between particles does not violate the theory of relativity.[181]

However, although the deep level of interconnected wholeness exists at the absolute, ultimate level of the process of reality, and permeates the 'lower' explicate levels, this does not invalidate the apparent independence of the substructures which have their 'explicate' mode of existence within the context of the undivided wholeness of the universe. When these substructures are perceived as being completely separate and fragmentary they may lose their full meaning, but they still function as independently functioning sub-units. Within the context of the whole, however, functioning substructures, or 'aspects', function as parts which clearly contribute to the functioning of the whole. In *Wholeness and the Implicate Order* Bohm wrote concerning this:

... in its totality, the holomovement is not limited in any specifiable way at all. ... *the holomovement is undefinable and immeasurable*. ... each theory will abstract a certain aspect that is relevant only in some context, which is indicated by some appropriate measure. In discussing how attention is to be called to such aspects, it is useful to recall that the word 'relevant' is a form obtained from the verb 'to relevate' ... 'to lift up'. We can thus say in a particular context that may be under consideration, the general modes of description that belong to a given theory serve to *relevate* a certain content ...[182]

And in *The Undivided Universe* we read:

The essential features of the implicate order ... that the whole universe is in some way enfolded in everything and that each thing is enfolded in the whole. However, under typical conditions of ordinary experience, there is a great deal of *relative* independence of things, so that they may be abstracted as separately existent, outside of each other, and only externally related. ... the explicate order, which dominates ordinary 'common sense' experience as well as classical physics, appears to stand by itself. But actually this is only an approximation and it cannot be understood apart from its ground in the primary reality of the implicate order, i.e. the holomovement.[183]

Thus we see that the existence of 'relevated' independently functioning substructures can be treated, for most purposes, as self-contained areas of discourse, with independent and self-enclosed modes of analysis and description, within the absolute context of wholeness. But, in Bohm's view, each substructure only has meaning within the context of the totality of wholeness. However, it is perhaps more precise to say that each substructure only has *ultimate* meaning within the context of the totality of wholeness.

What better way to end this chapter than to point out what a wonderful analogy the Hindu vision of Indra's Net is for Bohm's vision of an undivided wholeness constituted by mutually interpenetrating parts. In this image the whole universe is portrayed as an interpenetrating multidimensional net of jewels, which may be thought of as representing the infinite sparks of interconnected consciousness which underlie the appearances of the phenomena world. Jewels are set at every intersection of the net and each jewel reflects the light reflected in all the jewels around it, and each of those jewels in turn reflects the light from all the jewels around them, and this multifaceted mutually reflective process is repeated infinitely. In this way, all phenomena – events, entities and sentient beings reflect and express the radiance of the entire universe. All of totality can be seen in each of its parts. This later became incorporated into the Hua-yen doctrine which views the entire cosmos as a single nexus of conditions in which everything simultaneously depends on, and is depended on by, everything else. This description is from the *Avatamsaka Sutra*:

> There is a wonderful net which has been hung by some cunning artificer in such a manner that it stretches out indefinitely in all directions. In accordance with the extravagant tastes of deities, the artificer has hung a single glittering jewel at the net's every node, and since the net itself is infinite in dimension, the jewels are infinite in number. There hang the jewels, glittering like stars of the first magnitude, a wonderful sight to behold. If we now arbitrarily select one of these jewels for inspection and look closely at it, we will discover that in its polished surface there are reflected all the other jewels in the net, infinite in number. Not only that, but each of the jewels reflected in this one jewel is also reflecting all the other jewels, so that the process of reflection is infinite.[184]

Bohmian Quantum Emptiness

In a talk titled 'The Implicate Order: A New Order for Physics', given at a conference organized by the Center for Process Studies, David Bohm made the following comments:

> In other words, the energy in empty space is immensely greater than the energy of matter as we know it. Therefore, matter in itself is a kind of ripple in empty space. Matter is a relatively stable and autonomous ripple in the emptiness. Those of you who have studied the theory of solid states may not find this notion of emptiness entirely unfamiliar. For example, in a crystal of very dense material at absolute zero, if the crystal is of perfect order, electrons go right through it as if nothing were there. The suggestion is then that emptiness is really the essence. It contains implicitly all the forms of matter. The implicate order really refers to something immensely beyond matter as we know it -- beyond space and time. However, somehow the order of time and space are built in this vacuum.[185]

It is worth highlighting the following passage from the above quote from Bohm:

> The suggestion is then that emptiness is really the essence. It contains implicitly all the forms of matter.

And then contemplate its close resonance with the memorable passage from the *Heart Sutra* (*Prajnaparamita Hridaya*). In the following formula the term 'form' refers to what science calls 'matter':

> Form is emptiness; emptiness also is form. Emptiness is no other than form; form is no other than emptiness.[186]

This tells us that the material world is a manifestation from a deeper immaterial realm of potentiality - emptiness (*sunyata*). The Buddhist practitioner and translator Karl Brunnhölzl says of this in his excellent book *The Heart Attack Sutra*:

> In fact, emptiness is the very reason that anything can appear at all because emptiness is the fundamental space, nonsolidity, and openness in which appearance, movement, functionality, and change are possible. If things were solidly and independently existent, nothing could ever appear newly or change.[187]

We briefly saw in the last chapter that many physicists in the days of early quantum discoveries were shocked at what they were discovering about the quantum level of reality, Heisenberg exclaiming at one point:

> Can nature possibly be as absurd as it seems to us in these atomic experiments?[188]

More recently physicist Nick Herbert, in his book *Quantum Reality: Beyond the New Physics,* has written, in a similar vein, that;

> Nothing exposes the perplexity at the heart of physics more starkly than certain preposterous claims a few outspoken physicists are making concerning how the world really works. If we take these claims at face value, the stories physicists tell resemble the tales of mystics and madmen.[189]

The shock on the part of physicists when originally discovering the nature of quantum reality has a forerunner event in the reaction of monks when the Buddha disclosed a similar revelation when, in the second turning of the wheel of Buddhist metaphysics, he began teaching the emptiness of all conditioned phenomena. The reason that Brunnhölzl titled his book on the *Prajnaparamita Hridaya* '*The Heart Attack Sutra*' is that monks are said to have collapsed with heart attacks when they heard that what appeared to be a 'real' world was actually an 'empty' illusion. And Brunnhölzl indicates the connection between the Buddhist emptiness teachings and quantum physics:

> In a sense, the teachings on emptiness have a lot of parallels with quantum physics. Quantum physicists tell us there is really no world out there. There is actually not much, if anything. They are still looking for *something*, because it sounds better and we do not have to be scared that there is really nothing at all to hold on to. When physicists talk about a quantum field, it almost entirely consists of space and some energy in it, not even particles. They may talk about "particles," but this term does not refer to any kind of substance anymore, just statistical probabilities of relationships. This is very much what emptiness is about, meaning that there is no single phenomenon whatsoever that exists independently on its own.[190]

And this assertion by a Buddhist practitioner is clearly endorsed by the facts of quantum field theory, the following passage is from Jonathan Allday's book *Quantum Reality: Theory and Practice*:

Our whole manner of speech ... rather naturally makes us think that there is some stuff or substance on which properties can, in a sense, be glued. It encourages us to imagine taking a particle and removing its properties one by one until we are left with a featureless 'thing' devoid of properties, made from the essential material that had the properties in the first place. Philosophers have been debating the correctness of such arguments for a long time. Now, it seems, experimental science has come along and shown that, at least at the quantum level, the objects we study have no substance to them independent of their properties.[191]

Because there is no substance at the level of quantum field theory the term 'particle' is dropped and the term 'quanta' is used, and these are 'objects which have properties but not substances'. As the quantum field is the ultimate level of analysis we must therefore conclude that the ultimate sphere of reality is 'empty' of substance.

Buddhist metaphysics was and is always aware of the fact that a functioning world of change and activity would be impossible without a grounding in a immaterial, malleable, fluid and flexible substratum. As Vlatko Vedral, Professor of Quantum Information Theory, tells us:

> Quantum physics is indeed very much in agreement with Buddhistic emptiness.[192]

Emptiness, or *sunyata* (sometimes written as pronounced - *shunyata*), is the Buddhist concept of a fundamental non-substantial 'empty' ground of potentiality which gives rise to the multitudinous productions within dualistic experience through the operation of an internal primordial activity of cognition.

In his book *Physics and Philosophy* the physicist and philosopher Bernard d'Espagnat, having reached the conclusion that physics is incapable of ever completely unveiling the nature of a quantum 'veiled' reality conceived of as existing separately and independently of consciousness, suggests that insights into the nature of reality might very well come from other directions among which he cites mysticism.[193] In particular he refers to Buddhist thought which:

> ...rejects the notion of a 'ground of things' and even lays stress on the opposite notion, the one of an 'absence of foundation' or 'emptiness.'[194]

The concept of *sunyata*, usually translated as 'emptiness', does not indicate nothingness. Emptiness is a term which indicates that all phenomena are 'empty' of 'inherent existence', 'ultimately established existence' or 'true existence', there are various synonyms for this notion. When the elements of the everyday world are rigorously analysed they are found to lack a solid inner independent core or essence. All the conditioned phenomena of the everyday world are actually 'empty' of real internal solidity. As Khentrul Rinpoche points out, in his book *The Great Middle Way: Clarifying the Jonang View of Other Emptiness*:

> Even in modern science, we can see that when we analyze material particles with technology, we cannot find anything substantial there. Eventually all matter dissolves back into the quantum field.[195]

In fact, the everyday seemingly solid material world is in the region of 99.9999999999999% empty space!

The lack of internal 'real' solidity, which Buddhism indicates through the attribution of the lack of 'inherent existence' (*svabhava*) in all conditioned phenomena, is one aspect of emptiness. Another aspect of the Buddhist concept of 'emptiness' is the fact that everything in the universe depends on causes and conditions; there is nothing, anywhere to be found in the manifested world which exists solely and independently on its own, completely cut off from interaction with anything else, existing solely by its own independent essence:

> ...all phenomena originate from infinite interdependent causes and conditions and thus lack any intrinsic nature ...[196]

When this analysis is taken down to the quantum level we arrive at the insubstantial realm of potentiality which Buddhism designates 'emptiness', and Bohm referred to as the ultimate level of the implicate order. Bohm, like *Yogacara, Dzogchen,* and other Buddhist perspectives, views both consciousness and matter as appearances arising from the potentialities latent within the grounding field of emptiness.

As Khentrul Rinpoche indicates, the primary meaning of the term 'emptiness' refers to the 'self-emptiness' of all relative conditioned phenomena, the lack of internal 'inherent existence', or its lack of *svabhava*, which is the Buddhist term for independent internal solidity. However, it is also important to comprehend that the ground of the 'empty' dualistic appearances of the process of reality is the nondual ultimate sphere of consciousness-potentiality which itself is *not* empty of luminous awareness. To assert that the ground of dualistic phenomena of the manifested world is empty in the same manner as those dualistic phenomena would actually amount to asserting a complete nothingness. This would be absurd, in actuality the luminous quality of awareness-consciousness is the non-empty nature of the process of reality. The "intellectual form of emptiness" which constitutes knowledge of the 'self-emptiness' of relative conditioned phenomena "could never abide within one's own experience as the realization that apprehends reality as bliss and luminosity".[197] In other words there must be a insubstantial non-empty ground of potentiality for the arising of the 'empty' dualistic appearances of manifested consciousness and matter.

The crucial issue here lies in the subtle meanings of terms such as 'emptiness', 'existence', and 'nonexistence' as they are employed within Buddhist metaphysics. Such terms are often used within Western philosophy with an extraordinary lack of concern for rigorous definitions, and also are used by physicists and philosophers who do not understand the entire context of Buddhist philosophy. For example, the physicist Carlo Rovelli has recently published his book *Helgoland*, with a cacophonous publicity campaign. In this book Rovelli misrepresents the implications of quantum physics, and displays ignorance of the Buddhist *Madhyamaka* philosophy of Nagarjuna. Thus Rovelli writes at the end of his chapter on Nagarjuna that:

> Nagarjuna teaches the serenity, the lightness and the shining beauty of the world; we see nothing but images of images. Reality, including ourselves, is nothing but a thin and fragile veil, beyond which ... there is nothing.[198]

This assertion is incorrect from a physics point of view, and also mistaken in terms of Buddhist philosophy!

Firstly, let's look at the physics perspective. The world of phenomena arises from quantum fields, not from 'nothing'. In this context it is worth briefly examining a controversy which was prompted by the claim by the theoretical physicist Lawrence Krauss, made in his book *A Universe From Nothing: Why There Is Something Rather Than Nothing*, that the entire universe could have emerged from 'nothing'. The physicist and philosopher of science David Albert rightly took Krauss to task for claiming that quantum fields are 'nothing'. Albert wrote in a New York Times Review of the book:

> The particular, eternally persisting, elementary physical stuff of the world, according to the standard presentations of relativistic quantum field theories, consists (unsurprisingly) of relativistic quantum fields. And the fundamental laws of this theory take the form of rules concerning which arrangements of those fields are physically possible and which aren't, and rules connecting the arrangements of those fields at later times to their arrangements at earlier times, and so on — and they have nothing whatsoever to say on the subject of where those fields came from, or of why the world should have consisted of the particular kinds of fields it does, or of why it should have consisted of fields at all, or of why there should have been a world in the first place. Period. Case closed. End of story. ... Relativistic-quantum-field-theoretical vacuum states — no less than giraffes or refrigerators or solar systems — are particular arrangements of *elementary physical stuff*. The true relativistic-quantum-field-theoretical equivalent to there not being any physical stuff at all isn't this or that particular arrangement of the fields — what it is (obviously, and ineluctably, and on the contrary) is the simple *absence* of the fields! [199]

'Eternally persisting' quantum fields are quite clearly not 'nothings' but are fields of potentiality for universes containing sentient beings to come into illusory 'existence'. Physicist Sean Carroll also tells us that:

> The world is made of *fields* – substances spread out through all of space that we notice through their vibrations, which appear to us as particles.[200]

So, presumably for some intellectual/academic agenda, Rovelli has decided to misrepresent the physics.

And, furthermore, probably due to ignorance, Rovelli also misrepresents Nagarjuna. Nagarjuna nowhere suggests that the ultimate nature of the process of reality is 'nothing', or 'nothingness'. Nagarjuna did express views which, when not fully understood, might appear to imply an ultimate nothingness:

> If the nonempty existed in the least,
> Then emptiness too could have some slight existence,
> But when there is no such thing as the nonempty,
> How could emptiness exist?[201]

This does look as if Nagarjuna indicates ultimate 'nonexistence'. However, Nagarjuna also asserted that:

> "It exists" is an eternalist view; "It does not exist" is an annihilationist idea. Therefore the wise one should not have recourse to either existence or nonexistence.[202]

So when one appreciates the full scope of Nagarjuna's perspective, he refutes ultimate 'existence' and ultimate 'nonexistence'. And in this view Nagarjuna is using the terms 'existence' and 'nonexistence' in the mode of the everyday 'conventional' world.

Nagarjuna's *Madhyamaka* presentation, however, is a partial view, opening the way to a deeper nondual 'reality'. This ground field of nondual luminous potentiality, underlying the phenomena of the everyday world, which has an experiential nature of bliss, is termed within the *Jonang-Zhentong* Buddhist tradition as the 'Great Emptiness':

> The ground of emptiness is taught as the great emptiness that is the profound way things are. Moreover, it is the ultimate emptiness of other. ... it has the nature of the limitless enlightened qualities of the dharmakaya ... it is not the emptiness which is merely established as nothing at all ... The ultimate is taught to be the pristine wisdom of the noble ones, the great emptiness, and the great nirvana.[203]

This description, which is fully consistent with the Buddha's original teachings, is from the *Mountain Doctrine* of the enlightened *Jonang-Zhentong* practitioner Dolpopa Sherab Gyaltsen. The term 'noble ones' refers to enlightened beings. The 'great emptiness' is also designated 'other-emptiness', the term '*zhentong*' indicates 'other-emptiness'. This is because it refers to the ultimate field of awareness-potentiality which in its own nature is 'empty' of all the fleeting appearances which arise from within its limitless potentiality. The ultimate field in its own nature is 'empty of other', empty of the transient appearances. The *nature* of the fundamental ground of emptiness is nondual, luminous awareness-potentiality, and it is in its own nature empty of 'other' dualistic appearances.

So, although *in its nature* the field of 'great emptiness' is empty of the transient dualistic 'empty' phenomena of the manifested world which arise from it, and within it, it is the source of dualistic phenomena. As we saw in a previous chapter such a positive view of the emptiness-field is also an aspect of the *Yogacara* perspective:

> That is a unique feature of the Yogacarin presentation of emptiness, because emptiness is normally understood as a complete negation or a completely negative term rather than something positive. Here, once subject and object are negated, emptiness, which is reality, is affirmed in its place. A short passage from the *Madhyantavibhanga* says, "Truly, the characteristic of emptiness is nonexistence of the duality of subject and object, and the existence of that nonexistence." "The existence of that nonexistence" is reality. Duality is removed, but emptiness itself is another kind of existence.[204]

In the context of quantum theory this ground field would correspond to a quantum field.

Bohm was able to arrive at a similar conclusion to that of Buddhist metaphysics, concerning the ultimate nature of reality, through his appreciation of the nature of the quantum realm. In *Wholeness and the Implicate Order* he wrote:

> Thus, as we have seen, the easily accessible explicit content of consciousness is included within a much greater implicit (or implicate) background. This in turn evidently has to be contained in a yet greater background which may include not only neuro-physiological processes at levels of which we are not generally conscious but also a yet greater background of unknown (and indeed ultimately unknowable) depths of inwardness that may be analogous to the 'sea' of energy that fills the sensibly perceived 'empty' space. Whatever may be the nature of these inward depths of consciousness, they are the very ground, both of the explicit content and of that content which is usually called implicit. Although this ground may not appear in ordinary consciousness, it may nevertheless be present in a certain way. Just as the vast 'sea' of energy in space is present to our perception as a sense of emptiness or nothingness so the vast 'unconscious' background of explicit consciousness with all its implications is present in a similar way. That is to say, it may be sensed as an emptiness, a nothingness, within which the usual content of consciousness is only a vanishingly small set of facets.[205]

And:
> So we are led to propose further that the more comprehensive, deeper, and more inward actuality is neither mind nor body but rather a yet higher-dimensional actuality, which is their common ground and which is of a nature beyond both. Each of these is then only a relatively independent sub-totality and it is implied that this relative independence derives from the higher-dimensional ground in which mind and body are ultimately one (rather as we find that the relative independence of the manifest order derives from the ground of the implicate order).[206]

Taking these quotes together, we can make the reasonable conclusion that, from a Bohmian perspective, the explicate realms of consciousness and matter emerge or derive from a deeper, common ground which Bohm sometimes refers to as 'emptiness'. According to Bohm: "matter is a relatively stable and autonomous ripple in the emptiness", and: "emptiness is really the essence", and: "the vast 'unconscious' background of explicit consciousness ... may be sensed as an emptiness".

Several of Bohm's significant insights are remarkably consistent with some central Buddhist insights contained within the *Madhyamaka* (Middle Way) Emptiness school of metaphysics, as well as Tibetan 'schools' of metaphysics that, according to the historically ascendant *Geluk* school, lead, in an ascending sequence of subtlety of understanding, to the *Madhyamaka*. The *Madhyamaka* is one of the central philosophical-metaphysical schools of analysis within Tibetan Buddhism. Its mode of analysis derives from the important second century Buddhist philosopher-practitioner Nagarjuna, who authored the remarkable book of de-constructive metaphysical analyses *Mulamadhyamakakarika*, the *Fundamental Verses on the Middle Way*. In this significant work Nagarjuna relentlessly deconstructs the 'things' and 'processes' of the everyday conventional reality in order to reveal their illusion-like 'empty nature':

> Whatever is dependently arisen
> Does not arise, does not cease,
> Is not permanent, is not extinct,
> Does not come, does not go
> And is neither one thing nor different things.[207]

Nagarjuna's seemingly paradoxical philosophical analysis lays bare the quantum 'empty' facade of the everyday world. The above verse, which will be analysed in detail shortly, tells us that, if we consider the phenomena of the everyday world to be comprised of independently 'real' enduring entities, then we are mistaken. This everyday world is Bohm's 'explicate' order, the realm of 'classical' physics, and the realm which Buddhism terms 'conventional reality'. This explicate world of dualistic experience is derived from the deeper 'empty' implicate order.

The term 'conventional reality' relates to the Buddhist notion of the 'two truths' or 'two realities'. The Buddhist doctrine of the two truths divides the process of reality into the spheres of the 'seeming' or 'conventional' or 'relative' (Skt. *samvrtisatya*), which are all translated terms which refer to the unenlightened mode of perception, and the 'ultimate' or 'absolute' dimension of reality (Skt. *paramarthasatya*). This introduces a fundamental distinction within our understanding of what is real, or what appears to be real, depending upon the mode of perception:

> Thus two kinds of world are seen:
> The one of yogins and the one of common people.
> Here, the world of common people
> Is invalidated by the world of yogins.[208]

The 'seeming' or 'conventional' or 'relative' mode of perception, which corresponds to the 'classical' realm of modern physics, is the way that the world of phenomena *appears* within the experiential continuums of embodied, and unenlightened, sentient beings, who are completely reliant upon 'physical' sense organs and ordinary mental structures of perception. Buddhism asserts the possibility of having a direct experience of ultimate reality which lies beyond the duality of ordinary perception in the everyday world. Such ultimate nondual perception, however, is achieved by very few determined practitioners who achieve extraordinary meditative abilities. The conclusions which Buddhist philosopher-practitioners reached through reasoning were put to the test empirically through meditation insight, although the kind of 'inner' empiricism involved is rejected, wrongly, within mainstream Western thought.

Within the various schools of Mahayana viewpoints, there are various outlooks on the nature of the process of reality which can be considered to be conceptual frameworks of perception on a graduated path of deepening understanding. Each view presents a provisional delineation of reality for exploration:

> Also the yogins, due to differences in insight,
> Are overruled by successively superior ones.[209]

The following is the fifth century Buddhist philosopher Vasubandhu's presentation of of a 'lower level' version of the two truths – on the one hand the 'conventional', 'seeming' or 'relative', and, on the other, the 'absolute' or 'ultimate':

> Things which, when destroyed or mentally dissected,
> Can no longer be identified by the mind,
> Such as pots or water, are relative;
> All else besides is ultimately existent.[210]

On this view, only indivisible elements of reality can be considered to be 'ultimate' elements of reality, a view roughly corresponding to late nineteenth century / early twentieth century atomic theory. Vasubandhu reasoned that the apparently internally solid entities of everyday life must have internal structure to be able to change. If we assert that the internal 'essence' of a substance is continuous and unchanging then there can be no variation at all within the internal 'stuff' of such a substance. This means that there can be absolutely no weak points because the substance internal to the entity we are considering does not vary in any aspect, its essence is completely uniform, without variation or weak points. In this case it does not make sense for any material entity either to break or wear away, there would be no place for a rupture to occur. Here is the actual reasoning given by Vasubandhu:

> The change that these conditioned phenomena undergo over time is reasonable only if they are subject to a form of disintegration in which they arise and pass away with each moment; this phenomena is not reasonable if entities remained in an unchanging state.[211]

In other words, if material entities were solidly continuous, with no internal momentary changing structure, they simply could not deteriorate over time in the manner that they actually do. In a very general way, Vasubandhu prefigured later quantum insights, which apply at an even deeper level. Bohm, for example, tells us:

> What is essential to this new model is that the electron is instead to be understood through a total set of enfolded ensembles, which are generally not localized in space. At any given moment one of these may be unfolded and

therefore localized, but in the next moment, this one enfolds to be replaced by the one that follows. The notion of continuity of existence is approximated by that of very rapid recurrence of similar forms, changing in a simple and regular way (rather as a rapidly spinning bicycle wheel gives the impression of a solid disc, rather than of a sequence of rotating spokes).[212]

Today we know that even atomic constituents flash in and out of 'existence', as Buddhists indicated two thousand years ago with the assertion of the centrality of impermanence, and therefore momentariness, as a central 'mark', or characteristic, of existence.

In the above quote Bohm uses the example of a "spinning bicycle wheel" that "gives the impression of a solid disc" to illustrate how the appearance of solidity is generated from the movement of much less substantial phenomena. This image was also employed within Buddhist metaphysics to make the same point:

> ... know that the world has no self-nature and has never been born, it is like a cloud, a ring produced by a firebrand, the castle of the Gandharvas, a vision, a mirage, the moon as reflected in the ocean, and a dream; that Mind in itself has nothing to do with discrimination and causation, discourses of imagination, and terms of qualification; that body, property, and abode are objectifications of the Alayavijnana, which is in itself above the dualism of subject and object; that the state of imagelessness which is in compliance with the awakening itself, is not affected by such changes as arising, abiding, and destruction.[213]

It is difficult for us to really appreciate the fact that the vast universe, with its impressive and overwhelming appearance of the apparently material world, is actually like "a cloud, a ring produced by a firebrand, the castle of the Gandharvas (illusory beings), a vision, a mirage, the moon as reflected in the ocean, and a dream" etched out of the *alayavijnana* (nondual ground consciousness) by the powerful forces of habit-energies echoing across vast time scales. But, as we discover in the work of Bohm, if we take quantum theory and quantum field theory seriously then such is the truth of things.

Within the variety of Buddhist traditions, each one of which can be considered to be a different approach, highlighting different aspects, within an overall worldview. The *Geluk* school, which was derived from the works of the important Buddhist philosopher-practitioner Tsongkhapa (1357–1419), gained ascendancy. The Geluk metaphysical perspective became prominent and then organised some other views it accepted as partially correct into a hierarchy of viewpoints leading towards its own view. This is the view of the 'self-emptiness', or 'lack of self-existence', of all phenomenon, which was asserted to be superior to all other views. This, of course, is a contentious claim, only accepted by hardcore Geluk practitioners. The *Jonang* 'other emptiness' school for example refutes the Geluk assertion because of the failure of the Geluk school to fully endorse the 'buddhanature' of the ultimate ground of the process of reality:

> ...the nature of phenomena - buddhanature - is eternally stable and changeless as the inseparable ground and result. It is the primordial wisdom endowed with all aspects, nondual with the dharmadhatu. Having a nature of pure lucidity ...[214]

It seems fair to say that the Geluk school in its philosophical mode primarily analyses the nature of 'physical' reality, and the 'empty' nature of concepts, rather than make claims regarding the experiential nature of the ultimate ground of the process of reality. Some other schools consider it important to indicate more fully the experiential nature of ultimate reality.

A systematic study of each of the Geluk presented tenet schools in succession is said to develop insight into the increasingly more subtle understanding of each viewpoint on the nature of reality. In this scheme, the understanding of the less subtle concepts and metaphysical view of the lower tenets is an effective stepping stone to gaining insight into the more subtle concepts and metaphysical view of the highest viewpoint, which is, according to the Geluk school, the *Madhyamaka* emptiness perspective.

The four schools of tenets are:

1. The Great Exposition School (Sanskrit: *Vaibashika*)
2. The Sutra School (Sanskrit: *Sautrantika*)
3. The Mind–Only School (Sanskrit: *Cittamatra*)
4. The Middle Way School (Sanskrit: *Madhyamaka*)

The two lower views, *Vaibashika* and *Sautrantika*, are 'realist' concerning ultimate constituents. According to *Vaibhašika* philosophy at the ultimate level everything consists of indivisible atoms of matter and indivisible moments of consciousness. But all seeming / conventional / relative things are conglomerates of atoms and moments of consciousness. The flow of consciousness consists of separate moments of consciousness. Such conglomerates and the flow of consciousness make up the conventional level of truth / reality, they are conventional, not ultimate, because they don't exist as they appear to. For *Vaibashika* and *Sautrantika* the ultimate sphere of reality consists of momentary pulses of 'existence' which function to produce the 'illusion' of continuity of objects and consciousness.

Cittamatra Buddhist philosophy denies the real existence of indivisible atoms of matter and indivisible moments of consciousness. All that the *Sautrāntika* and *Vaibhasika* schools consider to be really existing at the ultimate level, *Cittamatra* considers as existing only on the relative level. On the ultimate level only Mind exists. The Mind can be characterised by clarity, which is the capacity to manifest phenomena, and intrinsic awareness. It is also impermanent, existing moment by moment. This fundamental mind is asserted to manifest all reality out of itself and is aware of its own manifestations. The final Buddhist 'classical' school is the *Madhyamaka*, which asserts *sunyata* - the non-substantial emptiness of all phenomena, as being the ultimate truth of reality. Conventional, relative, seeming phenomena are deceptive appearances.

The Buddhist division of the process of reality into an 'absolute', or 'ultimate', level and a deceptive 'conventional', 'seeming', or 'relative' level prefigured in a remarkable way the discussion by Bohm at the end of his book *Causality and Chance in Modern Physics*, in the section 'Absolute verses Relative Truth: The Nature of Objective Reality':

> We shall now sum up the ideas developed ...we are led to understand nature in terms of an inexhaustible diversity and multiplicity of things, all of them reciprocally related and all in terms of a treatment of the implications of the nature of the qualitative infinity of nature, with regard to the problems of the absolute *vs* relative character of truth ... no particular kind of thing can be no more than an abstraction from this process, an abstraction that is valid within a certain degree of approximation ... such an abstraction evidently cannot represent the absolute truth; for to do this it would have to be valid without approximation, unconditionally, in all possible contexts, and

for all time. Hence, any particular theory will constitute an approximate, conditional, and relative truth.[215]

Thus we see that Bohm makes the distinction between 'absolute' and 'relative' truth in a manner consistent with Buddhist thought. And, importantly, Bohm goes on to say that this does not mean that there is absolutely no objective reality. If it were the case that there were absolutely nothing rational and coherent underpinning the functioning of reality, it would not be possible to explain or account for the fact that the process of reality does, at least most of the time, function coherently. As Bohm says:

> We may then ask the question, "Does the fact that any given theory can only be approximately, conditionally, and relatively true mean there is no objective reality?" To see that this is not so, it is only necessary to ask the further question of whether the behaviour of things is arbitrary. For example, would it be possible for us to choose the natural laws holding within a given degree of approximation and a particular set of conditions at will, in accordance with our tastes, or what we feel would be helpful for us in the solution of various kinds of practical problems? The fact that we cannot actually do this shows that these laws have an objective content, in the sense that they represent some kind of necessity that is independent of our wills and of the way we think about things.[216]

This does not rule out the possibility of describing some aspect of reality in apparently different ways, but generally different perspectives on the same aspect of reality will have some degree of coherency. An example of this is the fact that Schrödinger and Heisenberg produced seemingly very different accounts of quantum functioning; Heisenberg producing the matrix mechanics mathematical technique, and Schrödinger creating his wave-equation to describe the same phenomena. Only later did the physicist Paul Dirac show the two descriptions were equivalent in mathematical outcomes, although seemingly very different in mathematical appearance. As we saw in the previous chapter, Bohm was also able to produce a further mathematical reformulation which, again, was capable of a different metaphysical 'interpretation'. But this does not mean we can come up with any description we fancy on a whim, it is clear that there is a deeper coherency of functioning underlying the process of reality, although, according to Bohm, we may never know it fully from a purely intellectual perspective.

As we have seen, Bohm suggested that our knowledge of quantum phenomena requires that we give up the view that objects of the everyday world, and the sub-atomic constituents that are supposed to be their constituents, are self-contained independent entities; they are, rather, appearances which derive from a deeper 'implicate' level which Bohm suggests in various places can be designated as deriving from, or being of the nature of, 'emptiness'. And 'emptiness', or *sunyata*, is a concept used by Buddhist philosophers to indicate the 'empty', interdependent and insubstantial, inner nature of reality.

This version of the doctrine of emptiness is referred to as *rangtong* emptiness, which is the lack of 'own-nature', or the lack of 'self-nature', the lack of an inner substantial core of independent inherent nature within all phenomena. This *rangtong* version of emptiness is complemented by another version we have briefly looked at previously, which is less well-known in the West, called *zhentong*, which is 'other-emptiness', the view that the ultimate nature of the totality of the sphere of reality is empty of the temporary appearances which manifest within it. The Buddhist scholars Klaus Dieter Mathes and Michael R. Sheeny, in their introduction to their book *The Other Emptiness* write concerning *rangtong* emptiness:

> It is this lack of independent, locally determined building blocks of the world that allows in Madhyamaka the Buddhist axiom of dependent origination. In other words, rangtong emptiness is the a priori condition for a universe full of open dynamic systems. The union of dependent origination and emptiness - the inseparability of appearance and emptiness - sets the ground for philosophical models of interrelatedness that are increasingly used in attempts to accommodate astonishing observations being made in the natural sciences, such as wave-particle duality or quantum entanglement.[217]

Here Mathes and Sheeny focus on how the phenomenon of quantum entanglement, whether known explicitly as in modern physics, or implicitly as in Buddhist metaphysics, can be viewed as the central ingredient in the *rangtong* view of emptiness. In the rest of this chapter we shall be concerned with *rangtong* emptiness, in a later chapter we shall see how the Bohmian worldview embraces both the *rangtong* and *zhentong* perspectives.

The *rangtong* aspect of the doctrine of emptiness is that all phenomena lack 'inherent existence', which is to say that no phenomenon can be a completely independent, self-sufficient and self-enclosed entity or event; everything is interdependent with everything else in a web of interpenetration. It is this assertion of a deep level of interconnectedness between all phenomena which prefigures quantum entanglement. As we can see, this *rangtong* metaphysical perspective resonates with the Bohmian quantum worldview in a remarkable way. Bohm also tells us that everyday objects and atoms appear to be independent and completely separate entities, but in reality, at a deeper level, everything is entangled with all other things in a universal web of interconnection.

The second century Buddhist philosopher-practitioner Nagarjuna indicated roughly two thousand years ago:

> Whatever is dependently arisen
> Does not arise, does not cease,
> Is not permanent, is not extinct,
> Does not come, does not go
> And is neither one thing nor different things.[218]

Here Nagarjuna indicates that if we consider something to be a really independently existent separate self-enclosed entity, then it should not depend for its very existence on other entities. It should exist on its own, completely sufficient on its own substance. This would mean that it would have an immutable core of independent substantial reality, which in Madhyamaka terminology is called *svabhava*, 'own-nature' or 'inherent existence'.

It is vital to hold in mind that the kinds of paradoxical analysis that is central to Madhyamaka applies to putative self-existent entities which are considered to be 'inherently existing' things, which are supposed to be changeless entities independent of all other entities. Although such entities do not exist, they are initially assumed to exist by the Madhyamaka in order to refute them. This tactic is used in order to press home the truth of the 'empty' nature of the everyday world through a logical analysis. As Mathes and Sheeny point out: "a universe full of open dynamic systems" requires *rangtong* emptiness in order to function. Bohm describes such a universe:

> Evidently, this principle of structure is universal.... living beings are in a continual movement of growth and evolution of structure, which is highly organized (e.g., molecules work

together to make cells, cells work together to make organs, organs to make the individual living being, individual living beings a society, etc). Similarly, in physics, we describe matter as constituted of moving particles (e.g. atoms) which work together to make solids, liquids, or gaseous structures, which similarly make larger structures, going on up to planets, stars, galaxies, galaxies of galaxies, etc. Here, it is important to emphasize the essentially dynamic nature of structation, in inanimate nature, in living beings, in society, in human communication, etc. (e.g. consider the structure of a language, which is an organised totality of ever-flowing movement).[219]

Such a universe requires both relative separation and creative interconnection, both of which are aspects of Bohm.s view of 'emptiness'.

In order to get a deeper appreciation of the Madhyamaka analysis it is worth examining the reasoning in the previous quote from Nagarjuna in greater detail. As we do so keep in mind that this is not meant to be just a quick intellectual exercise, the analysis is used to generate a deep understanding of emptiness which is meditated upon:

> Whatever is dependently arisen
> Does not arise, does not cease, ...

An inherently existing entity (which is an entity with '*svabhava*' = 'inherent existence') would have to be changeless because an object with an internal changeless essence obviously cannot change. It follows, therefore, that such an object, by definition, could not 'arise' from something else simply because it cannot come into or go out of existence, this would be a change!

Here we find the importance of the notion of 'dependent origination', which is a hallmark of 'emptiness' (*sunyata*). Anything that is 'dependently originated' is 'empty' because it is dependent and therefore not self-sufficient. What is being asserted here is that if something arises dependently then it does not arise as an inherently existent entity and neither can it cease as an inherently existent entity because if it were to cease then it could not have been an inherently existent entity in the first place! It is only a temporary *appearance* of an entity which is dependent on other causes and conditions.

Inherently existent entities cannot cease by definition. Something which arises on the basis of something else cannot be given credence as being a 'real' inherently existent thing because it has arisen in dependence on something else, so it is not self-powered, it depends on something else. It follows that this illusory 'thing', that we might think has come into inherent being has not actually arisen because it's not actually there as an inherently existent entity! It cannot, inherently, cease because there is nothing inherently existent to cease.

The next line is:

> Is not permanent, is not extinct, ...

It cannot be permanent because it appeared to arise in the first place, although it did not actually arise as an inherently existent thing, and because of this it cannot become extinct because there was never anything inherently existent to become extinct!

> Does not come, does not go...

Something which has not come into existence (as an inherently existent thing) can not come or go!

> And is neither one thing nor different things.

It cannot be one inherently existent thing because it has arisen and so is not an inherently existent thing in the first place. It cannot be different things because the analysis would apply to each of those things in turn and, anyway, it has appeared as being in the guise of one thing. The Madhyamaka begins, then, with a complete demolition of the notion of ultimate 'things' and ultimate thinghood, which it denotes by the term 'inherent existence', intrinsic existence' or 'own-nature' (*svabhava*).

The primary reason for the paradoxical precision of this kind of Madhyamika analysis is to completely convince a practitioner, at a deep level, through intellectual analysis and meditation, that the entire everyday world is an illusion-like appearance deriving from a deeper level. This must become, for a Buddhist Madhyamika practitioner, a living experience of everyday life. In Bohmian terms, of course, the Madhyamaka analysis, is deconstructing the explicate everyday world to reveal its implicate origins.

The paradoxical philosophical deconstruction performed by Madhyamaka Buddhism anticipates metaphysical insights of modern physics. Consider the Madhyamaka analysis described previously in the context of the following analysis from Bohm:

> For example, any localizable structure may be described as a world tube ... Inside this tube ... a complex process is going on... It is not possible consistently to analyse movement within this tube in terms of 'finer particles' because these, too, would have to be described as tubes, and so on ad infinitum. Moreover, each tube is brought into existence from a broader background or context ... while eventually it dissolves back into the background ... Thus, the 'object' is an abstraction of a relatively invariant form. That is to say, it is more like a pattern of movement than like a solid separate thing that exists autonomously and permanently. ... Particles are then to be regarded as certain kinds of abstraction from the total field, corresponding to regions of very intense field (called singularities). As the distance from the singularity increases ... the field gets weaker, until it merges imperceptibly with the fields of other singularities. But nowhere is there a break or a division. Thus, the classical idea of the separability of the world into distinct but interacting parts is no longer valid or relevant. Rather, we have to regard the universe as an undivided and unbroken whole. Division into particles, or into particles and fields, is only a crude abstraction and approximation. Thus, we come to an order that is radically different from that of Galileo and Newton – the order of undivided wholeness.[220]

Of course, Bohm is not the only physicist to offer such insights, although he does investigate the philosophical implications in much greater detail. The physicist Lee Smolin, for example, tells us that:

> Newtonian physics ... gives rise to the illusion that the world is composed of objects. ... But relativity and quantum theory each tell us ... no, better, they scream at us, that our world is a history of processes.[221]

And Jonathan Allday, in his book *Quantum Reality: Theory and Philosophy*, points out that quantum entanglement:

> ...presents us with a philosophical challenge, one that threatens to pick away at our notion of what a 'thing' is.[222]

So, we see that modern physics has completely validated the Buddhist Madhyamaka assertion that all phenomena are empty of ultimate independent substantial existence.

We looked at the phenomenon of entanglement and nonlocality, with the implication of necessary universal interconnection, in the previous chapter. In the Process Studies talk Bohm pointed out that:

> The original atomic theory had rigid bodies of some sort, but rigid bodies are not possible after Einstein. Let's say that a particle is made of smaller bodies -- of subparticles. Each of the subparticles, if it is extended, will meet the same problem as a rigid body. Therefore, a particle cannot be made of extended subparticles. Now then, what if it is made of particles with no extension at all, such as points whose tracks in space-time can be represented by lines? You will find that the fields around these point particles are infinite, leading to inconsistencies such as infinite mass and infinite charge and so on (especially in quantum mechanics).[223]

Quantum theory in general throws the idea of independent 'real' 'things' into question. In the above quote from Bohm the reason that "rigid bodies are not possible after Einstein" is, as Bohm explains:

> One view is that a particle is some extended structure. Now if I make a space-time diagram of a particle at rest whose boundaries are given by two lines and then suddenly accelerate it to another velocity, I see that if I push on one side of the object it immediately responds on the other side. However, in Einstein's views of relativity, this is not permitted. An impulse or a signal cannot be carried faster than the speed of light. Consequently, you cannot have a rigid or extended body in relativity.[224]

So Bohm indicates that a rigid application of Einstein's relativity theory rules out the possibility of 'rigid' extended particles. Particles can only be appearances, 'ripples in emptiness'.

This analysis, although it is not exactly the same, is reminiscent of, related to, and has a similar outcome to, the 'partless particle' analysis within the Madhyamaka discourse. Some other Buddhist schools (*Vaibhasika* and *Sautrantika*) assert that the entities of the physical world are comprised of assemblies of indivisible particles. The Madhyamaka analysis, however, undermines this viewpoint with a further analysis of the possibility of the existence of a 'partless particle'. Consider a putative 'indivisible particle'; it would need to be partless, otherwise it would be divisible conceptually at least, and there would need to be further reason why the parts could not be taken apart. Now consider how a 'partless particle' could be involved in the construction of the material world. A constructive partless particle might possibly be connected to six other particles which are located around the central particle in the six principle directions:

> If six particles join it simultaneously,
> This infinitesimal particle would have six parts.
> If all six together are partless,
> Then also their aggregation would be just an infinitesimal particle. [225]

If a particle is able to club together with other particles in order to produce the world of extension and experienced solidity then quite obviously it would need to connect in some way with the other particles. In other words the central particle would need to be connected to the others through its faces and this entails that any particle which can play a role in creating a world must have parts, which means that the 'partless' particles must have 'parts'! If particles were truly 'partless' then there is no way they could club together to produce a world with an appearance of solidity and extension! Thus the notion of partless particles which construct a material reality falls to pieces, so to speak!

The world cannot be constructed from ultimate entities devoid of parts, because, if these particles were devoid of parts then the world would simply collapse because it is impossible to produce extension from an extensionless entity:

> If you insist that this is truly so
> (Though it must also face the other particles),
> How is it that earth and water
> And all other things extend – or maybe they do not?[226]

And so:

> If you say that sides that face
> The different particles are different,
> How come the finest particle is one:
> A single entity devoid of parts?[227]

Furthermore, if the world really is substantial then it must be constructed from an ultimate substance, or substances, and this must imply the existence of ultimate particles of some sort. If this is the case then when we repeatedly divide up particles, we are supposed to arrive at a final particle which is the fundamental building block. If we find that particles are infinitely divisible into their parts then we can only stop when we reach a partless particle. But if this particle is truly partless then, as we have seen, it could not take part in the construction of the world because in order for a partless particle to connect to another partless particle it is necessary for one part of a partless particle to connect to a part of another partless particle!

So particles, if they existed as entities, must be infinitely divisible, each particle itself having parts. This means that each particle cannot be an inherently existent fully substantial indivisible entity because each particle is made up of its parts. This lack of inherent existence cascades down through the levels of assumed particles, arising from division, without end, thus demonstrating that there is simply nothing that can be found at the base which can provide a solid foundation:

> The particle, it's proved, does not exist inherently.
> And therefore it is clear that ... substance and the rest,
> The many things proposed by ours and other schools,
> Have no intrinsic being.[228]

This is called the reasoning of 'freedom from unity and multiplicity' which is said to open the door of emptiness. If it is not possible to demonstrate an ultimately unitary particle which exists independently, with no dependence upon its parts, then there can be no inherent unity. If there is no unitary entity to enter into the construction of a multiplicity there can be no inherent multiplicity either. Thus, we see that the Madhyamaka analysis completely undermines the idea of fully independent and substantial atomic entities making up fully independent and substantial entities in an external material world.

This Madhyamaka analysis completely accords with Bohm's conclusions:

> A particle is not a substance. A substance would be self-generated and self-maintained. But subsistence merely means that it depends on something else to be maintained. Democritus's original idea was that the atoms were substances -- self-maintaining and eternal. But now we are saying that particles are subsistants and not substances. This fits the facts of modern physics, because as I have just said, all particles can be created and destroyed and transformed, and so on. Therefore,

there is no sign that they are independent substances. We will say that particles are orders in the holomovement, which have the character of subsistence, a certain repetitiveness, stability, and so on.[229]

The observation by Bohm that a "substance would be self-generated and self-maintained" projects us immediately into the world of the Madhyamaka analysis, because this is a significant part the definition of an inherently existent substance used by the Madhyamaka. It is also part of the Madhyamaka deconstruction of the possibility of such a 'substance' actually to exist in reality.

The following Madhyamaka deconstruction of the notion of 'self-causation', which, again, echoes passages from Bohm we have looked at, uses the 'Diamond Slivers' reasoning, which is given this appellation because like a diamond it is said to be indestructible, and is able to cut through wrong views concerning the nature of reality. This reasoning focuses on the possible modes of causal processes in an inherently-existent universe, and shows their impossibility. The form of the reasoning has the logical structure of what is termed the Madhyamaka 'tetralemma' (*Catuṣkoṭi*), the fourfold logical paradoxical analysis:

> Neither from itself nor from another,
> Nor from both,
> Nor without a cause,
> Does anything whatever, anywhere arise.[230]

The refutation of production from self is aimed at a school of philosophy extant at the time (2nd century) which claimed that all phenomena were actually manifestations of a permanent primal cosmic substance. Therefore any entity, such as a sprout, was asserted to be produced from its own nature. This viewpoint, production from own-nature, however, produces some rather absurd consequences. When reading through these absurd consequences it should be kept in mind that these logical detonations are also used as meditation topics for direct experience of emptiness. Directly seeing the impossibility of real substantial substances and causal processes transforms perception of reality.

If an entity which already exists reproduces itself (a situation which corresponds to Bohm's "self-generated" substance, which he also refutes), then the reproduction is pointless because the entity already exists. Furthermore, if 'self-production' were an inherent feature of an entity, which would mean that it is a essential inner necessity of the ultimate nature of the entity, then once it had reproduced itself it would necessarily have to do so again, because self-production would be an inherent and essential feature of the entity. This must lead to an infinite, never-ending, sequence of reproductions. A seed, therefore, would never get around to producing a sprout because it would be too busy reproducing itself! Also, obviously, if the seed and its sprout, the cause and the effect, were the same and yet the one produces the other then the seed and the sprout should appear to be exactly the same which is absurd, because the seed is supposed to be the cause of the sprout. Effects are usually apprehended when the cause has ceased, but if the effect and the cause are identical then the effect should cease as soon as the cause does. Analysis of the logical implications of the idea of 'self-production' reveals the absurdity of the view.

The assertion that entities are produced from causes which are 'other' is the more usual view, a view which is derived from observation of the processes of the 'common-sense', conventional, everyday world. The idea that this view is impossible, then, can come as quite a shock. But the reasoning is unimpeachable:

> If something can arise from something other than itself,
> Well then, deep darkness can arise from tongues of flame,
> And anything could issue forth from anything.
> For 'nonproducer,' like 'producer,' is an 'other.'[231]

It is essential to bear in mind all the time when following Madhyamaka reasoning that the entities involved must always be viewed through the lens of 'inherent existence'. The Madhyamaka analysis is applied to the manner in which a fully-substantial, inherently-existent, Newtonian-type universe of fixed entities, which only interact externally, would have to function. From this perspective it is not possible to have gradations of 'otherness'. Something is either an 'other' or it is not. There are no in-between states of 'otherness'; this is a consequence of the lens of inherent existence and is entailed if we consider that things are inherently existent. So if we say that production is from something completely 'other' than what is produced, a rice seed being 'other' to a rice plant for instance, then it is also the case that a barley seed is equally 'other'. As there cannot be gradations of 'otherness' in an 'inherently existent' world, both the rice seed and the barley seed are equally 'other' to a rice plant. So, if both a barley seed and a rice seed are equal in otherness, both of these, or neither, must be capable of producing the rice plant. If both are capable then anything can indeed issue forth from anything. Such a world would be completely chaotic.

The third possibility considered in the Nagarjuna analysis is: "Nor from both", i.e. is it possible that things are caused by both self and other? This is refuted by employing both of the above refutations. The assertion of both cannot be correct because both possibilities have been refuted. Finally, if things arise without any cause, then things should be produced or caused randomly, which would essentially be the same as anything arising from anything. This, however, is not observed in the world.

The Madhyamaka reasonings, then, deconstruct any notion that there are inherently 'real' substantial causal processes involving substantial inherently existent entities to be found in reality. This is not to say, however, that the *appearance* of such processes does not occur because, quite obviously, they do. But these appearances are 'conventional' manifestations within a seeming, illusion-like reality. The Madhyamaka does not deny the *appearance* of the everyday world, which it terms 'conventional reality'. The everyday world does appear, very convincingly, to be made up of fully independent material entities. For ordinary beings the appearance of macroscopic reality is a 'seeming' reality that obscures the actual ultimate nature in which all phenomena are 'empty' and illusion-like. Ordinary beings, however, believe such a 'seeming' reality to be ultimately 'real'. This division of reality into two levels maps directly onto the dichotomy between the quantum level and the experiential macroscopic, 'classical' level of the everyday world.

The profound understanding that all phenomena have no independent substantial core of reality is repeatedly demonstrated within the Madhyamaka analysis. The seeming reality of the everyday world is taken as the ground from which the analysis begins, a thorough analysis, however, reveals repeated signs that point towards the insubstantial ultimate nature:

> These phenomena are like bubbles of foam ...'
> Like illusions, like lightening in the sky,
> Like water-moons; like mirages.[232]

This is not to say, of course, that there is absolutely nothing; but rather there is nothing substantial to be found in the manifestation of the seeming play of appearances, appearances which arise from the deeper level of 'emptiness'. In Bohm's terminology the appearances of the 'explicate order' arise from the functioning of the deeper level of the 'implicate order'.

The Madhyamaka also gives a precise definition of the existential configuration of emptiness. The ultimate nature of all phenomena, according to the philosopher-practitioners of the Madhyamaka (*Madhyamikas*), is that:

> Its character is neither existent, nor non-existent,
> Nor both existent and non-existent, nor neither.
> Madhyamikas should know true reality
> That is free from these four possibilities.[233]

This may be thought be be a bizarre irrelevant piece of Eastern mystical word-play, but, as we shall now see, it turns out to be *precisely* the existential configuration of quantum fields.

The Italian physicist Giancarlo Ghirardi, in his book *Sneaking a Look at God's Cards: Unraveling the Mysteries of Quantum Mechanics,* refers to the existential possibility configuration for a 'quantum chair', used as a convenient classical-level example, chairs of course do not exist at the quantum level. Ghirardi uses this object as a classical level example in the same manner that Schrödinger used his famous cat. Ghirardi begins by looking at the mathematical equation for a superposition of a quantum object which might be in one of two places, but actually is in a state of quantum superposition. In the following equation $1/\sqrt{2}$ is a number such that $1/\sqrt{2}$ multiplied by itself equals $1/2$, or one-half:

> ...according to the formalism the chair can be found ... in a state analogous to that of the photon above:
>
> $$|\,?\,> = 1/\sqrt{2}\,[|\text{there}> + |\text{here}>]$$[234]

This indicates that the quantum possibility for 'classical' manifestation of the position is spread between 'here' and 'there'. And Ghirardi then comments:

> What meaning can there be in a state that makes it illegitimate to think that our chair is *either* here or in some other place? ... only potentialities exist about the location of the chair, potentialities that cannot be realized, unless we carry out a measurement of position? How can it be understood that, attached to these potentialities, is a *nonepistemic* probability that in a subsequent measurement of position the chair will be found here or there (which is equivalent to

asserting that, before the measurement was carried out, the chair could be **neither here nor there, nor in both places, nor in neither place**)?[235]

The italicised word 'nonepistemic' is emphasised (in the original text) because the situation of 'hovering' between the four possibilities of existence is not a matter of our lack of knowledge, which would be 'epistemic'. *The quantum equation indicates the ontological condition of the quantum entity. It is important to comprehend this fact, quantum fields do, in reality, hover between four 'extremes of existence'.*

The following is from science writer Marcus Chown's book *The Never-Ending Days of Being Dead*, which contains entertaining elucidations of cutting edge physics:

> So, what of a water droplet that hovers half in existence and half out of existence? It goes without saying that nobody has actually seen such a schizophrenic water droplet ... Where does the quantum weirdness go.[236]

Here Chown is making the point that at the quantum level water molecules *can* hover "half in existence and half out of existence," it is only because there are a vast number of molecules in a water droplet that this "quantum weirdness" is eradicated. Indeed, it is the capacity to hover half in and out of existence that allows molecules to hold together, and this capacity to hover in semi-existence allows the entire universe to hold together. As the science writer Michio Kaku tells us:

> The reason why molecules are stable and the universe does not disintegrate is that electrons can be in many places at the same time. .... electrons can exist in parallel states hovering between existence and non-existence.[237]

It is because the electrons which encircle the atomic nucleus are actually clouds of hovering potential existence, that they are capable of being in two places at once, at the same time as being one thing, thus they hold molecules together.

The reader really should get themselves a stiff drink, or a coffee, read the next section, which is in bold, carefully, then put the book down and ponder for a few minutes.

About two thousand years ago, Buddhist practitioner-philosophers, using everyday observation, logical analysis and meditation insights,

came to their central conclusion that the everyday world did not exist as it appeared. The ultimate nature of all phenomena, they asserted, was that:

> Its character is neither existent, nor non-existent,
> Nor both existent and non-existent, nor neither.
> Madhyamikas should know true reality
> That is free from these four possibilities.[238]

Furthermore, they also asserted that this existential configuration, EMPTINESS, the lack of inherent existence, is essential for the functioning of the process of reality and the universe:

> If things were not empty of inherent existence, nothing could function…It is their emptiness of inherent existence that allows everything to operate satisfactorily.[239]

Today quantum physicists have discovered that these metaphysical insights on the part of Buddhist philosophers are precisely correct. Now to repeat a previous quote:

> The reason why molecules are stable and the universe does not disintegrate is that electrons can be in many places at the same time. …. electrons can exist in parallel states hovering between existence and non-existence.[240]

Which is why the Madhyamika master Nagarjuna wrote:

> For those for whom emptiness is possible,
> Everything is possible,
> For those for whom emptiness is not possible,
> Nothing is possible.[241]

So, according to both modern quantum theory and Buddhist metaphysics the state of 'emptiness', which is the quantum hovering between existence and nonexistence, is required for the universe to function.

However, the physicist Jean Bricmont, who is sceptical of quantum-mystical type views, has complained that:

> When one encounters claims about modern physics having been foreshadowed by …Eastern traditions, the first question to ask is: where are the equations … ?[242]

The first point to make to undercut Bricmont's disingenuous suggestion is that the claim generally made is not that Buddhism anticipated the *equations* of quantum physics, but it foreshadowed the *metaphysics* revealed by quantum discoveries. However, it is remarkable that the precise Buddhist formula of the existential configuration of ultimate reality, codified by Tibetan Buddhist metaphysics, is clearly a fairly impressive foreshadowing of a fundamental mathematical description of the quantum realm. And the notion that this is a coincidence is highly implausible. One can imagine a Tibetan mystic snake-oil huckster using the claim 'All is One' as a sales pitch, but surely the idea that anyone would proclaim "all phenomena hover between existence, non-existence, both, and neither" as the banner-head for a mystical cult scam seems highly unlikely!

This Buddhist view of the existential-hovering situation of the ultimate realm has a great deal in common with Bohm's notion of the 'implicate order', the realm more akin to 'emptiness', which is not 'nothingness' but is more like an indeterminate pool of potentiality. The 'explicate order' of the conventional everyday world manifests from this pool of empty potentiality. As Bohm tells us:

> ...one finds, through a study of quantum theory, that the analysis of a total system into a set of independently existing but interacting particles breaks down in a radically new way. One discovers, instead, both from consideration of the meaning of the mathematical equations and from results of the actual experiments, that the various particles have to be taken literally as projections of a higher-dimension reality which cannot be accounted for in terms of any force of interaction between them.[243]

In *Wholeness and the Implicate Order* Bohm described part of his vision as follows:

> It is being suggested here, then, that what we perceive through the senses as empty space is actually the plenum, which is the ground for the existence of everything, including ourselves. The things that appear to our senses are derivative forms and their true meaning can be seen only when we consider the plenum, in which they are generated and sustained, and into which they must ultimately vanish.[244]

What Bohm calls the 'plenum' here is equivalent to the Buddhist Madhyamaka notion of 'emptiness'. The term *sunyata*, which is usually translated as 'emptiness', does seem, because of this translation, as being nothing more than a 'void', which is in fact another possible translation. However, the original meaning of the term *sunya*, which is the Indian origin of the concept of zero, is 'the swollen', in the sense of an egg of potentiality which is about to burst into manifestation. The term 'emptiness' indicates the absence of any manifested thing, but the term also indicates the ground of all possible manifestations.

Bohm continues his description:

> This plenum is, however, no longer to be conceived through the idea of a simple material medium, such as an ether, which would be regarded as existing and moving only in a three dimensional space. Rather, one is to begin with the holomovement, in which there is the immense 'sea' of energy ... This sea is to be understood in terms of a multidimensional implicate order, ... while the entire universe of matter as we generally observe it is to be treated as a comparatively small pattern of excitation. This excitation pattern is relatively autonomous and gives rise to approximately recurrent, stable and separable projections into a three-dimensional explicate order of manifestation, which is more or less equivalent to that of space as we commonly experience it.[245]

Here we see that Bohm's holomovement is a movement of an "immense sea of energy" which is a " multidimensional implicate order". This is the 'plenum' which can be equated with the Madhyamaka universal ground of *sunyata*, which is emptiness. Out of this ground of empty potentiality the explicated "patterns of excitation" manifest as the experienced everyday world of the explicate order. Bohm later described this structure as follows:

> What is basic to the law of the holomovement is ... the possibility of abstraction of a set of relatively autonomous subtotalities. ... This operation will in general have these three key features:
>
> 1. A set of implicate orders.
>
> 2. A special distinguished case of the above set, which constitutes an explicate order of manifestation.

3. A general relationship (or law) expressing a force of necessity which binds together a certain set of the elements of the implicate order in such a way that they contribute to a common explicate end ...[246]

In the next chapter we will explore the mechanisms that operate within emptiness to determine how "elements of the implicate order" come to produce a manifested explicate order.

A Matter of Unfolding Mind?

The Dzogchen / Yogacara
Ground Consciousness
Implicates Bohm's
Implicate Order

The notion of the 'implicate order' is central within Bohm's worldview, and the notions of 'enfoldment' and 'unfoldment' are central concepts for understanding the functioning of the implicate order. Bohm explained in *Wholeness and the Implicate Order*:

> We proposed that a new notion of order is involved here, which we called the *implicate order* (from a Latin root meaning 'to enfold' or 'to fold inward'). In terms of the implicate order one may say that everything is enfolded into everything. This contrasts with the explicate order now dominant in physics in which things are unfolded in the sense that each thing lies only in its own particular region of space (and time) and outside the regions belonging to other things.[247]

Here Bohm indicates that the information that gives rise, by the operation of a process of 'unfoldment', to the everyday 'explicate order', has been previously 'enfolded' in some way into the 'implicate order'. He also suggests that physics at the time he was writing was more concerned with mechanisms within the explicate order, and neglected research into mechanisms taking place within the implicate order. In this chapter we will investigate the kind of mechanisms which could be involved in the processes of 'enfoldment' and 'unfoldment'. We shall look into evidence from physics, and both Eastern and Western metaphysics.

This process of enfoldment into the implicate order, and unfoldment from the implicate into the explicate order is clearly a center-stage aspect of the holomovement as it is referred to frequently:

> This is the *implicate* or *enfolded* order. In the enfolded order, space and time are no longer the dominant factors determining the relationships of dependence or independence of different elements. Rather, an entirely different sort of basic connection of elements is possible, from which our ordinary notions of space and time, along with those of separately existent material particles, are abstracted as forms derived from the deeper order. These ordinary notions in fact appear in what is called the explicate or unfolded order, which is a special and distinguished form contained within the general totality of all the implicate orders.[248]

And:

> Now, the word 'implicit' is based on the verb 'to implicate'. This means 'to fold inward' (as multiplication means 'folding many times'). So we may be led to explore the notion that in some sense each region contains a total structure 'enfolded' within it. It will be useful in such an exploration to consider some further examples of enfolded or implicate order. Thus, in a television broadcast, the visual image is translated into a time order, which is 'carried' by the radio wave. Points that are near each other in the visual image are not necessarily 'near' in the order of the radio signal. Thus, the radio wave carries the visual image in an implicate order. The function of the receiver is then to explicate this order, i.e., to 'unfold' it in the form of a new visual image.[249]

Here Bohm suggests that potentialities that are enfolded within the implicate realm are unfolded, perhaps by sentient beings acting as 'unfolders', into the experiences of the 'classical' dualistic world in a similar way to a radio set unfolding a sequence of images and sounds from a radio signal. The original content is 'unfolded' by tuning to the carrier frequency.

Another analogy Bohm used, perhaps more relevant to his vision of the holomovement, is that of a hologram:

> To indicate a new kind of description appropriate for giving primary relevance to implicate order, let us consider once again the key feature of the functioning of the hologram, i.e., in each region of space, the order of a whole illuminated structure is 'enfolded' and 'carried' in the movement of light. Something similar happens with a signal that modulates a radio wave .... In all cases, the content or meaning that is 'enfolded' and 'carried' is primarily an order and a measure, permitting the development of a structure. With the radio wave, this structure can be that of a verbal communication, a visual image, etc., but with the hologram far more subtle structures can be involved in this way ... [250]

The analogy of a hologram is extremely illuminating (!) in the elucidation of the processes of enfoldment and unfoldment. In *Science, Order, and Creativity* Bohm and F. David Peat indicate that the holograph (hologram) is the most appropriate analogy:

A better analogy to the behavior of an electron, for example, can be obtained by considering a holograph, which is a photographic record of light waves that have been reflected from an object. In normal photography a lens is used to focus light from an object, so that each small section of the object is reproduced in a small section of the photographic plate. In holography, however, the photographic record made by laser light does not in fact resemble the object but consists of a fine pattern of interference fringes. Each portion of the plate now contains information from the whole of the object. When similar laser light is used to illuminate the plate, the light waves emerging from it resemble those that originally came from the object. It is therefore possible to see, in three dimensions, an image of the original object. What is particularly significant, however, is that even if only part of the plate is illuminated, an image of the whole object is still obtained. This is because light from every part of the object is enfolded within each region of the plate. In normal photography, information is stored locally, but with the holograph it is stored globally. As successively smaller regions of the holograph are illuminated, the images as a whole are not lost. Instead fine detail becomes progressively more difficult to resolve. This global property of enfoldment of information and detail has something in common with both fractal and Fourier orders.[251]

So we see that there are two phases. Firstly, the production of the hologram involves the use of beams of coherent laser light. One beam reflects off the object and then interacts with a main beam, this produces a interference pattern which is recorded on a photographic plate. To 'unfold' the image from the holographic image, the hologram, it requires another coherent light beam to be shone on it. Thus there is an 'enfoldment' of the original image into the hologram, and then the subsequent 'unfoldment' of the original image from the hologram. One significant feature of the holographic image on the hologram is that any portion of the hologram contains the entire original image. If a small portion of the hologram is used to unfold the original image it will be blurred. The sharpness improves as more of the hologram is used. Bohm suggested that the way in which information is enfolded within the implicate order is like a hologram, and therefore its 'order' is different to the 'order' of the explicate order. In particular, as Bohm indicates: "Points that are near each other in the [explicate] image are not necessarily 'near' in the order of the [implicate order]".

This leads us to the issue of exactly what kind of mechanism might be involved in the processes of 'enfoldment' and 'unfoldment'. In *The Undivided Universe* Bohm and Hiley explain that:

> ... the notion of enfoldment is not merely a metaphor, ... it has to be taken fairly literally. To emphasize this point, we shall therefore say that the order in the hologram is *implicate*. The order in the object, as well as in the image, will then be unfolded and we shall call it *explicate*. The process ... in which the order is carried from the object to the hologram will be called *enfoldment* or implication. The process in which the order in the hologram becomes manifest to the viewer in an image will be called *unfoldment* or *explication*.
>
> .....
>
> Since all matter is now analysed in terms of quantum fields, and since the movements of all these fields are expressed in terms of propagators, it is implied by current physics that the implicate order is universal ... we can never have the same field point twice ... all properties that are attributed to the field have to be understood as relationships in its movement ... Whatever persists with a constant form is sustained as the unfoldment of a recurrent and stable pattern which is constantly being renewed by enfoldment and dissolved by unfoldment. When the renewal ceases the form vanishes.[252]

At first look this presentation may seem strange. Why should a particle, or 'stable pattern' be renewed by 'enfoldment', which 'enfolds' back into the implicate order, and why should the 'unfoldment' of a new particle dissolve it? We find further elucidation on previous pages. In the following the term 'wave' refers to a wave of quantum potentiality unfolding or enfolding, as the case may be, within the implicate order:

> Waves from each point unfold. But at the same time waves from many points are enfolding to give rise to a new wave front. So in the totality, the one process includes both enfoldment and unfoldment. It is only when we focus on a part that we are led to talk of these as distinct.

So quantum waves of potentiality are both enfolding into, and unfolding out of, the implicate order. Particles which make up the appearance of a 'classical' everyday reality are momentary flickers of explicated 'existence' which flicker into appearance, and as they do so they enfold back into the implicate order as, at the same time, they are dissolved by the next flicker of apparent 'existence'. As Bohm wrote in his article 'A New Theory of the Relationship of Mind and Matter':

> ...all things found in the unfolded, explicate order emerge from the holomovement in which they are enfolded as as *potentialities*, and ultimately they fall back to it. They endure only for some time, and while they last, their existence is sustained in a constant process of unfoldment and re-enfoldment, which gives rise to their relatively stable and independent forms in the explicate order.[253]

In his book *Mind, Matter and the Implicate Order*, the theoretical philosopher Paavo Pylkkänen elucidates on this:

> Bohm says that things *emerge* from the holomovement. But this is not "something out of nothing" emergence or creation. Instead, Bohm assumes in an Aristotelian fashion that there exist *potentialities* in the holomovement. A potentiality for him is an "enfolded order" that "actualizes" when it unfolds to the explicate order. A thing that has been actualized ... then *endures*, but only for some limited period of time ... While a thing endures, it does not have a continuous existence as a particle-like entity. Instead its existence is sustained in a constant process of unfoldment and re-enfoldment. Because such a process has *recurrence*, this gives rise to the relatively stable and independent form that we call a "particle". If you like, the "particle" is a recurring *phase* of an underlying process of unfoldment and enfoldment.[254]

This understanding raises the issue of what kind of 'stuff' the potentialities within the holomovement originate from. With regard to this issue the following observation by quantum physicist Henry Stapp is pertinent:

> The evolving quantum state, although controlled in part by mathematical laws that are direct analogs of the laws that in classical physics govern the motion of 'matter', no longer represents anything substantive. Instead, the evolving quantum

state would represent the 'potentialities' and 'probabilities' for actual events. Thus the 'primal stuff' represented by the evolving quantum state would be idealike in character rather than matterlike ... quantum theory provides a detailed and explicit example of how an idealike primal stuff can be controlled in part by mathematical rules based in spacetime.[255]

Thus, the quantum 'primal stuff' moving within the holomovement can be thought of as made up of a vast, possibly infinite realm, of "idealike" "potentialities". As we have seen, in their book *The Grand Design*, Stephen Hawking and Leonard Mlodinow tell us:

...the universe doesn't have just a single history, but every possible history, each with its own probability;[256]

And:

In this view, the universe appeared spontaneously, starting off in every possible way. Most of these correspond to other universes .... sometimes called the multiverse concept...[257]

In other words, the universe comes into existence as an infinite 'field', or 'sea', of potential histories. Such a 'multiverse of potentialities' view is clearly consistent with both Stapp's proposal and also Bohm's holomovement perspective. The multiverse contains infinite implicate future histories of the universe which are awaiting unfoldment into 'explicate' manifestation as the future 'unfolds'.

In *The Undivided Universe* Bohm and Hiley (B&H) compare their own view with the Many-Worlds interpretation, which asserts that all potential worlds are equally 'real' and there is no one privileged 'classical' world, and B&H say that:

...our interpretation ... gives a simple and coherent account of why the large scale world of common experience should be essentially classical.[258]

So we can conclude that Bohm's holomovement perspective takes as a background the presence of an implicate field, or fields, of infinite potential 'worlds' which await unfoldment into the explicate order. The later Bohmian Undivided Universe perspective, in contrast to the Many Worlds viewpoint, asserts that a privileged, because experienced, 'explicate' world unfolds from the implicate multiverse background of potentiality.

Bohm and Hiley criticize the Everett Many Worlds perspective because of its unfounded assumption that mind and consciousness can be simplistically reduced to purely quantum-material type processes which are completely described by the mathematical description of quantum functioning, which is designated as "Hilbert space" in the following quote:

> [Everett] not only assumes that the physical universe can be described completely in terms of Hilbert space, but he seems to imply that the same is true for mind, which he regards as being in essence just awareness and memory as he has defined them. It must be emphasized however that this in itself is a highly speculative assumption with very little evidence behind it. ... Even if we accept the as yet unproven assumption that memory can be explained the way that Everett does, it does not follow that this could be done for the whole of mind...[259]

The completely unsubstantiated, illogical, in fact Stapp refers to this view as "irrational",[260] and actually contrary to significant evidence, belief on the part of Everett, that the primarily experiential phenomena of mind can be just assumed to be reduced to, and included in, the mathematical description of the 'physical', indicates the operation of a thoroughly unscientific mindset within physics, which still permeates the quantum age. This mindset involves the assumption that consciousness is nothing special, its just 'physical' stuff doing its stuff so to speak!

Contrary to this completely "irrational", dogmatic and unscientific subtle quantum-materialist bias, Bohm and Hiley assert that quantum discoveries suggest that mind permeates the nature of the entire universe. Thus in *The Undivided Universe* they assert that:

> It is thus implied that in some sense a rudimentary mind-like quality is present even at the level of particle physics, and as we go to subtler levels, this mind-like quality becomes stronger and more developed. Each kind and level of mind may have a relative autonomy and stability. One may then describe the essential mode or relationship of all these as *participation*...[261]

This observation is clearly consistent with John Wheeler's view of the "participatory universe":

> The universe does not 'exist, out there,' independent of all acts of observation. Instead, it is in some strange sense a participatory universe.[262]

The nature of the "participation", however, is subtle, it is not the case that human beings can quantumly choose their material reality with beams of consciousness at will, any more than they can choose the laws of nature on a whim!

Bohm considered, in a very Buddhist-like perspective, that a universe, like our own, is the result of the coming together of even deeper movements of subtle energy-potentiality:

> …let us consider the current generally accepted notion that the universe, as we know it, originated in what is almost a single point in space and time from a 'big bang' that happened some ten thousand million years ago. In our approach this 'big bang' is to be regarded as actually just a 'little ripple'. An interesting image is obtained by considering that in the middle of the actual ocean (i.e., on the surface of the Earth) myriads of small waves occasionally come together fortuitously with such phase relationships that they end up in a certain small region of space, suddenly to produce a very high wave which just appears as if from nowhere and out of nothing. Perhaps something like this could happen in the immense ocean of cosmic energy, creating a sudden wave pulse, from which our 'universe' would be born. This pulse would explode outward and break up into smaller ripples that spread yet further outward to constitute our 'expanding universe'. The latter would have its 'space' enfolded within it as a special distinguished explicate and manifest order.[263]

And then this new universe itself becomes a holomovement of implicate-explicate orders operating as an explicating movement of implicate orders:

> Rather, one is to begin with the holomovement, in which there is the immense 'sea' of energy described earlier. This sea is to be understood in terms of a multidimensional implicate order, … while the entire universe of matter as we generally observe it is to be treated as a comparatively small pattern of excitation. This excitation pattern is relatively autonomous and gives rise to approximately recurrent, stable and separable projections into a three-dimensional explicate order of manifestation, which is more or less equivalent to that of space as we commonly experience it.[264]

But now we have the question of exactly how any of the various potentialities get activated, or 'chosen', to become experienced as part of the explicate order. According to Pylkkänen:

> It is particularly important to note Bohm's emphasis on the incompleteness of existence, for without it there would be no room at all for creativity in his concept of reality, as the unfoldment would be nothing but the realization of pre-existing enfolded potentialities. In contrast, incompleteness leaves room for the creation of new potentialities and new types of unfoldment ...[265]

Here Pylkkänen seems to suggest the possibility of completely new potentialties being created. This, however, would be a creation from 'nothing' and is not supported by quantum theory of any flavour. According to the Feynman sum-over-histories approach alluded to by Hawking and Mlodinow, all possible possibilities exist as potentialities at the dawn of time, so new ones could not be created ex-nihilo! Measurements cannot create completely new quantum potentialities, rather they trigger the selection of one of the available potentialities. Creativity, on this view, would be based on being able to jump out of established quantum paths in order to perceive novel potentialities in distant branches of the multiverse of potentialities.

Creativity resides in the manner of selection of pre-existing potentialities, and, as we shall see, the scope of this creativity is constrained, but not completely compelled, by selections which have been made in the past. To see how this works we can consult the insightful discussion between the biologist Rupert Sheldrake and Bohm, 'Morphic Fields and the Implicate Order'. The innovative biologist Rupert Sheldrake has proposed that biological inheritance and development is driven by 'morphic resonance' within a 'morphic field':

> The idea is that there is a kind of memory in nature. Each kind of thing has a collective memory. ... And how that influence moves across time ... is given by the process I call morphic resonance. It's a theory of collective memory throughout nature. What the memory is expressed through is the morphic fields, the fields within and around each organism. The memory processes are due to morphic resonance.[266]

For Sheldrake, genes are not the central feature of inheritance, although, of course, they are involved. However, according to his theory of morphic resonance the primary determining aspect of biological morphogenetic structuring lies deeper at the quantum level wherein there reside morphogenetic fields. According to Sheldrake morphogenetic fields are quantum:

> ...*probability structures* that depend on the statistical distribution of previous similar forms. The probability distributions of electronic orbitals described by solutions of the Schrödinger equation are examples of such probability structures, and are similar in kind to the probability structures of the fields of morphogenetic units at higher levels.[267]

According to Sheldrake, then, morphogenetic fields are internal structuring aspects of the implicate order. This was the subject of his discussion with Bohm, here are selected highlights :

Sheldrake: The developing organism would be within the morphogenetic field, and the field would guide and control the form of the organism's development. ...

Bohm: ... But from the point of view of the implicate order, I think you would have to say that this formative field is a whole set of potentialities, and that in each moment there's a selection of which potential is going to be realized, depending to some extent on the past history, and to some extent on creativity.

Sheldrake: But this set of potentialities is a limited set, because things do tend toward a particular end point. I mean cat embryos grow into cats, not dogs. So there may be variation about the exact course they can follow, but there is an overall goal or end point.

Bohm: But there would be all sorts of contingencies that determine the actual cat.

Sheldrake: Exactly. Contingencies of all kinds, environmental influences, possibly genuinely chance fluctuations. But nevertheless the end point of the chreode would define the general area in which it's going to end up. ...

Bohm: Each moment will therefore contain a projection of the re-injection of the previous moments, which is a kind of memory; so that would result in a general replication of past forms, which seems similar to what you're talking about. ...

Sheldrake: So this re-injection into the whole from the past would mean there is a causal relationship between what happens in one moment and what subsequently happens?

Bohm: Yes, that is the causal relation. When abstracted from the implicate order, there seems to be at least a tendency, not necessarily an exact causal relationship, for a certain content in the past to be followed by a related content in the future.

The discussion continues to explore various facets of the idea that repetitions of events in the past within the explicate order will strengthen the potentialities within the implicate order for those events to reoccur at a later point in time. The fundamental perspective which emerges is the understanding that there is a primary mode of functioning which indicates that potentialities which have been activated in the past are strengthened as potentialities for future activation, although there is also a limited degree of potentiality for creative novelty to occur.

Lee Smolin, in his book *Time Reborn*, agrees with this kind of quantum viewpoint with his *'principle of precedence'*:

> ... a principle stating that repeated measurements yield the same outcome ... Such a principle would explain all the instances in which determinism by laws work but without forbidding new measurements to yield new outcomes, not predictable from knowledge of the past.[268]

And Smolin quotes a similar insight from the philosopher Charles Sanders Pierce:

> All things have a tendency to take habits. For atoms and their parts, molecules and groups of molecules, and in short every conceivable real object, there is a greater probability of acting as on a former like occasion than otherwise. This tendency itself constitutes a regularity, and is continually on the increase. In looking back into the past we are looking towards periods when it was a less and less decided tendency.[269]

In his discussion with Sheldrake, Bohm observed that:

> If we can bring in time, and say that each moment has a certain field of potentials (represented by the Schrödinger equation) and also an actuality, which is more restricted (represented by the particle itself); and then say that the next moment has its potential and its actuality, and we must have some connection

between the actuality of the previous moments and the *potentials* of the next—that would be introjection, not of the wave function of the past, but of the actuality of the past into that field from which the present is going to be projected. That would do exactly the sort of thing you're talking about. Because then you could build up a series of actualities introjected that would narrow down the field potential more and more, and these would form the basis of subsequent projections. That would account for the influence of the past on the present.[270]

Thus, we see that Bohm is clearly aware that his proposals naturally lead towards a 'presence of the past', or 'principle of precedence', quantum metaphysics.

At a later point in the discussion Sheldrake points out that:

If, however, you start using psychological language, and you start talking in terms of thought, then you've got a handier way of thinking of the influence of the past, because with mental fields you have memory. And one can extend this memory if one thinks of the whole universe as essentially thought-like, as many philosophical systems have done. You could say that if the whole universe is thought-like, then you automatically have a sort of cosmic memory developing. There are systems of thought that take exactly this view. One of them is a Mahayana Buddhist system—the idea of the Alayavijnana, store consciousness, is rather similar to the idea of cosmic memory.[271]

Here Sheldrake suggests that the implicate order, or implicate orders, and therefore the holomovement in general, should be considered to be more correctly characterised as being of the nature of thought, or mind, rather than matter. And, furthermore, he points to a connection with a Mahayana ('Great Vehicle') Buddhist metaphysical system which has as its central concept of a universal 'store consciousness', which is called the *Alayavijnana*, sometimes translated as 'ground-consciousness'.

The *Alayavijnana*, ground-consciousness, is described in the *Lankavatara Sutra,* which is an exposition of the Consciousness-Vehicle / Mind-Only (*Yogacara-Chittamatra*) Buddhist psycho-metaphysical perspective, and which later become a prominent Zen text. This sutra expounds the manner in which the operations of the dualistic world, which involves the *appearance* of matter and the play of the dualistic consciousnesses (*vijnana* – each sense organ is considered to have its own consciousness all of which are collected together by a mental consciousness) comes into being. Both the apparently material world and dualistic consciousnesses are considered to be ultimately unreal and illusory (because of not ultimately existing) productions out of an infinite field of potential energy-awareness, produced through the internal operation of 'habit-energy'.

The nature of the fundamental Mind-stuff itself is nondual energy-awareness, and the operations of habit-energy within it produces the 'illusions' of the dualistic world. A part of this passage has been presented previously, but it is worth a second look:

> Mahamati, there are some Brahmans and Sramanas who assume something out of nothing, saying that there exists a substance which is bound up in causation and abides in time, ... [but, on the other hand] Mahamati, there are some Brahmans and Sramanas who recognising that the external world which is of Mind itself is seen as such [i.e. mistakenly seen as 'real' external 'substance'] owing to the discrimination and false intellection practised since beginningless time, know that the world has no self-nature and has never been born, it is like a cloud, a ring produced by a firebrand, the castle of the Gandharvas, a vision, a mirage, the moon as reflected in the ocean, and a dream; that Mind in itself has nothing to do with discrimination and causation, discourses of imagination, and terms of qualification; that body, property, and abode are objectifications of the Alayavijnana, which is in itself above the dualism of subject and object; that the state of imagelessness which is in compliance with the awakening itself, is not affected by such changes as arising, abiding, and destruction.[272]

The first sentence indicates that there are philosophers who embrace a materialist view, who believe that there are 'real' independent bits and pieces of "substance which is bound up in causation and abides in time", and this substance is assumed to have been magically created from nothing in the big bang. There seem to be physicists who embrace such a view around today! Others, and Bohm is in this set, conceive of the process, or holomovement, of the universe to be of the nature of Mind.

As pointed out previously, it is difficult for us to really appreciate the fact that the vast universe, with its impressive and overwhelming appearance of materiality is actually like "a cloud, a ring produced by a firebrand, the castle of the Gandharvas (illusory beings), a vision, a mirage, the moon as reflected in the ocean, and a dream" etched out of the "quantum dream stuff", as the physicist Wojciech H. Zurek[273] refers to it, of the *alayavijnana* ground-consciousness by the powerful forces of habit-energies echoing across vast time scales. But such metaphors are appropriate for Bohm's later quantum perspective of the 'holomoving' interconnected Totality.

It is worth noting here that in this passage from a Buddhist *Yogacara* - Mind-Only and Zen text we find the claim that: "the world has no self-nature and has never been born, it is like a cloud, a ring produced by a firebrand". In the last chapter we covered the reason why some Buddhist schools of metaphysics assert that the world "has never been born", this is because it actually does not inherently exist in a substantial manner in the first place so could not have been born! We also came across the following quote from Bohm, reiterating, in another equivalent metaphor, the 'ring produced by a firebrand' analogy:

> The notion of continuity of existence is approximated by that of very rapid recurrence of similar forms, changing in a simple and regular way (rather as a rapidly spinning bicycle wheel gives the impression of a solid disc, rather than of a sequence of rotating spokes).[274]

In his writings and discussions Bohm seems to avoid fully positive ontological attributions for the nature of the implicate order(s), but he is explicit, as we have seen, that he does rule out the idea that the any of the various layers of reality are fully and solidly 'material'. Thus, in *Wholeness* he wrote:

> To obtain an understanding of the relationship of matter and consciousness has, however, thus far proved to be extremely difficult, and this difficulty has its root in the very great difference in their basic qualities as they present themselves in our experience. This difference has been expressed with particularly great clarity by Descartes, who described matter as 'extended substance' and consciousness as 'thinking substance'. Evidently, by 'extended substance' Descartes meant something made up of distinct forms existing in space, in an order of extension and separation basically similar to the one that we have been calling explicate. By using the term

> 'thinking substance' in such sharp contrast to 'extended substance' he was clearly implying that the various distinct forms appearing in thought do not have their existence in such an order of extension and separation (i.e., some kind of space), but rather in a different order, in which extension and separations have no fundamental significance. The implicate order has just this latter quality, so in a certain sense Descartes was perhaps anticipating that consciousness has to be understood in terms of an order that is closer to the implicate than it is to the explicate.[275]

Here we see that the original definitions of mind, or consciousness, and matter used within Western philosophy, which pretty much match the definitions in Buddhist philosophy, rule out one of them transforming into the other. They are defined in a kind of mutual opposition: non-thinking extended-in-space matter-stuff cannot magically transform into immaterial thought-stuff, and vice-versa. Furthermore, physics has now reached the level beneath the appearance of the material world, the level of quantum field theory. And as the physicist Jonathan Allday, in his book *Quantum Reality: Theory and Philosophy,* tells us, quantum fields are 'empty' of substance. He writes:

> Now, from a philosophical point of view, this is rather big stuff. Our whole manner of speech ... rather naturally makes us think that there is some stuff or *substance* on which properties can, in a sense, be glued. It encourages us to imagine taking a particle and removing its properties one by one until we are left with a featureless 'thing' devoid of properties, made from the essential material that had the properties in the first place. Philosophers have been debating the correctness of such arguments for a long time. Now, it seems, experimental science has come along and shown that, at least at the quantum level, the objects we study have no substance to them independent of their properties.[276]

However, there is no logical reason why a deep layer of quantum field thought-stuff cannot transmute into the *appearance* of matter-type stuff. Such a transformation regularly occurs in dreams!

The fact that Bohm tells us that "consciousness has to be understood in terms of an order that is closer to the implicate than it is to the explicate" indicates that the implicate order is of the nature of mind and not matter. This is not to say that the implicate order is made up of exactly the same kind of 'stuff' as 'moves' through a person's mind when they consider making a cup of tea, its nature is obviously more profound than that! The nature of individualised consciousness must be derived from the nature of the ultimate nature of the implicate order, but the experiential quality of individualised consciousness is, for most sentient beings, toned down in its radiance and power, so to speak.

According to Bohm:

> ...matter and consciousness can both be understood in terms of the implicate order. We shall now show how the notions of implicate order that we have developed in connection with consciousness may be related to those concerning matter, to make possible an understanding of how both may have a common ground.[277]

Bohm does not explicitly state the precise nature of his "common ground". However, it should be clear that if matter and consciousness emerge from the common ground of the implicate order, and the nature of the implicate order is "closer" to the nature of consciousness than it is to matter, then the implicate order must be some kind of energetic field which has the potentiality to produce pulses of matter-like events as well as a field of consciousness which experiences those matter-like events. And it is "closer" of the nature of consciousness than it is to the apparently material world.

The term 'matter-like events' is used because, firstly, as we have seen previously, Descartes' notion of classical-type extended matter is antithetical to the definition of consciousness, so the two could not co-exist as potentialities within a common field, and, secondly, it coincides precisely with Bohm's own discourse. As Pylkkänen tells us:

> ... the building blocks of the Bohmian universe are moments, and are thus not permanent. Yet the essential idea of his ontology is the idea of the implicate order: each moment contains within it all the other moments in some way. It is the stability of the explicate order which makes each unfoldment, and a kind of "presence of the past", possible.[278]

In such a "Bohmian universe" Cartesian type matter, which is actually what the word 'matter' refers to in general discourse, has no place. Such 'matter' is self-enclosed independent extended 'stuff', the kind of stuff which has no room for any kind of "presence of the past" in the Bohmian sense. Pylkkänen continues:

> According to Bohm, it is a very fundamental feature of the mode of the universe that it consists of moments. A moment has a limited duration as a moment; but it can continue its existence, at least in some sense, as an unfoldment, or trace in future moments. ... The explicate order thus provides a kind of continuity of existence and dependence of the present moment upon the past moments of the universe ... Without the explicate order there would be no way for past moments to exist in the present moment, and the world would lose its structure.[279]

As can be clearly comprehended from this characterisation, the understanding is that there is continuity within the explicate order because of the process of enfoldment into the implicate order, followed by a future unfoldment of another 'moment', the nature of this future 'moment' being conditioned by the natures of moments in the past. This is central. As Bohm says:

> The quantum field contains information about the whole environment and about the whole past, which regulates the present activity ... in much the same way that information about the whole past and our whole environment regulates our own activity as human beings, through consciousness.[280]

Bohm describes his notion of 'moment':

> In certain ways this notion is similar to Leibniz's idea of monads, each of which 'mirrors' the whole in its own way, some in great detail and others rather vaguely. The difference is that Leibniz's monads had a permanent existence, whereas our basic elements are only moments and are thus not permanent. Whitehead's idea of 'actual occasions' is closer to the one proposed here, the main difference being that we use the implicate order to express the qualities and relationships of our moments, whereas Whitehead does this in a rather different way.[281]

According to the internet Stanford Encyclopaedia of Philosophy:

> The ultimate expression of Leibniz's view comes in his celebrated theory of monads, in which the only beings that will count as genuine substances and hence be considered real are mind-like simple substances endowed with perception and appetite.[282]

And:

> Whitehead's ultimate ontology—the ontology of 'the philosophy of organism' or 'process philosophy'—is one of internally related organism-like elementary processes (called 'actual occasions' or 'actual entities') in terms of which he could understand both lifeless nature and nature alive, both matter and mind ... [he] claims that not only our perception, but our experience in general is a stream of elementary processes of concrescence (growing together) of many feelings into one— "the many become one, and are increased with one"[283]

So, as we can see, the comparisons that Bohm himself makes indicates that his notion of a 'moment' can only be of the nature of experience, or mind. As Bohm indicated:

> It follows, then, that the explicate and manifest order of consciousness is not ultimately distinct from that of matter in general. Fundamentally these are essentially different aspects of the one overall order. This explains a basic fact that we have pointed out earlier – that the explicate order of matter in general is also in essence the sensuous explicate order that is presented in consciousness in ordinary experience.[284]

A 'moment' is a dualistic event wherein there is an experience of either an internal mental event, or a perception of a pulse of apparent 'external' materiality. Such a momentary dualistic event occurs primarily as an enfoldment of a previous 'moment' which then conditions and gives rise to the unfoldment of the next 'moment' in question. The view that all conditioned phenomena are of a momentary nature is shared by all schools of Buddhist metaphysics. The fact of momentariness is required by the central insight for Buddhist philosophy and spirituality of the universality impermanence within the conditioned world.

The implicate order, then, must be:

1) Essentially nondual and undivided, whereas the explicate order is dualistic, divided into subject-object experience. This subject-object experience is a flow of 'moments' each of which are conditioned by the collectivity of moments which have preceded the current moment.

2) Closer to the nature of consciousness, which means it must be of the nature of a deep level of Mind-energy.

The nature of this undivided common deep level of Mind should not be taken to be exactly like the 'stuff' of the minds of sentient beings as they manifest in the cogitations of everyday life. The individual 'minds' of sentient beings are dramatically stepped down in quality and power from the energy-potentiality of what is termed in Dzogchen Buddhism 'Primordial Mind'. The following passage comes from the chapter 'How Samsara and Nirvana Originated from the "Basis" as "the Appearances of the Basis"' from the book '*The Practice of Dzogchen*:

> The primordial purity of the original basis transcends the extremes of existence and non-existence, and it is the great transcending of (the objects of) conception and expression. As the essence ... (of the basis) is primordially pure, it transcends the extreme of existence, eternalism, and it is not established as the phenomena of things or characteristics. As the nature (of the basis) is spontaneously accomplished, it transcends the extreme of non-existence, nihilism, and it is present as the purity, the ultimate nature ... of emptiness-clarity, as the nature of the primordial Buddha, as the state of changeless ultimate body (*Dharmakaya*), as non-existent either as samsara or nirvana, and as the self-arisen great intrinsic wisdom which is present from primordial time like space. ... The primordial purity, the basis, is present (in the mode of) essence [entity], nature [character], and compassion [power]. The essence is the ceaselessness of the changeless intrinsic wisdom, and it is called the nature of "the youthful vase body"....
>
> Having broken the shell ... of the "youthful vase body," the primordial basis of the originally pure inner ultimate sphere, by the flow ... of the energy/air of primordial wisdom, the self-appearances of the intrinsic awareness flash out ... [285]

Here in Dzogchen psycho-metaphysics, which has been to a large extent discovered through profound meditational states of direct investigation of the deep levels of awareness, we find a description of the insubstantial ground of reality which corresponds to quantum field theory, although from a subjective perspective. The techniques of meditation developed within Dzogchen, and in all Buddhist traditions, enable practitioners to be aware of profound levels of reality, levels that Bohm called 'implicate orders', which are more or less hidden to ordinary consciousness.

There are several significant points in this Dzogchen quote. The first is the reference to the fact that "the original basis transcends the extremes of existence and non-existence". In Bohmian terms the 'original basis' corresponds to what Bohm, in his later thinking called the 'super-implicate order', the deepest level at which particles enfold and unfold. The super-implicate order will also be at the level of quantum fields wherein particles manifest and are destroyed. Such quantum fields, then, organize the momentary 'existence' of 'particles' and their subsequent 'non-existence'. Thus we see that the Dzogchen claim concerning the transcendence of "the extremes of existence and non-existence" as being a primary condition of the ultimate ground conforms to what is known about quantum fields. In the previous chapter we have seen that the existential configuration of quantum fields conforms to the nature of 'emptiness':

> Its character is neither existent, nor non-existent,
> Nor both existent and non-existent, nor neither.
> Madhyamikas should know true reality
> That is free from these four possibilities.[286]

The Dzogchen text is reiterating this view, which is now validated within quantum theory.

The next significant point from the Dzogchen quote is that the ultimate nature is described as:

> ...emptiness-clarity, as the nature of the primordial Buddha, as the state of changeless ultimate body (*Dharmakaya*), as non-existent either as samsara or nirvana, and as the self-arisen great intrinsic wisdom which is present from primordial time like space...

The term "emptiness-clarity" relates to Bohm's notion of the common ground within the implicate order from which the 'explicate order' derives. 'Emptiness' indicates the lack of any explicit manifestation, its existential configuration is "neither existent, nor non-existent, nor both existent and non-existent, nor neither". 'Clarity' refers to the cognizant-luminous capacity for explicit awareness, although it is not, in its nondual potentiality, explicit itself, unless made so through advanced meditation practices. This is an important point. Even though Bohm had many discussions with Jiddu Krishnamurti, he did not practice advanced meditations. Even the meditations advocated by Krishnamurti did not approach the depth of meditation practices of advanced Buddhist practitioners who can control their mind-streams to the level at which they are to be able to control their death process.

The "primordial Buddha" is the experiential aspect of pure expanse of utterly uncontaminated mind-energy at the very base of the process of reality, which is also *"the state of changeless ultimate body (Dharmakaya - body of potential phenomena)"*. At this level there is neither *samsara* - which is the repeated round of rebirth within the dualistic spheres of the process of reality, or *nirvana* - which is the relinquishing of *samsara*. The "self-arisen great intrinsic wisdom" refers to the fact that the ultimate 'wisdom', which is a direct experience of the ultimate mind-energy of reality, arises naturally as an interior facet of the process of reality because it is intrinsic to the very nature of the process of reality. A Tibetan Buddhist adept tells us that:

> Self-arisen means self-liberated. Any phenomena whatsoever is the true nature of the energy of mind. Therefore, it is self-arisen and self-liberated. It is like a wave in the ocean. Where does the ocean's wave come from? The ocean. And when the wave dissolves, what does it dissolve back into? The ocean. Both when it arises and when it dissolves, the wave is the ocean itself. Similarly, when any phenomena arises, it is the true nature of mind itself. And when it is liberated, it is the true nature of mind itself.[287]

The same applies to the direct wisdom which is a direct experience of the nondual mind-energy of the process of reality, it is not produced by anything other than the nature of reality itself. Because all phenomena are of the nature of the "energy of mind", they are ultimately nothing other than this "energy of mind", and a direct experience of this energy is "self-arisen great intrinsic wisdom".

The above quote concerning phenomena arising from a mind-energy ocean can be compared with Bohm's observation that we have previously surveyed:

> The new form of insight can perhaps best be called *Undivided Wholeness* in *Flowing Movement*. This view implies that flow is, in some sense, prior to that of the 'things' that can be seen to form and dissolve in this flow. One can perhaps illustrate what is meant here by considering the 'stream of consciousness'. This flux of awareness is not precisely definable, and yet it is evidently prior to the definable forms of thoughts and ideas

which can be seen to form and dissolve in the flux, like ripples, waves and vortices in a flowing stream. As happens with such patterns of movement in a stream some thoughts recur and persist in a more or less stable way, while others are evanescent. The proposal for a new general form of insight is that all matter is of this nature: That is, there is a universal flux that cannot be defined explicitly but which can be known only implicitly, as indicated by the explicitly definable forms and shapes, some stable and some unstable, that can be abstracted from the universal flux. In this flow, mind and matter are not separate substances. Rather, they are different aspects of one whole and unbroken movement.[288]

And here we find another resonance with the passage from the Dzogchen text we are contemplating in the context of Bohm's ideas. In the final section of the above Bohm passage he indicates that his view of the holomovement is that of a "flow" of a fundamental "flux of awareness" wherein "mind and matter are not separate substances". Compare this perspective with the Dzogchen view that:

> The essence is the ceaselessness of the changeless intrinsic wisdom, and it is called the nature of "the youthful vase body".... Having broken the shell ... of the "youthful vase body," the primordial basis of the originally pure inner ultimate sphere, by the flow ... of the energy/air of primordial wisdom, the self-appearances of the intrinsic awareness flash out ... [289]

The "youthful vase body" is elucidated:

> ...a common metaphor in the Dzogchen teachings is that of a youthful image or 'body' enclosed within a vase. This signifies the dharmakaya in which all qualities are present but not visible from the outside. The body is described as youthful to indicate that these qualities are pure and pristine, untainted by samsara, and immune from birth and death. Chökyi Drakpa says: "Since the essence of the dharmakaya is beyond birth and death, it is described as youthful, and since there is a clarity that comes from its knowing aspect, it is called the vase body."[290]

So, it is the "flow ... of the energy/air of primordial wisdom" which breaks the 'vase' which contains the infinite potentialities of the *Dharmakaya,* which is the sphere of all potentialities, and as a consequence the dualistic "self-appearances of the intrinsic awareness flash out". The resonance with Bohm's description of the functioning of the 'holomovement' is clearly apparent.

According to the text *Fundamental Mind: The Nyingma View of the Great Completeness*:

> From its factor of luminous self-effulgence, it is called "self-arisen pristine wisdom." And due to its not changing in any aspect, it is called "fundamental mind." In other texts it is called "fundamental cognition" and "natural mind of clear light." From the viewpoint of its immutability, it is called "mind-vajra" ['vajra' = diamond-nature] since it does not undergo any change. The mind-vajra pervades wherever space is present, and thus this basal mind of clear light is called "that endowed with the space-vajra pervading space." Though it is taught with such synonyms, all of them are not different in fact from only the nondual sphere of reality and pristine wisdom, the noumenon of the mind, the ultimate mode of subsistence, the vajra-like mind of enlightenment itself. Therefore, although it is called the "sphere of reality," it is not to be understood as a mere empty sphere but as emptiness endowed with all supreme aspects, without any conjunction with or disjunction from luminosity. Though it is called "self-arisen," it is to be understood not as a compounded awareness-endowed with marks, a subject realizing emptiness within a division of object and subject-but as having a luminous nature without even a particle of any mark to be designated as compounded.[291]

The 'self-arisen' luminous-mind nondual ground of of the process of reality, which when activated within a mind-stream is "enlightenment itself", and it is also the intrinsic nature of the implicate realms of the holomovement.

The process of reality must self-arise as a play of illusory appearances precisely because, ultimately, existence does not exist. That is to say fully manifested changeless existence cannot come into being precisely because this requires a change of non-being to being. But, for the same reason, i.e. the impossibility of substantial eternal changeless existence, neither can 'non-existence' fully 'exist'. In other words, a complete and absolute lack of anything whatsoever, a complete absence of even absence, would have the same existential status as changeless existence, and cannot exist for the same reason! Therefore, a momentary dream state of illusion-like appearances, hovering between existence and non-existence naturally arises, with its own internal self-disclosing 'wisdom', which is a direct knowledge accessed in deep meditation.

In his meditation manual *Minding Closely: The Four Applications of Mindfulness*, the Dzogchen practitioner and teacher Alan Wallace describes the *alayavijnana*, the ground or substrate consciousness. He describes how, by using *shamatha* or focused meditation, a practitioner can experience the nature of a deep level of nondual awareness which lies beneath the everyday moving mind:

> Everyone's individual psyche is unique, like a snowflake. Your psyche is built from the experiences of this lifetime and is influenced by previous lifetimes, genetic dispositions, parenting, cultural values, and language, which make your psyche and everyone else's absolutely unique. But if we melt any snowflake, its fundamental ingredient is simply water. Similarly, when you or anyone "melts" the psyche by using shamatha, and it settles back into the substrate consciousness from which it arose, then the three traits that you or anyone will find, regardless of genetic and cultural background, are that the substrate consciousness is blissful, luminous, and nonconceptual.[292]

Wallace also indicates the quantum source of this level of consciousness. Consciousness arises from the same 'emptiness' of space as do apparently 'material' particles:

> Quantum field theory includes very elegant theoretical systems and experimental methods to probe and characterize the nature of space. My undergraduate work in physics was focused on the energy that is implicit in the essence of space itself, called the "zero-point energy." When Paul Dirac (1902-1984) mathematically integrated special relativity and quantum mechanics into quantum field theory, the concept of space was altered radically. In classical physics, space is inert - simply a location in which things can happen. In general relativity, space becomes far more interesting because it can be warped by massive objects. In quantum field theory, the very nature of empty space is characterized by the zero-point energy. Besides containing ordinary matter, space can contain energy in thermal, gravitational, electromagnetic, and other forms. When all such

matter and energy is removed, what remains is the zero-point energy: the energy of empty space. The very nature of space can be thought of as an equilibrium, symmetry, or homogeneity - the same in every direction. But circumstances can break this symmetry, causing virtual particles to emerge spontaneously from "empty" space. A virtual electron or another elementary particle might be detected, but it will rapidly vanish with little effect. Other more durable phenomena also emerge from empty space, and we call them particles and fields. According to quantum field theory, all particles of matter and fields of energy, virtual and real, are simply configurations of empty space. From galaxies to wristwatches to dark matter and energy, everything emerges from and consists exclusively of configured space. Everything eventually dissolves back into space. Whether phenomena are ephemeral or durable, quantum field theory describes their common ground as the nature of space.[293]

The same must be true of the immaterial qualities of consciousness, there is nowhere else for this experiential qualitative continuum of awareness to arise from. The following quote from Stapp's book *Mind, Matter and Quantum Mechanics* is appropriate here:

> The physical world thus becomes an evolving structure of information, and of propensities of experiences to occur ... The new conception essentially fulfils the age old philosophical idea that nature should be made out of a kind of stuff that combines in an integrated and natural way certain mind-like and matter-like qualities, without being reduced to either classically conceived mind or classically conceived matter.[294]

In fact, given that all quantum fields are immaterial, it would seem to be quite reasonable to suppose that the fundamental nature of the realm of quantum fields is energetic potentiality with an internal quality of mind-awareness-energy, or Primordial Mind.

Within the Yogacara / Dzogchen perspective, the *alayavijnana*, the ground or store consciousness has a fundamental function of being the nondual ground which, like Bohm's implicate order, is able to retain a memory of what has occurred in the past and triggers future occurrences of a similar nature. According to the Buddhist worldview all actions performed by all unenlightened beings, including seemingly neutral perceptions, cause repercussions. *Karma-vipaka*, action and resultant effect, action and feedback, is the universal process of cause and effect which operates on all levels of reality, including the appearance of a material world, *karma-vipaka* is not limited to only the moral dimension. This means that there is a dimension of the operation of karma which is involved in the manifestation of what we perceive as an external 'material' reality:

> ...since beginningless time we have been perceiving sights, sounds, smells, tastes and bodily sensations and these perceptions have been creating imprints or latencies in the ground consciousness. Habituation of having experienced a certain visual form will create a latency for that very form. Eventually, that latency will manifest from the ground consciousness as a visual form again, but it will be perceived as external to ourselves.[295]

And:

> ...the mind is the principle creator of everything because sentient beings accumulate predisposing potencies through their actions, and these actions are directed by mental motivation. These potencies are what create not only their own lives but also the physical world around them. All environments are formed by *karma*, that is actions and the potencies they establish. The wind, sun, earth, trees, what is enjoyed, used, and suffered-all are produced from actions.[296]

The resemblance of this Buddhist perspective to the following insight from physicist John Wheeler is unmistakable and remarkable:

> Directly opposite to the concept of universe as machine built on law is the vision of *a world self-synthesized*. On this view, the notes struck out on a piano by the observer participants of all times and all places, bits though they are in and by themselves, constitute the great wide world of space and time and things.[297]

This insight formed the basis for Wheeler's idea that the universe was self-created through an internal mechanism of quantum level 'self-perception'. Wheeler also indicated that he thought that all the phenomena of the process of reality originates ultimately from Mind or a fundamental level of consciousness:

> Where does Space-Time come from?
> Is there any answer except that it comes from consciousness?[298]

This observation echoes a similar one made by Max Planck, a founding father of quantum theory:

> All matter originates and exists only by virtue of a force... We must assume behind this force the existence of a conscious and intelligent Mind. This Mind is the matrix of all matter.[299]

Such views clearly fit with the Dzogchen / Yogacara perspective.

The *alayavijnana* is described by Walpola Rahula as:

> The deepest, finest and subtlest aspect layer of ... consciousness. It contains all the traces and impressions of past actions and all ... future potentialities.[300]

The Yogacara perspective describes the world of psychophysical embodiment as being comprised of seven consciousnesses, which are driven by, and emerge from the *alayavijnana*. The seven consciousnesses which are other than the ground consciousness, which is the eighth consciousness, comprise firstly the five basic sense consciousnesses which are associated with the faculties of sight, hearing, smell, touch and taste, and then the sixth is the mental consciousness. The seventh consciousness is a deluded awareness which conceives the psychophysical continuum to be an independent and separate entity, rather than a momentary sub-process located within a greater totality. The first six consciousnesses are all based on direct perceptions and are therefore temporary. The deluded layer of the mental process is aware of an apparent continuity amidst the flow of temporary impressions, causing a sense of continuity of a 'self'. This deluded mind aspect is known as the *klistamanas* and it overlaps with the store-consciousness, which is the most fundamental part of the mind process that stores the perceptions of every experience in the mental continuum. This creates the illusion that there is a permanent 'self'. According to Buddhism, of course, the perception of a permanent, unchanging 'self' is an illusion.

This perspective, again, matches up with that of Bohm, although it is far more detailed in its description of the functioning of embodied consciousness. First we have the notion that human beings (and so all sentient beings) are ultimately interconnected with the entire universe:

> Ultimately, the entire universe (with all its 'particles', including those constituting human beings, their laboratories, observing instruments, etc.) has to be understood as a single undivided whole, in which analysis into separately and independently

existent parts has no fundamental status.[301]

But, nevertheless, human beings also function as if they are independent substructures or subunits which seem to have a degree of independence and separation:

> We may begin by considering the individual human being as a relatively independent sub-totality, with a sufficient recurrence and stability of his total process (e.g., physical, chemical, neurological, mental, etc.) to enable him to subsist over a certain period of time. In this process we know it to be a fact that the physical state can affect the content of consciousness in many ways. ... Vice versa, we know that the content of consciousness can affect the physical state ... This connection of the mind and body has commonly been called psychosomatic (from the Greek 'psyche', meaning 'mind' and 'soma', meaning 'body'). This word is generally used, however, in such a way as to imply that mind and body are separately existent but connected by some sort of interaction. Such a meaning is not compatible with the implicate order. In the implicate order we have to say that mind enfolds matter in general and therefore the body in particular. Similarly, the body enfolds not only the mind but also in some sense the entire material universe. (In the manner explained earlier in this section, both through the senses and through the fact that the constituent atoms of the body are actually structures that are enfolded in principle throughout all space.)[302]

When reading passages such as this from Bohm's works it is important to keep in mind that terms such as 'matter', 'material', 'body', and so on, do not indicate Cartesian-type matter. The "entire material universe" is in fact an *appearance* of material reality, more akin to congealed mind-potential-energy, rather than independent, self-enclosed, 'real' material 'stuff'. Thus the minds and material appearances, i.e. bodies, of sentient beings are 'sub-totalities' or 'sub-structures' which are ultimately interconnected with the whole.

In the following passage from Bohm we can detect a resonance with the Dzogchen / Yogacara worldview, wherein the apparent material 'explicate' world derives from a deeper mind-like energy field of potentiality:

> One reason why we do not generally notice the primacy of the implicate order is that we have become so habituated to the explicate order, and have emphasized it so much in our

thought and language, that we tend strongly to feel that our primary experience is of that which is explicate and manifest. However, another reason, perhaps more important, is that the activation of memory recordings whose content is mainly that which is recurrent, stable, and separable, must evidently focus our attention very strongly on what is static and fragmented. This then contributes to the formation of an experience in which these static and fragmented features are often so intense that the more transitory and subtle features of the unbroken flow ... generally tend to pale into such seeming insignificance that one is, at best, only dimly conscious of them. Thus, an illusion may arise in which the manifest static and fragmented content of consciousness is experienced as the very basis of reality and from this illusion one may apparently obtain a proof of the correctness of that mode of thought in which this content is taken to be fundamental.[303]

This understanding is remarkably reminiscent of Buddhist psycho-metaphysics. For example, Bohm refers to how the illusion of the material world is so overwhelming that it veils the deeper implicate levels of the process of reality. Within Buddhism it is pointed out that the appearance of 'inherent existence' (*svabhava*) within the ordinary world is so powerful that very few will notice that in fact material reality is actually 'empty'. As mentioned earlier, the material world is in the region of 99.99999999999999% empty space! The 'stuff' of reality only just fulfils a notion of what 'stuff' might be. For instance in his book *The Theory of Almost Everything* Robert Oerter tells us that:

> The laws of physics were saying that matter as we know it simply can't exist. It was time for some new laws of physics.[304]

And later on in the book we read:

> Is the quantum field real? True, it describes the motion of matter ...[305]

Is seems that there is something the matter with matter! The 'particles' which apparently comprise the 'material' world flash in and out of 'existence' from the immateriality of quantum fields.

In Buddhism this misunderstanding, the belief that the external world is really real, and the idea that a human being is a completely separate, independent, self-enclosed unit of reality, cut off and separate from the totality, is termed 'ignorance' (Sanskrit: *avidya* Pali: *avijja*). Ignorance is basically ignorance about the way that reality functions and what it amounts to. And the fundamental ignorance is ignorance of the fact that the entire process of dualistic 'reality' is a dream-like illusion driven by 'craving' for existence. And this craving gives rise to *dukkha*, which is suffering and dissatisfaction. The pervasiveness of *dukkha* in the dualistic world is the first of the Four Noble 'Truths' or 'Realities': (i) the pervasiveness of *dukkha* within conditioned dualistic existence; (ii) the origin of *dukkha* which lies in the existence of 'craving'; (iii) the possibility of the cessation of *dukkha*; (iv) the path to achieving cessation of *dukkha* by becoming enlightened. These 'truths' or 'realities' are called 'noble' truths (realities) because they are only *directly* seen to be 'true' by 'noble' beings, i.e. enlightened beings.

Enlightenment can be thought of as the activation of a level of consciousness which brings the interconnection of the totality of Bohm's holomovement into everyday consciousness. This does not mean that everyday consciousness is completely turned off, enlightened beings are quite capable of turning on a light and cooking breakfast, but their everyday consciousness exists within the context of direct perception of the universal ground consciousness, the implicate order(s), which 'exist' as the background of everyday consciousness.

In the following passage, which we have seen previously, taken from the commentary to the *Diamond Sutra* by the contemporary Chinese Buddhist teacher Hsing Yun, the 'floating' momentary appearances of dualistic everyday world are compared to 'dust' floating within the 'clarity of perfect awareness', the term *'lakshana'* indicates momentary 'characteristics' or 'signs,' which we may interpret as activated 'moments' of dualistic experience:

> Dust clouds the metaphorical pool of enlightened awareness. ... Lakshana rush into the mind and appear before it like clouds of dust-like lakshana; impure intentions are based on deluded visions of dust. Dust clouds the mind on all levels; matter is dust, illusion is dust, and thoughts and perception also are dust. Only the Tathagata sees the 'vast realm of emptiness' in which all of this floats in the clarity of perfect awareness.[306]

Tathata - 'suchness' - is a Buddhist term for the ultimate nature which manifests to a practitioner's awareness when the true nature of the seeming-realities of dualistic reality are penetrated and seen to be illusory. Thus a practitioner who has achieved the ultimate realisation of *tathata*, the ultimate 'thusness' or 'suchness' of the process of reality, is called a *tathagata*, which literally means 'one who has gone to thusness.'[307]

In the following passage the term 'Kun-gZhi' is the Tibetan term for the *alayavijnana*, the ground consciousness:

> As the universal ground (*Kun-gZhi*) is the root of *samsara*, it is the foundation of all the traces, like a pond. As the *Dharmakaya* (ultimate body) is the root of *nirvana*, it is the freedom from all the traces, and it is the exhaustion of all contaminations... In the state of clear ocean-like *Dharmakaya*, which is dwelling at the basis, the boat-like universal ground filled with a mass of passengers – mind and consciousness and much cargo, karmas and traces – sets out on the path of enlightenment through the state of intrinsic awareness, *Dharmakaya*.[308]

Thus we see that the *alayavijnana* is likened to a boat, filled with a mass of sentient beings, which is coursing through the uncontaminated and clear "intrinsic awareness" of the ultimate *Dharmakaya*. Here again, we can see the connection with Bohm's notion of human beings as 'sub-totalities' or sub-structures within the process of the overall holomovement. But whilst the alayavijnana/holomovement is in operation the majority of sentient beings are unaware that ultimately the nature of the process of reality is that of the ultimate Dharmakaya, all else is actually an illusion. Up until the point when buddhahood is achieved and the Dharmakaya is realised, the holomovement operates through the alayavijnana, which determines the possible paths taken by sentient beings according to their karmic traces.

In the introduction to *Adorning Maitreya's Intent: Arriving at the View of Nonduality*, Buddhist translator Christian Bernert writes concerning the *alayavijnana*, which he translates as the 'all-base consciousness', that:

> Without the existence of the consciousness, it would be difficult to account for the continuity of experience, rebirth, and the maturation of karma, in either the distant future or a future life. ... what is the link between an action carried out in the past and its karmic result in the future? According to Yogacara, it is the all-base consciousness that enables this

process. It is termed "basis of all" because as the source of the other seven consciousness, it is the foundation of all experience. One of its main function is as a "storehouse," containing seeds (*bija*) and latencies (*vasana*) of all actions. Every karma, every deed of body, speech, and mind, leaves a mark or imprint on the most subtle mental continuum. These imprints are like seeds in the sense that they contain the potential for a future experience ...[309]

In this passage there is a reference to rebirth, which is a central feature of Buddhist psycho-metaphysics. It is important to understand the meaning of the notion of rebirth in the context of a worldview that also asserts that there is no fixed 'self' which could possibly exist from one moment to the next, let alone pass from one body to another from one lifetime to the next!

The important point that needs to be appreciated is that within Buddhism all phenomena are asserted as impermanent and momentary and therefore there is only ever an appearance of persistence, things appear to be mostly unchanged from moment to moment. This applies to the sense of 'self' within all sentient beings. Because of the appearance of this continuity, the apparent continuity of various 'mental factors' and an internal sense of continuity from moment to moment, the appearance of persistence is taken as being a 'real' independent 'self' which is absolutely unchangeable. However, in reality this appearance of being an unchanging 'self' is actually based on the fact that, in general, subtle momentary change is not noticed, and gross changes are generally gradual in nature.

Furthermore, every sentient being does have a tiny degree of freedom to determine actions at any point in time, even though there is no permanent unchanging 'self' which makes decisions. Because all the energetic factors which make up a psycho-physical continuum function through interconnected dependent arising from one moment to the next, decisions made in any one moment will affect the qualities of consciousness, and the possibilities for decisions, at later points in time. Without such a mechanism of a degree of freedom, any spiritual path would be impossible. Because of this, although consciousness is momentary, the fact that there is a interconnection between moments, consciousness functions, and is experienced, as if there is a continuity of a 'self'. This 'self', however, changes in a continuous manner from moment to moment, each moment conditioning future moments.

At death enlightened beings, who have relinquished the deep seated 'clinging' (*upadana*) for embodiment, can blissfully dissolve into the nondual ground of the process of reality, unless they choose to be reborn. When unenlightened beings die, however, they give rise to a subtle 'rebirth consciousness' which carries karmic potentialities into a future rebirth and actively seeks to re-embody itself because of the quality of clinging generated during their lifetimes. This rebirth consciousness is like a 'clinging' quantum morphogenetic field carrying 'seeds' of potentiality derived from past actions. From the point of view of ordinary everyday consciousness the rebirth-consciousness is 'unconscious', and for most people, a person being comprised of a continuity of consciousness-energy taking a continuity of rebirths, their previous lives remain completely unconscious because the upper levels of gross consciousness do not have access to the rebirth level of awareness, which lies within the ground-consciousness (*alayavijnana*).

Physicist Henry Stapp has written a paper entitled *Compatibility of Contemporary Physical Theory with Personality Survival* in which he addresses this issue in the context of quantum theory. In the usual understanding, according to Stapp, a 'reduction' or 'collapse' of a quantum wavefunction of potentiality results in a psychophysical event. In other words there is a subjective experience which has an objective content. This corresponds exactly to the *Yogacara* view we have looked at. However, Stapp tells us that it is also possible for purely mental events, without a gross 'physical' aspect to take place:

> ...a natural resolution of the problem of biocentrism leads to a relaxing of the notion that all reduction events must be psychophysical events possessing both mental and physical components. That natural resolution of the biocentrism problem is to allow, in addition to the psychophysical reduction events that dynamically connect our human thoughts to the physically described world around us, reduction events that involve only physical properties. ... An analogous possibility exists on the mental side. William James drew attention to "the fantastic laws of clinging" that allow a stream of conscious thoughts, with its ever-changing intermingling of related ideas, to hang together like a persisting entity. If there were purely mentalistic laws of

clinging, then in our normal streams of consciousness these mentalistic laws could be acting in coordination with the physical laws of clinging, to produce the coordinated streams of consciousness that we experience.[310]

Stapp, like Buddhism, suggests that such "laws of clinging" must be a primary aspect of the process of reality, underlying both physical and mental manifestations of the dualistic world. In a Bohmian world the implicate order is primary, and in this case the mental laws of clinging are primary, creating the material appearances. As Stapp indicates:

> This line of thought suggests that the mental laws of clinging could be the more basic, and that they could create the physical aspects....[311]

With this insight into the central importance of 'mental clinging' Stapp, following William James, rediscovers a key psycho-metaphysical insight that the Buddha discovered and elucidated two and a half thousand years ago; that *upadana*, 'mental clinging' to dualistic experience, is a primary force in the process of reality. As Stapp points out:

> If the reduction events need not always be dual in character, but can sometimes be purely mental or purely physical, and if events of each pure kind can, under appropriate conditions, cling together by virtue of their own dynamical laws, then it would seemingly become possible for the mental and physical aspects of a living person to go their separate ways upon the death of the physical body. [312]

The body, of course, now disintegrates. But it is entirely reasonable to consider that the 'clinging' structure of mentality underlying this process, which according to Buddhism is not a fixed 'soul' but a developing energetic-psychic structure, would not dissipate but, as Buddhist psycho-metaphysics asserts, seek further rebirth. And, according to Buddhism, rebirth can be into a physical body, which is reincarnation, or into a purely mental realm. Such a view is entirely consistent with Bohm's worldview.

The perspective outlined here can be referred to as the Quantum-Mind-Only worldview. The mechanism involved, which accounts for the transformation of quantum mind-potentiality into *the appearance* of matter, is *karmic resonance.* All actions and perceptions leave potencies within a deep level of collective mind called the *alayavijnana*, or ground-consciousness, a level of the process of reality which can be shown to correspond to the realm of quantum emptiness or potentiality. When these potencies are activated through being combined with potencies within the mind-streams of vast numbers of other sentient beings, a subsequent resonant inter-subjective creation of a shared material environment comes into being. This description of the process of reality, including the production of the inter-subjective illusion of the material world, involves the mechanism of karmic cause and effect and quantum resonance, the carrying forward and subsequent inter-subjective activation of potentialities within a deep collective mind-stream. When the subjective potentialities resonate together in a reinforcing manner due their overall similarity, then the collective experiential solidity of the apparently independent material world emerges. From this perspective the 'objective' world of apparent material reality is an inter-subjective creation on the part of all sentient beings who have ever 'existed' and 'exist' within the universe. Within this collective dream-like manifestation, which manifests within the potentialities of the overall Mind-Energy-Potentiality of the process of reality, which Bohm calls the 'holomovement', each relatively independent sentient being have their own 'clinging' quantum-mental continuity of mind-energy, which manifests as individuated consciousness. It is this continuity of 'clinging' consciousness-energy which continuously takes rebirth.

Within both *Yogacara-Vijnanavada* (Consciousness-Basis-Way) and *Dzogchen* (Great Perfection or Completion) Buddhist metaphysical perspectives, the primordial ground is not only conceived of as a field of 'empty' potentiality (which is to say 'empty' of any particular manifestation), it is also asserted as having the fundamental and inseparable function of cognition. The ground of the universe is an infinite pool of potentiality and awareness, or empty-cognizance, which must create the infinite 'illusions' within the dualistic experiential realm because of its fundamental nature of awareness has the impetus to explore its own nature through cognitive activity. Herbert V, Guenther, in his book on Dzogchen metaphysics *The Matrix of Mystery* explains this 'pristine' cognitiveness of the fundamental 'matrix':

> What this term refers to derives directly from the self-excitatoriness (*rang-rig*) of the field as the universe of and for experience, and as such denotes a sensitivity and alertness that makes cognition possible as such on every

level of the biosphere. This pristine cognition has a self-referential intentionality of atemporal primordiality...[313]

Thus we are returned, within a Buddhist context, to Wheeler's vision of the universe as a 'self-synthesized' universe, or the Dzogchen 'self-excitatory universe', which comes into being through an infinite web of internal self-perceptions.

The only way that the universe could 'unfold' from within itself in this manner is if the ground contained both the potentialities and the cognitive mechanism of perceptual 'unfoldment' within its own nature. As Guenther explains:

> In Dzogchen thought there is the additional factor of intelligence which inheres in the very dynamics of the universe itself, and which makes primordiality of experience of paramount importance. The atemporal onset of this unfoldment occasions the emergence of various intentional structures...[314]

It is worth noting that Guenther explicitly uses Bohm's term 'unfoldment' in his explanation of *Dzogchen* psycho-metaphysics. And, as Bohm pointed out:

> We can say that human meanings make a contribution to the cosmos, but we can also say that the cosmos may be ordered according to a kind of 'objective' meaning. New meanings may emerge in this overall order. That is we may say that meaning penetrates the cosmos, or even what is beyond the cosmos. For example there are current theories in physics that imply that the universe emerged from the 'big bang'. In the earliest phase there were no electrons, protons, neutrons, or other basic structures. None of the laws that we know would have had any meaning. Even space and time in their present well-defined form would have had no meaning. All of this emerged from a very different state of affairs. The proposal is that, as happens with human beings, this emergence included the creative unfoldment of generalized meaning.[315]

Each sentient being is an individualized structure of experiential meaning-values embodied within individualised consciousness, each sentient being embodies a fundamental evolutionary impetus to maximise the overall meaning value of the individualized meaning-matrix, the final endpoint being enlightenment, wherein the limited awareness of a sentient being dissolves into its universal source, which is the ultimate nondual meaning-field.

Guenther describes the beginning phases of the evolution of the manifested and materialized world of dualistic experience from the 'evolutionary zero point' according to the Dzogchen worldview as follows:

> It is excitatory intelligence that provides the necessary programming information for initiating a dramatic unfolding process (the big bang) tending towards ever greater degrees of complexity (the evolving universe) while simultaneously, throughout all its phases, retaining the intelligence that initiated the process. When this big bang occurs, the surging of intelligence-qua-isotropic radiation develops a special envelope-like structuring of radiation field...The unitary process as an envelope-like structure which results from this surging of intelligence is termed the meaning-saturated field as pristine cognitiveness. [316]

At this level of development there is a cascade of quantum templates of meaning-manifestation, levels of quantum downward evolution from the nondual zero point, levels that Bohm termed 'implicate orders'. Each implicate order enfolds a new level of meaning evolution in a quantum descent into apparent materiality, and this descent requires the materialization of sentient beings as carriers of individualized awareness of a particular locus of meaning-awareness. Thus Bohm tells us that:

> Later, with the evolution of new forms of life, fundamentally new steps may have evolved in the creative unfoldment of further meanings. That is, we may say that some evolutionary processes occur which could be traced physically, but we cannot really understand them without looking at some deeper meaning which was responsible for the changes. The present view of the changes is that they are random, with selection of those traits that were suited for survival, but that does not explain the complex, subtle structures that actually occurred. [317]

Here Bohm indicates the serious shortcomings of the materialistic and mechanistic view of the evolutionary process enshrined in the materialistic vision of the Darwinian evolutionary process.

In contrast to the now completely unacceptable, and debunked, vision of the lifeless magically producing life through the blind chaotic mechanical churning of mindless bits and pieces of inert matter, Bohm suggests that evolution must be driven by an intentionality which acts towards the manifestation of life through increasingly more materialized levels of quantum potentiality, a quantum potentiality which has an internal meaning-function, or force-for-meaning. Evolution, according to Bohm, must essentially be an intentional quantum process by which subtle quantum structures cascade down to less subtle levels to eventually become fully apparently 'materialized' meaning-structures. This is close to Planck's 1944 claim that "Mind is the matrix of all matter".[318] And this is an observation which is worth contemplating alongside the following distillation from the fourteenth century Tibetan Buddhist masterpiece of the Buddhist Jonang 'Other-Emptiness' school: *The Mountain Doctrine: Ocean of Definitive Meaning: Final Unique Quintessential Instructions* by Dolpopa Sherab Gyaltsen:

> I am called the matrix of attributes....
> I am called the pure matrix....
> The essence of ... of cyclic existence
> Is only I, self-arisen.
> Phenomena in which cyclic existence exists
> Do not exist-even particles-
> Because of being unreal ideation.[319]

Dolpopa's exposition is devoted to a lengthy and comprehensive elucidation of the nature of the 'matrix of phenomena', a fundamental Buddhist concept which is analogous to Bohm's holomovement. When the above fragment is unravelled and explicated from within its own context, it turns out to be a spectacular Buddhist treatment of Bohm's quantum worldview.

In the terminology of Dzogchen the creative 'force' referred to by Planck is the 'excitatory intelligence' which acts through the subtle or 'implicate' quantum levels towards manifestation on an apparently materialized level. This life-force creates 'envelopes' which are quantum demarcation structures which designate boundaries within the cognitive process of materialization. These quantum structures mark out areas of differentiation between the activity of subjective cognition and the projected stabilized cognized objects; and in this way the 'pristine cognitiveness' hides its unitary nature in an imaginational field of activity, a field of activity within which the possibilities for the evolution of sentient beings and the collective environments shared by the various varieties of sentient beings takes shape.

This dramatic psycho-metaphysical perspective is articulated within the Buddhist Dzogchen tradition in texts such as *You Are the Eyes of the World*, composed by the remarkable fourteenth century philosopher-yogi Longchenpa:

> Listen, because all you beings of the three realms
> Were made by me, the creativity of the universe,
> You are my children, equal to me.
> Because you and I are not separate,
> I manifest in you.[320]

Again, we can find echoes of this Dzogchen view within the work of Bohm:

> One must then go on to a consideration of time as a projection of multidimensional reality into a sequence of moments. Such a projection can be described as creative, rather than mechanical, for by creativity one means just the inception of new content, which unfolds into a sequence of moments that is not completely derivable from what came earlier in this sequence or set of such sequences. What we are saying is, then, that movement is basically such a creative inception of new content as projected from the multidimensional ground. In contrast, what is mechanical is a relatively autonomous sub-totality that can be abstracted from that which is basically a creative movement of unfoldment. How, then, are we to consider the evolution of life as this is generally formulated in biology? First, it has to be pointed out that the very word 'evolution' (whose literal meaning is 'unrolling') is too mechanistic in its connotation to serve properly in this context. Rather, as we have already pointed out above, we should say that various successive living forms unfold creatively.[321]

As mentioned previously, Bohm's perspective, and quantum physics and metaphysics in general, clearly demonstrates the falsity of materialist-mechanistic Darwinism. In contrast, Bohm emphasizes the 'creativity of the universe'.

In *Science, Order, and Creativity* Bohm and F. David Peat write:

> If creative intelligence originates in the infinitely subtle depths of the generative order, which is basically not in the order of time, then it follows that the discussion of creative intelligence must bring in this timeless order in a fundamental way. This order must be considered all at once, rather than in an order of succession. ... In terms of the implicate order, it is clear that if the flow were only from the subtler to the more manifest, then it would reduce to a purely timeless order ... Such an order could in a certain sense be intensely creative. But if what happens in one moment would not be related to the next moment, such creativity would resemble an arbitrary series of kaleidoscopic changes with little total meaning. Moreover, more manifest levels would have no autonomy in relation to the subtler levels. A more meaningful kind of creativity can be obtained by relating the eternal order to the time order, and by allowing the more manifest orders to have some degree of relative autonomy.[322]

Here we see that in Bohm's perspective there is both a downward creativity from the "eternal order", as well as a necessary 'presence of the past' "time order" which operates within the manifest, or explicate, order.

According to Longchenpa:

> Out of the state of pure and total presence, the impetus for everything
> From which come the five great elements whose very being is this state,
> I, the creativity of the universe,
> Arise as teacher, in five forms of pure and total presence.[323]

These "five teachers," which are generated by the "creativity of the universe which fashions everything",[324] are earth, water, fire, wind and space, in other words all the basic factors which make up the dualistic world of experience. And:

> If I [the state of pure and total presence which is the creativity of the universe] did not exist, you would not exist.
> When you do not exist, the five teachers [i.e. the dualistic and material world of experience] also do not come about...[325]

It is intriguing to compare these observations with some of Wheeler's 'quantum-mystical' musings, such as:

> Yes, oh universe, without you I would not have been able to come into being. Yet you, great system, are made of phenomena; and every phenomena rests on an act of observation. You could never even exist without elementary acts of registration such as mine.[326]

Here Wheeler indicates the same self-referential creativity, wherein the creativity of the fundamental field of the universe gives rise to the sentient beings who are themselves participants in the creativity of the universe. This is the activity of the primordial consciousness that activates what Longchenpa calls the "majestic creativity [of the universe] which fashions everything."[327]

According to Guenther:

> In the human context, intelligence reaches into man's life as his spirituality, constituting itself as human subjectivity. The latter, therefore, is not an immutable essence; rather it is a product of an overall evolutionary force moving in an optimizing direction, thereby enabling the subject to transcend itself by overcoming its limited domains. This force is felt as giving meaning to man's life and is experienced as having existential significance.[328]

In the Buddhist Dzogchen worldview, which is fully in accord with modern physics, and is also remarkably consistent and resonant with Bohm's ideas, we have an inspiring vision of the universe as a meaning-machine, or meaning-organism, using sentient beings both as creative agents and also agents of transcendence reaching towards ever greater vistas of universal meaning-values. This perspective indicates a universal directedness towards ever more universal modes of experience within consciousness, the ultimate experience being 'enlightenment', which is full awareness of the expanse of the Totality.

This view resonates vibrantly with views Bohm advanced in an important discussion with Renee Weber, '*Meaning as being in the implicate order philosophy of David Bohm*'. In this discussion Bohm makes the following striking comments:

> Meaning is being.[329]

And:

> Being and knowing are inseparable.[330]

These profound and unconventional insights are part of an extended discussion which is worth careful consideration, Here is the start of the discussion:

> **Weber:** You are more and more interested in meaning, so can we explore what meaning is ...?
>
> **Bohm**: ... it is the essential feature of consciousness, ... meaning is being as far as the mind is concerned.
>
> **Weber:** *Is* meaning being?
>
> **Bohm**; Yes. A change of meaning is a change of being. If we say consciousness is its content, therefore consciousness is meaning. We could widen this to a more general kind of meaning that may be the essence of all matter as meaning.[331]

It is important to understand the deeply novel perspective, from a modern Western point of view, that Bohm is proposing here. Bohm has suggested that the realms of consciousness and the appearance of matter derives from a deeper unified common ground. In the above statements Bohm now indicates that this common ground is fundamentally of the nature of meaning, perhaps we may call this common ground a 'quantum meaning field':

> The universe is supposed to have started from this big bang. We might say that is the formation of a certain meaning and a certain structure of meaning which unfolds. There could be other universes, within this sea of infinite energy. Let's look at basics: meaning, energy, matter and, ultimately, self-awareness. Meaning infuses and informs energy, giving it shape and form. Now a certain form is matter, which is energy which has stabilized into a regular form, more or less stable, with some independence. But there must be a meaning that is behind it. ... it would be a formative cause, a field of meaning. ... meaning runs through the implicate order as well as the explicate order, at all levels. ... and would apply at all levels of implication, inner and outer. ... This is a universe that is alive ... and somehow conscious at all its levels.[332]

According to Bohm, "consciousness is meaning" and also "the essence of all matter is meaning". This would seem to suggest that Bohm considers that meaning may underpin both consciousness and the appearance of matter. Here is another insight from the discussion with

Weber:

> **Weber**: ... is there meaning in the non-human world, in the world of nature, and in the universe as a whole?
>
> **Bohm**: That's what I am proposing; not only that there is meaning to it, but rather *that it is meaning*...[333]

Thus, Bohm indicates that meaning suffuses the entire holomovement; the meaning-field unfolds into more explicit meanings as it unfolds through implicate orders towards the explicate level. The whole process itself, however, is suffused with degrees of the existential quality of meaning. Furthermore, according to Bohm the universe is "somehow conscious at all its levels".

In the article 'Soma-Significance and the Activity of Meaning' Bohm emphasizes that he considers that 'meaning' is a primary quality internal to the universe. According to Bohm 'meaning' can be considered to be the most fundamental aspect of reality because it enfolds the other aspects of consciousness and 'matter', and it can also enfold itself, which is to say that it is possible to have multiple levels of meanings; higher level meanings can relate together meanings on levels beneath it, and so enfold them into a unity. So because the qualitative aspect of 'meaning' enfolds all three aspects, including itself, it is the fundamental aspect of experience:

> ... meaning refers to itself directly and this is in fact the basis of the possibility of that intelligence which can comprehend the whole, including itself. On the other hand, matter and energy obtain their self-reference only indirectly, first through meaning.[334]

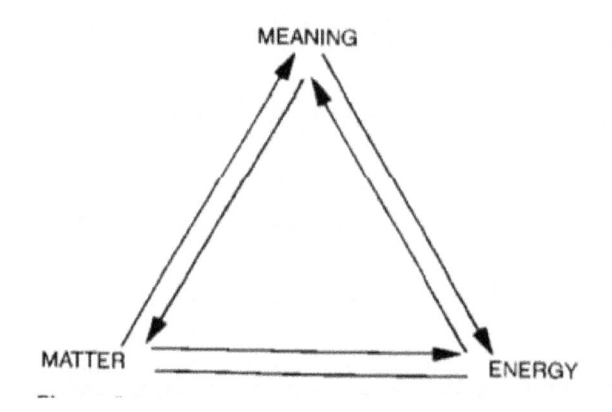

The foundational existential qualitative dimension of Meaning makes all comprehension and understanding possible. And, it seems that according to Bohm, Meaning also underlies the creation of the appearance of the material world. From a Buddhist perspective, it would be more correct to assert that the operation of meaning-functionality is a primary aspect of the inner nature of awareness-consciousness. The quality of 'meaning', which results from the functioning and unfoldment of consciousness, is able to refer to itself and thus can act as a basis for a self-referring, self-creating and self-perceiving universe. Bohm indicates that meaning is a self-referring function that is intrinsic to reality; it is the inner quality of the universe:

> Rather than ask what is the meaning of this universe, we would have to say that the universe *is* its meaning. ... And of course, we are referring not just to the meaning of the universe for us, but its meaning 'for itself', or the meaning of the whole for itself.[335]

Without this inner quality of meaning being intrinsic to the universe from the start the universe could never mean anything, to itself or to anything within it. The function of meaning, then, can be looked at as the central source of the experiential polar aspects of mind and matter. Matter is an appearance of objective meaning to mind, and individuated consciousness, or awareness, is the ground of subjectively experienced meaning and the objective appearance of matter.

In a dramatic moment Bohm declares:

> We do not know how far the self-awareness would go, but if you were religious, you would believe it in the sense of God, or of something that would be totally self-aware.[336]

Within *Yogacara* and *Dzogchen* Buddhism this nondual "totally self-aware" ground field, from which all manifestation originates, is the *Dharmakaya*, which has the internal aspect of 'suchness' - *Tathata* or pure Meaning-Being;

> ...suchness is undifferentiable, which refers to the true nature of the mind abiding without change and interruption as being similar in type from sentient beings up through buddhas. This nature of the mind is taught through many names and examples. In the sutras, it is referred to as prajnaparamita ['perfection of transcendent wisdom'], ultimate reality, the true end, the basic nature, the unchanging perfect nature, the nature of phenomena,

mind as such, emptiness, and so on. In the mantrayana it has many synonyms such as primordial protector, connate wisdom, great bindu, natural luminosity, and Mahamudra ['the Great Seal']. This pure luminous nature is obscured by cloud-like adventitious stains, which arise simultaneously with it, like film on gold, and consist of the consciousnesses that manifest as the dualistic appearances of apprehender and apprehended. They are given many names, such as alaya-consciousess, dependent nature, mistakenness of the seeming, and the ground of the latent tendencies of ignorance. ... [By practice] ... the basic nature is realized perfectly and the adventitious stains eradicated, which leads to the manifestation of the buddha heart. This is called "dharmakaya." Since this natural luminosity was primordially never tainted by stains, there is nothing to be removed in it – the stains are fabricated and adventitious, and therefore the basic element is empty of them.[337]

And, here is another Dzogchen description, this is from the inspirational work *Buddhahood without Meditation* which contains Dzogchen psycho-metaphysical doctrines revealed to the nineteenth century Dzogchen master Dudjom Lingpa:

> The cause is the ground of being as basic space which is pristinely lucid and endowed with the capacity for anything whatsoever to arise. The condition is a consciousness that conceives of an I. From the coming together of these two, all sensory appearances manifest like illusions. In this way, the ground of being as basic space, ordinary mind that arises from the dynamic energy of that ground, and the external and internal phenomena that constitute the manifest aspect of that mind are all interlinked like the sun and its rays.[338]

This indicates the operation of an infinitely fertile universal potentiality-ground which can bring into being an extraordinary appearance of a vast 'material' universe containing infinite varieties of consciousnesses, all of which inhabit an individualized field of meaning-values. Each sentient being is an individualized structure of meaning-values which are not static but partake of a fundamental evolutionary impetus to maximize the overall meaning-value of the individualized meaning-matrix.

The very first glimmer of manifestation occurs when the 'pristine lucidity' or 'awareness', which is an innate aspect of the ground of being and the holomovement, stirs with the conception of an 'I'. In other words this first stirring of cognition, a movement towards a perceiver and perceived, is the first movement within the quantum ground of 'emptiness and cognizance' in the direction of manifestation. It is the first quantum field creation operator. Following this, a dramatic manifestation takes place within which the apparently material world arises as a container for the individuated consciousnesses which will evolve through the evolution of sentient beings inhabiting the material container:

> Yet regarding the sensory appearances of the outer world as a container, the animate beings contained therein, and the objects manifesting in between... when you go to sleep, the outer sensory appearances of the inanimate universe, the outer container, the animate beings contained therein, and the objects manifesting in between as the five kinds of sensory stimuli dissolve into the space of the unconscious blankness of the ground of all ordinary experience, just like the artifices of a magical illusion collapsing in basic space.[339]

And embodied consciousness is actually a direct manifestation of the infinite power of the energetic awareness of the ground of reality; it is in fact *the innate glow of the ground of being*:

> When the true face of the ground aspect of buddhahood - a state of purity and mastery of the ground of being - is obscured by the nonrecognition of awareness, ... timeless awareness - the innate glow of the ground of being - subside into an inner glow whose radiance is directed outwardly ...[340]

The implications of this are profound. Just about all sentient beings are entirely unaware that their everyday consciousness is actually 'the innate glow of the ground of being' and because of this their awareness is completely directed into the external environment. Furthermore, such unaware beings think that the apparently material world is a 'real' material world; a 'real' and ultimately mindless material world that somehow magically gives rise to an aspect of manifestation – consciousness – that has little import beyond enabling a lumbering bit of organic matter to survive for a paltry few years on a hunk of rock hurtling through space. Most human beings do not realize that if they turn their consciousness around and, using precise meditation techniques, explore the deeper levels of their embodied consciousness they can discover that embodied consciousness is indeed 'the innate glow of the ground of being'. And the experiences that can be produced in this way will confirm the Dzogchen view that this is perhaps why the universe manifested – so that embodied consciousness can come to know the extraordinary qualities of the 'glow of the ground of being'.

What is enlightenment? It is the direct nonconceptual understanding of the ground of Being by the fundamental cognizant aspect of the ground of Being itself. In other words enlightenment occurs when the ground of Being fully and directly and nonconceptually cognizes, comprehends and understands its own nature through the agency of a sentient being. This is brilliantly explained in the excellent Dzogchen text *Wonders of the Natural Mind* by Tenzin Wangyal Rinpoche. The ground of Being is characterized within Dzogchen as an 'empty' energy field of potentiality which has an internal spontaneous cognizant quality. The field of potentiality is designated 'emptiness' and the internal spontaneous cognizant quality is designated 'luminosity' or 'clarity'. Tenzin Wangyal Rinpoche writes:

> Who then understands emptiness? There is the self-understanding of emptiness by emptiness itself, by the clarity aspect of emptiness that enables understanding by direct perception. Understanding is not separate from emptiness. Emptiness understands itself and illuminates itself, ... Herein lies the inseparability of emptiness and clarity; self-understanding is self-clarity or self-awareness.[341]

Quantum & Buddhist Views of Free-Will

Bohm, Henry Stapp, Michael Mensky & The Buddhist Spiritual Path

Bohm presented his view on the subject of 'free will' in his essay 'Freedom and the Value of the Individual'. Here Bohm tells us that:

> ...ultimately *everything* is participating creatively in the action of the totality. For example in its grossest levels, this creative participation consists of continuing to re-create its past forms, with modifications, in a way that is approximately mechanical. ...Such creation of a sustained but ever changing existence of matter of the grosser [mechanical] levels opens the way for the action of higher levels of creativity, such as life and mind.[342]

Here we see that Bohm considered that the mechanisms of the enfolding into, and unfolding out of, the implicate levels of potentiality, thus producing the relatively stable explicate world of apparent materiality, eventuality creates the "higher levels of creativity" associated with "life and mind". The lower levels function more mechanically whereas the higher levels associated with mind have a greater, although still limited, degree of freedom.

However, although Bohm indicates that the higher realms of individualized minds, especially in the human realm, have the possibility of making free-willed decisions, he also tells us that there are deep obstacles preventing decisions actually being truly free, although decision makers may think they are free. As Bohm points out:

> Freedom has been commonly identified ... with *free will* or with the closely associated notion of *freedom of choice*. ... the basic question is: Is will actually free, or are actions determined by something else ... such as our heredity constitution, our conditioning, our culture, our dependence on the opinions of other people ...[343]

But, of course, the obstacles cited by Bohm here are not blocks to the metaphysical possibility of free will in principle. They are factors that interfere with the proper functioning of free will. It is quite clear that Bohm considers that free will is a very real capacity within the mental make-up of human beings, but he also considers that it is very rarely a significant feature of most people's decision making processes. As Bohm says:

> ...if one does not have correct knowledge of the consequences of one's action, freedom of will and choice have little or no meaning.[344]

Lee Nichol points out in the introduction to Bohm's essay:

> It is Bohm's view that in general we do not exhibit genuine free will, and thus do not rise to the original definition of "individual" - one who is undivided. A systematic limitation on free will springs from a fundamental misunderstanding of knowledge itself ... knowledge is typically a mechanical projection from the past. As such, suggests Bohm, it is incapable of meeting the rich and complex nature of the living present - the very milieu in which true free will would necessarily act.[345]

Here it is clear that free will is accepted by Bohm as a capacity internal to the process of reality at its higher levels of embodied mind, but that its true operation is generally undermined by interfering and obstructing mental knowledge-structures deriving from the past.

Clearly, the possibility of true free will, which can be developed through mental development, is a definite feature of the Bohmian worldview. As Pylkkänen points out:

> ...it is important to remember that Bohm assumes that the holomovement is incomplete and in a constant process of unfoldment. Such unfoldment is not only the actualization of pre-existing potentialities. This leaves room for genuine creativity ...But note that Bohm thinks that recurrence and stability have an extremely important role to play in the universe, for they make it possible for there to exist relatively independent sub-totalities. ... we might think of an individual human being as such a "relatively independent sub-totality". Relative independence means that the individual has, in principle, some freedom.[346]

According to Bohm's perspective, the holmovement consists of several aspects. Firstly, there is the downward unfolding, through increasingly more stabilized, or 'materialized', levels of implicate orders. These are layers of potentiality that become more stabilized as they descend towards the explicate order of the experienced everyday world. At the same time there is the time order of the 'presence of the past' movement from past to future. In this aspect there is a continuous enfolding of activity into the implicate orders, followed by subsequent unfolding, in a manner that establishes a relative stability within the everyday explicate experiential world. Events occurring at any point are, as we have seen, conditioned by events in the past, thus establishing a stability of recurrence of similarity. This establishes the appearance of a stable 'material' world'. Eventually, the process of unfoldment gives rise to relatively stable 'sub-totalities' of individualized consciousness, these are the sentient beings who come into existence as aspects of the holomovement that are actually directly aware, to various degrees, of the explicate productions of the holomovement. Human beings are the most explicitly self-aware sentient 'sub-totalities', although very few achieve full self-awareness. According to Bohm, true creative use of free-willed freedom of choice requires a heightening of human self-awareness.

According to Bohm:
> Given that a human being may be creative when his or her consciousness arises directly from the "timeless" holomovement, we come to another question: is the creative human being merely an instrument or a projection of the creative action of the totality? Or does one act from one's own being independently? I suggest that this is the wrong question, as it presupposes a separation of the human being from the totality, which I have denied at the very outset of this enquiry. A better question is: can we be free to participate in the creativity of the totality at a level appropriate to our own potential?[347]

Here Bohm emphasizes that human beings are never fully separate from the totality, and obviously they could not be absolutely separate. They are, however, "sub-totalities" with a degree of independence and freedom, without this they could hardly be expected to participate in the "creativity of the totality". Physicist Henry Stapp makes a similar point, in a more theological context, a little more emphatically when he asserts that the quantum universe is:

> ...concordant with the idea of a powerful God that creates the universe and its laws to get things started, but then bequeaths part of this power to beings created in his own image, at least with regard to their power to make physically efficacious decisions on the basis of reasons and evaluations.[348]

If we use the equivalence of Stapp's "powerful God" with Bohm's implicate levels of the holomovement, we can see that Stapp's view is concordant with Bohm's, and we may take Bohm as implying that the process of unfolding through implicate levels into the explicate can be viewed as bequeathing a tiny part of the creative power of the holomovement as a glimmer of free will within the mind-streams of all human beings.

Bohm is a little less enthusiastic than Stapp because he is setting the scene for his assertion that most humans are not capable of utilizing the available freedom because of the restrictive influence of past conditioning. However, this does not alter the fact that freedom to some extent can be developed through mental development. This view closely corresponds to the Buddhist perspective which asserts that ultimate and absolute freedom is attained, through spiritual development, in 'enlightenment'. This can be conceived of as the process through which the limited mind of a human practitioner 'sub-totality' dissolves the limiting confines of the sub-totality structure to become co-extensive with the totality.

The issue of 'free-will', of course, is of great importance for notions of the possibility of spiritual practice which can lead to enlightenment. If there were absolutely no free will then it would be very difficult to explain how anyone could possibly endeavor to follow an ethical code such as is required by Buddhism, 'ethics' would be a matter of determinism, rather than spiritual effort. One possible avenue of explanation that could work in the case of a completely deterministic universe would be to claim that the universe itself is directed towards enlightenment and therefore sentient beings at a naturally occurring point in development would, as a matter of the causal nature of the universe itself, develop the desire to follow a spiritual path and therefore follow a moral code. In a sense there is some truth in this, because according to the Buddhist worldview the universe is directed towards enlightenment and all sentient beings have the opportunity to choose a spiritual path in order to become enlightened.

However, Buddhism requires a degree of free-will, Buddhism is clear on the issue of free-will and determinism, as Kyabje Kalu Rinpoche indicates:

> It is very important to understand clearly that although karma conditions our experiences and actions, we still enjoy a certain measure of freedom – what would be called free will in the West – which is always present in us in varying proportions.[349]

This assertion that there is a limited degree of free will which is "always present in us in varying proportions", is clearly fully in line with Bohm's view. *Karma-vipaka* is the Buddhist technical term for the mechanism by which actions (*karma*), which include simple perceptions, leave traces of potentiality which may be triggered to produce effects (*vipaka*) of the same or similar kind at a future point in time. In the *Yogacara* Buddhist worldview it is the *alayavijnana*, or ground consciousness, which facilitates this process:

> The common characteristic of the alayavijnana is the seed of the receptacle-world' means that it is the cause of perceptions which appear as the receptacle world. It is common because these perceptions appear similarly to all who experience them through the force of maturation that is in accordance with their own similar karma.[350]

And:

> The ground consciousness is the foundation and location for mind because all karmic latencies are stored in the ground consciousness. A momentary visual consciousness instantly ceases (when the next instant appears). Similarly, a mental consciousness is created and ceases instantly; sometimes a mental consciousness does not appear at all. However, the latencies for the arising of these consciousnesses are contained within the ground consciousness. Thus we can remember a visual perception that occurred in the past; and remembering it, strengthens the latency.[351]

Even the appearance of the apparently material world is a karmic appearance because it has been created over vast time periods by the perceptions of uncountable numbers of sentient beings. This Buddhist viewpoint is completely consistent with Bohm's viewpoint, as well as John Wheeler's. In Bohm's perspective, like the Buddhist, the appearance of the material world is a more recurrent, stable and long-lasting phenomenon within which the more evanescent sentient beings have their being for a limited time period.

According to the Yogacara view, enhanced with a quantum perspective, a fundamental feature of consciousness is that even the tiniest movement of energy within the structure of consciousness leaves a trace within the ground consciousness *which increases the probability that the same movement of energy will occur at a later point in time*. This reinforcing process takes place at all levels of consciousness, including those deep structures of psychophysical embodiment not available to direct awareness. As the scholar and Buddhist monk Nyanaponika Thera tells us:

> Like any physical event, the mental process constituting a karmic action never exists in isolation but in a field, and thus its efficacy in producing a result depends not only on its own potential, but also on the variable factors of its field, which can modify it in numerous ways. [352]

Some karmic actions are freely-willed actions, although the degree of freedom involved overall will vary. Within Buddhism it seems that all karmic actions have at least a glimmer or residual trace of free-will involved precisely because karmic actions are *intentional* actions:

> The Buddha used the term *karma* specifically referring to volition, the intention or motive behind an action. He said that karma is volition, because it is the motivation behind the action that determines the karmic fruit. Inherent in each intention in the mind is an energy powerful enough to bring about subsequent results.[353]

Within Buddhism, even simple non-enlightened perceptions are intentional and karmic because they are driven by clinging to the process of reality as a real process, and this is also a clinging which sustains the fiction of an independent ego or 'self'. This belief in the reality of the appearance of the world and the independent existence of a personal 'self' is called *avijja* or ignorance, which is the first link in the Buddhist twelvefold path of dependent origination of the samsaric, unenlightened, cycle of continued rebirth in the explicate realm of Bohm's holomovement.

The Twelve Links of Dependent Origination (*Paticcasamuppada*) are expounded in the *Sammaditthi Sutta: The Discourse on Right View*[354] and the *Paticca-samuppada-vibhanga Sutta: Analysis of Dependent Co-arising*[355]. In the following there is additional information from Maurice Walshe's *Thus I have Heard* and the excellent *Buddhism for Dummies* by Jonathan Landaw and Stephen Bodian. The most natural, and generally accepted, way of understanding the operation of these links is as spanning three lifetimes, 1-2 relate to a previous life, 2-10 the present life (number 2 actually crosses between the previous and current life because it is the rebirth consciousness that carries the karmic [Pali: *kammic*] tendencies from one to the other), links 11-12 refer to the next life.

The twelve links are:

1. **Ignorance** (*avijja*): Ignorance is basically about the way that reality functions and what it amounts to. And the fundamental ignorance is ignorance of the fact that the entire process of dualistic 'reality' is a dream-like illusion driven by 'craving' for existence. And this craving gives rise to *dukkha*, which is suffering and dissatisfaction. The pervasiveness of *dukkha* in the dualistic world is the first of the Four Noble Truths or Realities: (i) the pervasive existence of *dukkha*; (ii) the origin of *dukkha* which lies in the existence of 'craving'; (iii) the possibility of the cessation of *dukkha*; (iv) the path to achieving cessation of *dukkha* by becoming enlightened. These 'truths' or 'realities' are called 'noble' truths (realities) because they are only *directly* seen by 'noble' beings, i.e. enlightened beings. Craving itself is actually a crucial element of the twelvefold chain of interdependent origination. This means that the twelvefold chain itself is part and parcel of the second noble truth, because it is a detailed and graphic depiction of the arising of *dukkha*, which is the first noble truth. Also, it is clear that part of the reason for the arising of *dukkha*, or suffering and dissatisfaction, is ignorance of *anatta* (absence of fixed and inherent personal self and lack of substantiality in phenomena) and its co-component *anicca* (the impermanence of all phenomena). Thus basically the ignorance indicated concerns the fundamental and essential nature of reality and its functioning. On the Wheel of Life image this is illustrated by a hobbling blind man, at the top and slightly to the right of image – the rest of the twelve links go clockwise around the 'wheel'.

2. **Karmic (Kammic) Formations** (*sankhara*): Because of ignorance, sentient beings perform intentional actions which have karmic (Pali-*kammic*) consequences, leaving traces upon the mind stream of the sentient being performing the acts, of body, speech and mind. The image on the wheel of life is of a potter fashioning a pot. The idea is that kammic actions fashion future potentialities. This link indicates the importance of free will in the process of reality as described by Buddhist psycho-metaphysics. Although in any one human lifetime a majority of the circumstances that are in that particular life are not under the control of the free will of the current life, many of these circumstances will be the result of free-willed choices and activities in previous lives.

3. **Consciousness** (*vinnana*) Karmic (*kammic*) actions condition the nature of the consciousness which is projected into a future life. The

image is of a monkey scampering down a tree, this represents the consciousness, with karmic/kammic traces, leaving one life in preparation for the next. This rebirth consciousness, which is a subtle consciousness carrying karmic traces will determine the nature of the next link.

4. **Name and Form / Mentality and Materiality** (*nama-rupa*): The image is that of two travelers in a boat, one of the travelers is 'form' (*rupa*), which is the body; the other is 'name' or 'mentality' (*nama*). Thus the body and mind, the psychophysical embodiment that one has in any lifetime depends on previous lifetimes.

5. **Six sense media / bases / gates** (*sal-ayatana*) Represented by an empty house with six windows; the six gates (ear, eye, nose, tongue, touch, and mental sense) are the bases for experience, although there is no one inside, there is no fixed and permanent 'self' (*anatta*).

6. **Contact** (*phassa*): This link is represented by a man and woman embracing and kissing and engaging is sexual activity, this link is the coming together of senses and sense objects which then leads to the next link.

7. **Feeling** (*vedana*). Represented by a man with an arrow in his eye; feelings may be pleasant, unpleasant or neutral.

8. **Craving** (*tanha*) Represented by someone drinking alcohol. Pleasant experiences produce a craving for more of them. Unpleasant experiences produce cravings to be rid of them.

9. **Clinging / grasping** (*upadana*): Represented by a monkey snatching a fruit; this is a deep, instinctual grasping at existence which is conditioned by endless lifetimes of habitual grasping. This grasping often becomes viscerally desperate at the time of death and conditions the leap into a future life.

10. **Becoming** (*bhava*): Represented by a pregnant woman; here is the beginning of the next lifetime.

11. **Birth** (*jati*): A woman giving birth.

12. **Aging and Death** (*jara-maranam*): Represented by someone carrying a corpse. From the moment we are born we are on the way to dying again.

This sequential description of the *samsaric* cycle of ultimately dissatisfying existence is actually the detailed account of the generation of the first 'noble truth' or reality of *dukkha*. The twelvefold cycle of links is driven at its core by 'craving' (*tanha*) and 'grasping' (*upadana*); which are the two crucial factors of embodiment which are etched deeply and unconsciously into the psycho-physicality of embodiment.

Wheel of Life

As the Buddha said at the beginning of the *Dhammapada* "All phenomena are preceded by mind", indicating that all free willed intentions leading to actions have effects at all levels of the process of reality. *Karma-vipaka* is the universal law of action and maturation, cause and effect, which operates at all levels of reality, including the creation of the potentialities, or seeds, within the ground consciousness which mature or manifest as experiences of a supposedly external 'material' reality:

> Space, earth, wind, sun,
> The oceanside, and waterfalls
> Are aspects of the true, internal consciousness
> That appear as if being something external.[356]

According to the Yogacara account of the process of reality, the vast experiential web of reality is a resonant interactive field of primordial quantum mind-energy which is fundamentally of the nature of awareness-consciousness. It is the amplificatory mechanism of the universal *karmic* cause and effect process within the fundamental implicate orders of reality that creates the appearances of the dualistic world. In this characterisation of the quantum process, the appearance of the 'classical' world of experience and materiality is generated through a continuous web of rapidly repeated perceptions on the part of countless numbers of sentient beings over vast time-scales. These sentient beings are Bohm's consciousness-embodying 'sub-totalities' cycling within the holomovement.

The Buddhist philosopher William Waldron describes this fundamental aspect of the Yogacara account of the functioning of reality as being driven by 'self-grasping' which is the deep instinctual habit within all sentient beings to crave individuated experience. Waldron describes this as a linguistically recursive process; however the 'linguistic' levels operative within the Yogacara account of the process of reality operate deep within the psychophysical structure of embodiment, directly structuring and determining the potentialities for manifestation of future experience at deep psychophysical levels:

> ...this linguistic recursivity, which colours so much of our perceptual experience, including our innate forms of self-grasping, now operates unconsciously ... and ... these processes are karmically productive at a collective level as well as individual level – that is they create a common 'world'.[357]

This constitutes an unconscious 'intersubjective feedback system' and therefore:

> ...it is the unconscious habits of body speech and mind to which we are habituated that give rise, in the long term and in the aggregate, to the habitats we inhabit, the 'common receptacle world' we experience all around us.[358]

Although this formulation attributes the creation of the 'common receptacle world' to the unconscious habits that 'we' have become habituated to (over countless lifetimes) it is important to understand that this is an inter-subjective process that involves both deep non-individuated quantum implicate levels of the universal process of Bohm's holomovement as well as infinite acts of free-willed perception and activity.

Henry Stapp is another significant physicist who has come to the conclusion that quantum discoveries require that we accept the reality of free will. Stapp is one of the few current physicists who worked with some of the founding fathers of quantum theory, and he tells us that:

> We live in an *idealike* world, not a matterlike world. The material aspects are exhausted in certain mathematical properties, and these mathematical features can be understood just as well (and in fact better) as characteristics of an evolving idealike structure. There is, in fact, in the quantum universe no natural place for matter. This conclusion, curiously, is the exact reverse of the circumstances that in the classical physical universe there was no natural place for mind.[359]

And within this idealike quantum universe, according to Stapp, there is a mechanism which he calls 'the two-way quantum psycho-physical bridge':

> ...the connection between physical behaviour and human knowledge was changed from a one way bridge to a mathematically specified two-way interaction that involves *selections* made by conscious minds.[360]

And:

> ...each such choice is *intrinsically meaningful*: each quantum choice injects meaning ...[361]

And:

> ...the quantum universe tends to create meaning: the quantum law of evolution continuously creates a vast ensemble of forms that can act as carriers of meaning... [362]

So for Stapp, like Bohm, 'meaning', and therefore consciousness, are *intrinsic* aspects or qualities of the process of reality. Stapp tells us that:

> According to the orthodox interpretation, these interventions are probing actions *instigated by human agents who are able to 'freely' choose which one, from among the various probing actions they will perform.* ... The concept of intentional actions by agents is of central importance. Each such action is intended to produce an experiential feedback.[363]

As we have seen, within the Buddhist perspective the mechanism of *karma-vipaka*, which corresponds to Bohm's similar mechanism of enfoldment and unfoldment, is the central mechanism which drives the process of the wheel of existence within the realm of the dualistic world; actions and subsequent effects make up the process of dualistic experience, which is what Bohm calls the 'explicate' order. It is remarkable that this spiritual-metaphysical doctrine finds a deep resonance at the level of the microscopic foundations of the process of reality within the discoveries of quantum physics.

As Stapp tells us, the evidence of quantum theory, a theory which places *actions* at the center of the ontological structure of reality:

> ...upsets the whole apple cart. It produced a seismic shift in our ideas about both the nature of reality, and the nature of our relationship to the reality that envelops and sustains us. The aspects of nature represented by the theory are converted from elements of *being* to elements of *doing*. The effect of this change is profound: it replaces the world of *material substances* by a world populated by *actions*, and by *potentialities* for the occurrence of the various possible observed feedbacks from these actions.[364]

In his important book *Mindful Universe* Stapp analyses how the 'orthodox' John von Neumann quantum perspective relates to the issue of free will:

> In John von Neumann's rigorous mathematical formulation of quantum mechanics the effects of these free choices upon the physically described world are specifically called 'interven-

tions'... These choices are 'free' in the sense that they are not coerced, fixed, or determined by the physically described aspects of the theory. Yet these choices, which are not fixed or determined by any law of orthodox contemporary physics, and which *seem to us* to depend partly upon 'reasons' based on felt 'values' definitely have potent effects upon the physically described aspects of the theory. These effects are specifically described by the theory.

Nothing like this effective action of mind upon physically described things exists in classical physics. There is nothing in the principles of classical physics that requires, or even hints at, the existence of such things as thoughts, ideas, and feelings, and certainly no opening for aspects of nature not determined by the physically describable aspects of nature to 'intervene' and thereby influence the future physically described structure. In fact, it is precisely the absence from classical physics of any notion of experiential-type realities, or of any job for them to do, or of any possibility for them to do anything not already done locally by the mechanical elements, that has been the bane of philosophy for three hundred years. Eliminating this scientifically unsupported precept of the closure of the physical opens the way to a new phase of science-based philosophy.[365]

And in in his article *'Free Will'* Stapp writes that:

A criterion for the existence of human free will is specified: a human action is asserted to be a manifestation of human free-will if this action is a specific physical action that is experienced as being consciously chosen and willed to occur by a human agent, and is not determined within physical theory either in terms of the physically described aspects of nature or by any non-human agency.[366]

Stapp presents an account of how the "orthodox quantum mechanics that flows from John von Neumann's analysis of the process of measurement in quantum theory" leaves a "causal gap" which is closed by the presence of free will. As Stapp points out:

...the re-bonding [between mind and matter] achieved by physicists during the first half of the twentieth century must be seen as a momentous development: a lifting of the veil. Ignoring this huge and enormously pertinent development in basic science, and proclaiming the validity of materialism on the basis of an inapplicable-in-this-context nineteenth century science is an irrational act.[367]

As we have previously seen, in order to explain the nature of the 're-bonding' between mind and matter Stapp employs an interpretation of von Neumann's presentation of the process of the 'collapse of the wavefunction', which is the quantum process through which an actual experienced reality appears to emerge from the potentialities that are contained within the quantum wavefunction. He outlines his version of von Neumann's analysis as the following three stages of the quantum experimental process:

Process 1: The 'free choice' of the experimental setup, Heisenberg called this phase "a choice on the part of the 'observer'" constructing the measuring instruments and reading their recording. This choice is "not controlled by any known physical process, statistical or otherwise, but appears to be influenced by understandings and conscious intentions."[368]

Stapp calls this process 1 "a dynamical psychophysical bridge".[369]

Process 2: The deterministic quantum evolution of the potentialities within the Schrödinger wavefunction, this is the mathematical description of the development of the probabilities associated the potentialities within the quantum realm.

Process 3: This is what Paul Dirac called a "choice on the part of nature". It is the yes or no feedback from the experimental setup – yes, reality is this way or no, reality is not this way; Stapp indicates that complex questions can be reduced to yes-no choices.

This feature of quantum theory naturally accommodates the phenomenon of consciousness acting freely as a causally effective aspect of reality:

> ...the founders of quantum mechanics instituted a profound break with one of the basic principles of classical physics: they inserted the conscious experiences of human beings into the dynamical workings of the theory. Human beings were allowed, and indeed required, to act both as causally efficacious agents, and also as causally efficacious observers. ... Both of the two actions, the query and the feedback, are causally efficacious; they alter in different non trivial ways the physically described state of the universe.[370]

This situation, Stapp says, constitutes "an idea-based quantum triality", and:

> ...the dynamical structure of quantum theory contains certain causal gaps. In particular, the process-1 agent-generated choices of probing actions are determined, within the theory, neither by the physically described aspects of nature, nor by any non-human agency. Thus, within the framework of orthodox quantum mechanics, the process-1 probing actions are, according to the specified criterion, manifestations of human free will...[371]

Stapp has also pointed out that this situation applies not just in quantum experiments but also in everyday life.

Stapp's analysis here echoes a thought experiment presented by John Wheeler who concocted an intriguing illustration of the way in which the functioning of a 'participatory' universe would allow the process of participation to determine the nature of the future of the Universe on the basis of a kind of intersubjective party game of 'Twenty Questions' involving 'yes' or 'no' responses to 'probing' questions. The usual way the game is played is that one of a group of people leaves a room whilst the rest of the group agree on an object or concept. The other person now returns and tries to discover the chosen object or concept by asking questions to which the rest of the group may only reply 'yes' or 'no' Wheeler's version is that instead of the group inside the room agreeing on a definite object or concept they agree to allow any 'yes' or 'no' response which is consistent with what has gone before. Thus the result of the yes-no game is not pre-determined, but neither is it completely random and chaotic. Physicist Anton Zeilinger has suggested that "this is a beautiful example of how we construct reality out of nothing."[372] We must, however, be aware that the word "nothing" in this context is incorrect, the quantum 'game' operates on quantum potentialities within Bohm' holomovement.

In his 2017 book *Quantum Theory and Free Will: How Mental Intentions Translate into Bodily Actions,* in a section titled 'The Role and Importance of Free Will', Stapp indicates that:

> The linkage between the *free question* and the *random answer* ties the mental and physical aspects of things into a single cohesive dynamically evolving reality. In this evolution, the mental actions of observers play an essential role. ... Your "free

choices" of probing actions, combined with nature's responses, enter actively into the determination of your future psycho-physical states and the futures also of other observers of the system observed by you.[373]

However, there is a problem in Stapp's presentation of quantum free will scenario as developed to this point. In the above quote Stapp admits that at this point the free-willed activity in the Process-1 phase only sets the stage for 'nature' to give a random yes or no response. The degree of free will involved here is limited in its effectiveness.

In his article 'Quantum Interactive Dualism: An Alternative to Materialism' (and elsewhere) Stapp addresses, and closes this limitation by utilizing the evidence of the Quantum Zeno Effect. The nature of the quantum wavefunction, which describes the field of quantum potentiality, is such that when it is unobserved it will spread out, its probability distribution becoming increasingly smeared out over larger and larger areas. It is only when measured or 'observed' that it 'collapses' back into a momentary fixed position and, as soon as the observation ceases, the wavefunction will once more begin to smear out, becoming more and more spread out from the position it was observed at. Rapid observations, or measurements, at the quantum level therefore can pin a quantum 'particle' into a relative stability; this phenomenon is called the Quantum Zeno Effect.

This mechanism of fixing quantum reality through habitual rapid perception is not speculation, it has been demonstrated in quantum experiments. Physicist John Gribben describes an experiment carried out at the U.S. National Institute of Standards and Technology which employed the Zeno effect "to make the pot beryllium ions boil, and to watch it while it was boiling – which stopped it boiling."[374] The experiment involved getting beryllium ions to jump between quantum states in a given period of time. The experimenter then stopped this quantum jumping by constantly observing, using a laser, and thereby fixing the ions into the first state:

> The act of looking at the ion has forced it to choose a quantum state, so that it is now once again purely in level 1. The quantum probability wave starts to spread out again, but after another 4 milliseconds another peek forces it to collapse back into the state corresponding to level 1. The wave never gets a chance to spread far before another peek forces it to collapse back into level 1 [375]

This is an example of the quantum Zeno effect being consciously employed to manipulate and stabilize quantum reality through rapid perception.

Stapp employs this quantum effect, in the context of his von Neumann free will analysis, to produce a mechanism by which he considers the mind can interact with the brain and body through brain patterns that Stapp calls 'templates for action':

> The quantum "holding" effect is a rigorous consequence of the basic orthodox laws of quantum mechanics, supplemented by the assumption that mental effort can instigate a rapid repetition of a Process 1 action. In the present context the intentional act is associated with a macroscopic pattern of brain/body activity identified as a "Template for Action". This particular pattern of neural/brain activity is actualized by the 'Yes' response to the Process 1 intentional probing action. The succession of similar probing actions must occur rapidly on the scale of the natural time-scale of the template for action in order for the quantum Zeno effect to come into play and hold this template of action in place. The "Quantum Zeno Effect" can, in principle, hold an intention and the associated template for action in place in the face of strong mechanical forces that would tend to change the latter. This means that agents whose mental efforts can increase the rapidity of Process 1 actions would enjoy a survival advantage over competitors that lack this feature. Agents who possess this capacity could sustain beneficial templates for action in place longer than competitors who lack it. Thus the dynamical rules of quantum mechanics can endow conscious effort with the causal efficacy needed to permit its evolution and deployment via natural selection. Given this potentially strong causal effect of mental effort on brain activity, both the survival dynamics in the evolution of species and the trial and error learning in the life of the individual would tend to establish a positive correlation between the conscious intention associated with a Process 1 action and the functional effect of this action on the brains of the agents.[376]

Thus Stapp considers that the development of the capacity for minds to develop a mechanism for controlling bodily functioning through patterning the brain into "templates for action" that are controlled by focused intentionality is an aspect of the evolutionary process.

Stapp's proposal has significant evidential support. 'Neuroplasticity' is the term given for the brain's ability to 'rewire' itself in accordance with the changes in behaviour, physical activities, environmental changes and so on. Up until the 1970's the received wisdom within the psychological and neuroscientific academic world was that the brain was unable to change its structure much after very early infancy. As Jeffrey M. Schwartz and Sharon Begley point out in their excellent book *The Mind and The Brain: Neuroplasticity and the Power of the Mental Force*:

> The great Spanish neuroanatomist Santiago Ramon y Cajal concluded his 1913 treatise "Degeneration and Regeneration of the Nervous system" with this declaration: "In adult centres the nerve paths are something fixed, ended, immutable, Everything may die, nothing may be regenerated." Cajal based his pessimistic conclusion on his meticulous studies of brain anatomy after injury, and his gloomy sentiment remained neuroscience dogma for almost a century. "We are still taught that the fully mature brain lacks the intrinsic mechanisms needed to replenish neurons and reestablish neuronal networks after acute injury or in response to the insidious loss of neurons seen in neurodegenerative diseases," noted the neurologists Daniel Lowenstein and Jack Parent in 1999.[377]

However, this dogmatic 'mechanomorphic' (a term coined by the psychologist Abraham Maslow) fantasy was not shared by all investigators and philosophers. The significant psychologist William James wrote in his groundbreaking 1890 work *Principles of Psychology* that:

> Plasticity, then, in the wide sense of the word, means the possession of a structure weak enough to yield to an influence, but strong enough not to yield all at once. Each relatively stable phase of equilibrium in such a structure is marked by what we may call a new set of habits. Organic matter, especially nervous tissue, seems endowed with a very extraordinary degree of plasticity of this sort; so that we may without hesitation lay down as our first proposition the following, that the phenomena of habit in living beings are due to the plasticity of the organic materials of which their bodies are composed.[378]

This crucial insight by James foreshadowed recent crucial and spectacular discoveries regarding neuroplasticity.

We now know that neuroplasticity is a fact, it has been shown that controlling and manipulating the functioning of consciousness has a significant physical effect upon brain structure, it can mould brain structure. Furthermore, as the work of Henry Stapp in collaboration with the neuropsychiatrist Jeffrey M. Schwartz indicates, the mechanism involved is quantum in nature: the quantum zeno effect. This is, of course, what one would expect as the brain itself is nothing else than a classical manifestation of the more fundamental quantum realm. The evidence for the reality of neuroplasticity is now conclusive; as is demonstrated by Schwartz's excellent book *The Mind and the Brain*. Schwartz clearly draws the conclusion that the brain is moulded by the mind:

> Mind, we now see, has the power to alter biological matter significantly; that three pound lump of gelatinous ooze within our skull is truly the mind's brain.[379]

This has been demonstrated conclusively by experiment.

As we have seen, the notion that the minds and bodies of sentient beings, as well as the apparently external 'material' environments that they inhabit, are the result of a continuous habituation over vast time scales was central to the Yogacara view. The Buddhist philosopher William Waldron describes the Yogacara view of the process of reality as being "psycho-ontological" in the sense that according to the central metaphysical mechanism of *karma* – the cause and effect mechanism of the process of reality – it is the intentionality exercised by sentient beings which builds up the physical and organic structures of the worlds that they inhabit, including their own sequence of bodies over innumerable lifetimes.

The habitual patterns of experience and behaviour which are etched into the karmic fabric of reality are called '*sankharas*'. These are habitual patterns of experience, behaviour and organic form which echo as potentialities into the future like Sheldrake's notion of 'morphogenetic fields'. Waldron describes this 'psycho-ontological' process:

> These sankharas are thus formative influences which not only continuously condition our bodily forms, but also our intentional activities, the nature and direction of our mental and spiritual energies as well. That is, contoured by these banks, our stream of consciousness continuously flows with both the bubbling surface

of its swift, churning waters and the deeper, hidden currents flowing beneath its surface - both of which subtly yet continuously make their mark upon the contours of that very riverbed and its banks, scouring out pockets here, accumulating deposits there. Together, the river and riverbed constitute a continuous mutually conditioning relationship that has been built up by nothing more than the history of their own previous, continuous interactions. Sankharas built up from the past serve as the continuous basis for our current activities.[380]

The Yogacara psycho-metaphysical process of reality, within which the apparent ontology of the 'physical' world is built up via the operation of karma and intentionality, takes place over enormously long time periods:

> ...in the early Buddhist worldview the various kinds of bodies we inhabit, with their specific types of cognitive and sensory dispositions and apparati, are also built up over the course of countless lifetimes in the particular conditions of cyclic existence. The paths our continued embodied existence take are directed by the accumulated results of our past actions, which are continually reinforced - which increase and "grow"...[381]

Such a worldview, of course, is one involving 'plasticity' across lifetimes.

In the Buddhist worldview universes also pass through successive 'incarnations'. At the end of the first chapter we looked at some of the weak points in Bohm's 1952 pilot wave proposal which were highlighted by Stapp. One of these unsettling features is described by Stapp as follows:

> Bohm's model does however retain one feature of classical physics that can be regarded as objectionable ... This is the need for an arbitrary-looking choice of initial conditions. In particular, some definite initial position for each of the particles in the universe must be chosen.[382]

So how could we account for the initial positions of all the particles at the dawn of time, given that we do not want to appeal to an independent creator God putting them all in precise places before setting the universe in motion, so to speak?

It is difficult to see how this could be solved in Bohm's undeveloped 1952 pilot-wave proposal. However, there is an interesting link to

Buddhism if we consider the problem from Bohm's later perspective wherein the implicate fields serve as the common ground of consciousness and matter. The Buddhist metaphysical view of the way in which universes succeed one after the other, each one inheriting an initial configuration as a karmic echo from the one that has gone before, just as sentient beings also cycle through Bohm's holomovement, provides a mechanism for accounting for universal initial conditions.

As we have seen, according to the Buddhist worldview karmic causality operates at all levels of the process of reality. Many people in the West think that the mechanism of *karma-vipaka*, or action and future result, is only a moral mechanism. This, however, is not the case, *karma-vipaka*:

> ...is the theory of cause and effect, or action and reaction; it is a natural law, which has nothing to do with the idea of justice or reward and punishment. Every volitional action produces its effects or results.[383]

To reiterate, within Buddhist psycho-metaphysics it is understood that perceptions by sentient beings are karmically active and therefore condition future similar perceptions. The subjective structures of intentionality are embodied within the various types of sentient beings and operate to produce an intersubjectively created field of potentiality which is carried by the *alayavijnana*, the ground consciousness. And, even when sentient beings are absent, the field of the deepest level ground consciousness, which corresponds to the deepest levels of Bohm's implicate orders, remains active.

The resonant traces, or 'seeds', which are 'stored' in the ground, or 'store', consciousness, will later 'mature', or be activated into experiential reality. *Karma-vipaka* is the universal law of action and maturation, cause and effect, which operates at all levels of reality, including the creation of the potentialities, or seeds, within the ground consciousness, which later mature or manifest as experiences of an apparently external 'material' reality.

The *Yogacara* consciousness-ground-of-all perspective is quite clear that what we experience as the physical world is in fact built up through the operation of the collective karma within the fundamental consciousness-field which is the ground of reality:

> The entire world was created through latent karmic imprints. When these imprints developed and increased, they formed the

earth, the stones, the mountains, and the seas. Everything was created through the development or propagation of these latent karmic potentials.' ... 'How can all these external forms arise out of latent karmic imprints? All these mountains, oceans, the sun and moon are so solid and so vivid. How can they arise out of latent karmic imprints in the mind.' ... 'These things arise through the power and propagation of thought'.[384]

The fact that the process of karma is responsible for the appearance of the material world is little known or understood in the West even among Buddhists. But this understanding of the process of reality is clearly stated within the Buddhist worldview.

The Buddhist metaphysical account described above is further amplified and elucidated by some teachings regarding the formation of universes from the *Kalacakra Tantra,* which describes:

> ...vast world systems throughout infinite space. At any particular time, some world systems are arising, some abiding, some disintegrating, and others remaining dormant. In this view, there is no absolute beginning. There is simply the beginningless interplay of various factors that make world systems arise, abide, disintegrate and remain dormant.[385]

And, furthermore, the *Kalacakra Tantra* describes the state as it exists between the destruction of one universe and the karmic arising of the next as follows:

> During this time of emptiness the subtle particles ... exist as isolated fragments and are not in any conventional sense objects of the sensory powers of the eye and so forth. They are known as *empty particles*....[386]

So, to answer Stapp's query as to how particles can start out in an appropriate configuration for the generation of a coherent universe, this situation can be accounted for if such 'empty' particles, or particles of quantum potentiality, are generated as karmic echoes from one disintegrated universe into the next universe. Thus both sentient beings and entire universes can be viewed as cycling through Bohm's holomovement, driven by karmic intentional actions, both individual and collective.

Returning to the context of individual free-will; in his explorations in the topic of free-will Stapp addresses the apparently contrary evidence advanced by the neuroscientist Benjamin Libet of the University of California in San Francisco. Experiments carried out by Libet shocked many people because they seemed to show that human beings have less control over their brains than they thought. According to the results he obtained it seemed that a lot of what appears to be free will cannot be what it appears to be. Libet's experiments were interpreted as showing that free-will is an illusion.

Libet set up an experiment in which he was able to monitor when a brain appeared to be in a state of readiness to flex a finger. This was compared to the report by the experimental subject as to when they decided to perform the act. When he performed the experiment he found that the brain's preparation apparently occurs before the experimental subject reported having made a decision. In the introductory summary to his paper *Do We Have Free Will* Libet writes:

> Freely voluntary acts are preceded by a specific electrical change in the brain (the 'readiness potential', R.P) that begins 550 ms before the act. Human subjects became aware of intention to act 350-400 ms after RP starts, but 200 ms. before the motor act. The volitional process is therefore initiated unconsciously. But the conscious function could still control the outcome; it can veto the act. Free will is therefore not excluded. [387]

Despite the fact that Libet wrote that his work does not exclude free will some proponents of materialism made much of these results, claiming that it proves that conscious free-will is an illusion. Stapp, however, counters this claim. In his book *Mind, Matter and Quantum Mechanics*, for example, Stapp indicates that the interpretation of Libet's work as indicating that the brain automatically makes decisions independently of consciousness is based on ignorance of the complexities of the process at the quantum level:

> In the context of the Libet experiments the critical point is that according to the Heisenberg picture there must first be a separation, generated by the evolution in accordance with the deterministic equation of motion, of the physical state into parts representing several macroscopically distinct possibilities before the act of choosing one of these macroscopically distinct alternatives occurs. In the brain most of the processing activity is done at the unconscious level: the lower-level process first

prepares the distinctive alternatives, and the Heisenberg actual event then selects and actualizes one of them. Thus the delay found by Libet is demanded by this quantum-mechanical theory of consciousness.[388]

According to Stapp, then, the actual process by which a decision is made, taking Heisenberg's quantum mechanics as fundamental, requires that the brain divides the superposition of quantum possibility, within consciousness, into distinct alternatives before a choice is made by the chooser. This would mean that Libet's 'readiness potential' was not an indication that the choice has been made but that *the alternatives had been prepared ready for a choice to be made*. Taking into account the quantum mental separation of alternatives occurring prior to the decision, Libet's time gap is, required by Stapp's theory of the operation of free-willed quantum consciousness.

The quantum free will perspective advanced by Stapp is very close to the view proposed by the little known but hugely important Russian physicist Michael Mensky, a physicist and professor who worked at the Lebedev Physical Institute of the Academy of Science in Moscow. Mensky, although not a Buddhist himself, was deeply impressed by the fact that quantum discoveries were indicative of the validity of various claims within the world's spiritual traditions, and in particular Mensky was impressed by aspects of Buddhism which he considered to be metaphysically close to various quantum metaphysical insights.

In the early part of his book *Consciousness and Quantum Mechanics: Life in Parallel Worlds*, Mensky tells us that:

> Life is not the function of the body, and consciousness is not a function of the brain. Rather body is a realization of life, and brain is an instrument of consciousness.[389]

Mensky calls his quantum view of the life-process of reality the Quantum Concept of Consciousness (QCC) and also the Extended Everett Concept (EEC), which, as the name indicates, extends Hugh Everett's 'Many Worlds' quantum metaphysics to incorporate consciousness as an active ingredient. In the usual presentation of the Many Worlds proposal all the vast, perhaps infinite, possibilities within the universal quantum wavefunction 'exist' at the same time, and there are multiple copies of all sentient beings experiencing all the various possibilities, each 'self' being ripped apart into different universes at each moment of time. According to Mensky, however, there are not duplicate 'copies' of sentient beings being divided into different worlds, in reality each individual 'selects' a path through the multitude of worlds.

Mensky's quantum metaphysical perspective indicates that core features of some spiritual traditions, such as karma, rebirth and enlightenment, are consistent with, perhaps even indicated by, quantum discoveries. Furthermore, Mensky considers that spiritual goals are inherent within the process of life. He points out:

> The phenomenon of consciousness demonstrates mystical features that are experienced by some people. All religions and spiritual schools that [have existed] for thousands of years include the mystical component as a necessary part of their message ...[390]

Mensky says that his paradigm, involving quantum consciousness, is a "bridge between the natural sciences and humanities, between matter and spirit".[391] His approach "unifies natural sciences with the sphere of spiritual knowledge including religion".[392] Within Mensky's Quantum Concept of Consciousness (QCC) "science, philosophy and religion meet together",[393] and the QCC explains "numerous strange phenomena, which are described in ... mystical teachings, including religion and oriental philosophies".[394] The QCC shows that:

> ...mystical aspects characteristic of any religion not only are compatible with natural sciences, natural sciences (first of all their central part, quantum mechanics) is logically defective without the inclusion of the concept of consciousness with its mystical features.[395]

Furthermore, Mensky's quantum metaphysical worldview is consistent and supportive, perhaps enhances, Bohms's insights. These issues will be the focus of the next chapter, so only a brief indication will be given here, for the purposes of giving an insight into Mensky's quantum free will proposal.

Mensky's account of quantum 'free will' indicates that consciousness is a quantum field phenomenon associated with the "separation of alternatives". This means, exactly as Stapp suggests, the quantum field of the mind, when presented with a choice, organizes into a quantum superposition within which the alternatives corresponding to the choices become primary. Subsequently a free willed selection is made by the person concerned:

> ...the consciousness as a whole splits between the alternatives but the individual consciousness subjectively chooses (selects) one alternative.[396]

So, according to Mensky, when the individual quantum state of consciousness is in a superposition of possibilities with equal, or close to equal, probability weightings, due to the arising of the necessity for a decision to be made, quantum theory suggests an aspect of the individuated consciousness can have a direct effect upon the alternative possibilities for action, enhancing one of them in order to make the decision:

> If I wish to go to the right and actually go to the right, how (does) this happen? ... In the framework of EEC [Extended Everett Concept], if the modification of probabilities is assumed, free will is explained quite naturally. There are two alternatives: in one (of) Everett's world(s) I go to the right, in the other I go to the left. Both alternatives have non-zero probabilities. My consciousness modifies the probabilities, increasing the probability of the first alternative. As a result, with a high probability I go to the right. This is my free will.[397]

Thus, Mensky gives a precise account of the kind of quantum free will proposal also advanced by Stapp. A similar proposal has been made by physicist Jack Sarfatti:

> ...the self-measuring system self-organises its wavefunction so that it splits into non overlapping wave packets. That is the system creates its own measuring observable ... in a self consistent way and then chooses which channel to occupy. This is free will.[398]

And the physicist Amit Goswami, after considering the Many Worlds model, suggests a very similar view of the interaction of mind and an environmental wavefunction at a more universal context:

> Suppose that the parallel universes of the many-worlds theory are not material but archetypal in content. Suppose they are universes of the mind. Then, instead of saying that each observation splits off a branch of the material universe, we can say that *each observation makes a causal pathway in the fabric of possibilities in the transcendent domain of reality.* Once the

choice is made, all except one of the pathways are excluded from the world of manifestation.[399]

Again, we are presented with a quantum metaphysical viewpoint which resonates with Bohm's perspective as well as Buddhism. In Goswami's version the many-worlds are conceived of as possibilities which are etched, as it were, into the fabric of the objective wavefunction or the 'transcendent domain of reality'. This view is clearly analogous to Bohm's implicate order(s) and explicate order holomovement model. Only one of these possibilities can be actualised for any one sentient being; and Goswami considers that the one world which is actually 'materialised' out of the possibilities is a matter of 'choice' rather than complete necessity, a perspective which clearly resonates with the views of Stapp and Mensky.

Paavo Pylkkänen has written an article comparing the quantum free will views of Stapp and Bohm, 'Henry Stapp vs. David Bohm on Mind, Matter, and Quantum Mechanics', and this provides an excellent place to round off our discussion of quantum free will. According to Pylkkänen:

> ...the advantage of Bohm's mind-matter scheme is that it allows for a stronger possibility for conscious free will to influence quantum dynamics than Stapp's approach, in which human "free choice" merely sets the stage for the entry of the statistical choice "nature" makes, obeying the Born rule.[400]

However, Pylkkänen, presumably because of his enthusiasm for Bohm's worldview, does a disservice to Stapp's point of view. Although Stapp does take as his taking-off point the "free choice" that "sets the stage" for a statistical choice by 'nature', Stapp also develops beyond this starting point when he proposes the notion of a "template for action", which is a quantum brain structure which is activated by intentionality. In his book *Mindful Universe: Quantum Mechanics and the Participating Observer* Stapp writes that:

> In quantum theory the enduring states are vibratory states. ... Such states tend to endure as organized oscillating states, rather than quickly dissolving into chaotic disorder. I call by the name 'template for action' a macroscopic brain state that will, if held in place for an extended period, tend to produce some particular action. Trial and error learning, extended over the evolutionary development of the species and over the time of the individual agent, should have an effect of bringing into the agent's repertoire

of intentional process-1 actions the 'Yes-No' partitions such that the 'Yes' response will, if held in place for an extended period, tend to generate achievement of the intent. Successful living demands the generation through effort based learning of templates for action.[401]

Thus, one can consider the brain as being a quantum interface between the body and the experiences and activities of an individual's mind, including intentional states of mind. Intentional mind-states interact with the brain's quantum templates-for-action and through this mechanism the body can be controlled and motivated. This does not alter the fact that the vast majority of body processes are unconscious and automatic, of course, but it accounts for those aspects of embodied life which are amenable to intentional control.

When one fully understands Stapp's approach it is clear that it is, in fact, coherent with Bohm's viewpoint, although presented in a different guise. Pylkkänen elucidates Bohm's viewpoint as follows:

> Bohm did not want to reduce the human mind and consciousness to the quantum level, but he also wanted to avoid dualism. He suggested that the quantum ontology can be extended to include higher level fields, each influencing and being influenced by levels below, and that the human mind could be a part of such a hierarchy of levels of information associated with certain neural processes .... In this way, say, when I move my hand, the information content in my thought could act down the hierarchy all the way to the level of the quantum field, which latter could then control particles (e.g., in synapses or some other relevant "quantum sites" in the brain). Such effects could then be amplified to control macroscopic neural processes.[402]

So, it seems that the main significant difference between Bohm's approach and that of Stapp is Bohm's view that there is a hierarchy of implicate levels involved on the way down to the explicate level. Indeed, Stapp is critical of Bohm's penchant to implicate multiple implicate orders:

> Bohm certainly appreciated the need to deal more substantively with the problem of consciousness. He wrote a paper on the subject ... which ended up associating consciousness with an infinite tower of pilot waves, each one piloting the wave below. But the great virtue of the original pilot-wave model namely the

fact that it was simple and deterministic with cleanly specified solvable equations, became lost in this infinite tower.[403]

Bohm's 1952 pilot wave theory was very sparse and precise in terms of what the pilot wave actually did. It guided particles around the world, and there were no pilot wave hooks where consciousness could latch on and get involved to do some piloting on its own accord! Because of this Bohm later extended his metaphysical perspective in order to open a role for consciousness. Pylkkänen says of this:

> When Bohm reflected on the mathematical form of the quantum potential Q in the late 1970s, he noticed that Q only depends ... on the form or shape of the quantum field. This form typically reflects the form of the environment (e.g., whether one or two slits are open in a two-slit experiment). This suggests that the particle is not being pushed and pulled mechanically by the quantum field, but rather that the particle is able to respond to the form of the field, or is literally IN-FORMED actively by the information contained in the field. Note that this is information for the electron, not information for us. Thus, Bohm called this type of information active information. There exists potentially active information everywhere where the quantum potential is non-zero, while the information is actually active where the particle is ... How is this notion of active information relevant to the mind-matter relationship? Bohm noted that what is typical of mental phenomena is activity of form (as opposed to activity of substance). When we are reading a newspaper, we are abstracting the forms of the letters; we do not need to eat the paper. These forms are taken up by the nervous system and eventually give rise to an experience of meaning. The meaning, in turn, can be active ... Thus, there is at least an analogy between active information at the quantum level and active information in human subjective experience.[404]

Thus we see that Bohm used an analogy between the way he conceived of the pilot wave 'in-forming' the particles that it guides, and the way that the human mind functions, to speculate that a significant link can be made between these aspects.

So, the notion of the link between the mental realm of 'active information', and the 'active information' of the quantum potential of the pilot wave, indicates that intentional configurations of consciousness, created by thought, now, in Bohm's later perspective, directly hook into the implicate realms underlying the explicate order. On this view, Bohm asserts a subtle yet direct connection between active information of consciousness and active information within the quantum field underlying the pilot wave. Thus Bohm presents an avenue for, a limited, quantum free will which is consistent with the views of Stapp and Mensky.

In making this, very subtle, link Bohm sometimes used the analogy of music, suggesting that the quantum realm of the implicate and explicate orders and consciousness can be viewed from a perspective of flowing movement and unfoldment from the quantum implicate order :

> This activity in consciousness evidently constitutes a striking parallel to the activity that we have proposed for the implicate order in general. Thus ... we have given a model of an electron in which, at any instant, there is a co-present set of differently transformed ensembles which inter-penetrate and intermingle in their various degrees of enfoldment. In such enfoldment, there is a radical change, not only of form but also of structure, in the entire set of ensembles; ... and yet, a certain totality of order in the ensembles remains invariant, in the sense that in all these changes a subtle but fundamental similarity of order is preserved. In the music, there is, as we have seen, a basically similar transformation (of notes) in which a certain order can also be seen to be preserved.... there is a feeling of both tension and harmony between the various co-present transformations, and this feeling is indeed what is primary in the apprehension of the music in its undivided state of flowing movement.[405]

And:

> We see, then, that each moment of consciousness has a certain explicit content, which is a foreground, and an implicit content, which is a corresponding background. We now propose that not only is immediate experience best understood in terms of the implicate order, but that thought also is basically to be comprehended in this order. Here we mean not just the content of thought for which we have already begun to use the implicate order. Rather, we also mean that the actual structure, function and activity of thought is in the implicate order. The distinction between implicit and explicit in thought is thus being taken here

to be essentially equivalent to the distinction between implicate and explicate in matter in general.[406]

Given that Bohm considered that both the appearance of matter and consciousness had their origin within the common ground of the implicate order, it was natural that he also speculated that to a certain degree that consciousness, conceived as a primary internal aspect of implicate orders and the holomovement, can itself in-form the quantum field, and this would obviously apply to interactions between minds and brains. As Bohm indicated:

> One might then suggest that in intelligent perception, the brain and nervous system respond directly to an order in the universal and unknown flux that cannot be reduced to anything that could be defined in terms of knowable structures. Intelligence and material process have thus a single origin, which is ultimately the unknown totality of the universal flux. In a certain sense, this implies that what have been commonly called mind and matter are abstractions from the universal flux, and that both are to be regarded as different and relatively autonomous orders within the one whole movement.[407]

So, although mind and brain may appear as "relatively autonomous" at the explicate level, within the implicate order they are closely interconnected. It was this line of thinking led Bohm and his associates to propose the new primary category of the process of reality - 'active information'. Thus F. David Peat has suggested:

> It is proposed, in the spirit of open speculation, that science is now ready to accommodate a new principle, that of active information, that will take its place alongside energy and matter. Information connects to concepts such as form and meaning which are currently being debated in a variety of fields from biology and the neurosciences, to consciousness studies ... It may well provide the integrating factor between mind and matter.[408]

It seems clear, then, that Bohm considered that human beings had the capacity for free will through the ability to influence active information at the implicate level. However, it may also be said that Bohm did not explicitly provide a precise mechanism for the operation of quantum free will, whereas both Stapp and Mensky give credible and mutually consistent accounts of a viable quantum free will mechanism.

In his book *Mindful Universe* Stapp observed that:

> I once asked Bohm how he answered Einstein's charge that his model was 'too cheap'. He said that he completely agreed! ... in the last two chapters of his book with Hiley, Bohm goes beyond this simple model and tries to come to grips with the deeper problems ... by introducing the notions of implicate and explicate order. But those extra ideas are considerably less mathematical, and much more speculative and vague, than the pilot-wave model that many other physicists want to take more seriously than Bohm did himself.[409]

Here Stapp betrays his dislike of what he sees as Bohm's later intellectual move in the direction of "speculative and vague" quantum metaphysics. But, as we have seen in previous chapters Bohm's later *Undivided Universe* phase arises coherently as a logical metaphysical necessity from his initial 1952 pilot wave perspective, especially when the necessity of encompassing the phenomenon of consciousness is factored in.

In this context it is important to bear in mind that Bohm considered that a general non-mathematical worldview was a vital and necessary requirement, even though mathematics does provide greater precision:

> I would say that in my scientific and philosophical work, my main concern has been with understanding the nature of reality in general and of consciousness in particular as a coherent whole, which is never static or complete, but which is in an unending process of movement and unfoldment.

And:

> In discussing this whole, we begin with the general language which is used for description in physics. We may then be said to mathematize this language, i.e. to articulate or define it in more detail so that it allows statements of greater precision from which a broad range of significant inferences may be drawn in a clear and coherent way. In order that the general language and its mathematization shall be able to work together coherently and harmoniously, these two aspects have to be similar to each other in certain key ways, though they will, of course, be different in other ways (notably in that the mathematical aspect has greater possibilities for precision of inferences).[410]

And, as we have seen, Stapp is no slacker when drawing inspiring psycho-metaphysical conclusions from quantum insights:

> ...the person who recognizes himself to be an integral component of a universal process that selectively weaves waiting potentialities into dynamic new forms that create potentialities for still newer integrations should be inspired to engage actively and energetically in the common endeavour to enhance the creative potentialities in all of us.[411]

I think it is safe to suggest, based on the explorations of this chapter, that, rather than seeing the various proposals from Bohm, Stapp and Mensky regarding the implications of quantum theory for our understanding of the phenomenon of free will as being antagonistic, it is more appropriate to view them as mutually illuminating, elucidating and enhancing aspects of Bohm's holomovement. As Bohm wrote:

> It seems clear, then, that we are faced with deep and radical fragmentation, as well as thoroughgoing confusion, if we try to think of what could be the reality that is treated by our physical laws. At present physicists tend to avoid this issue by adopting the attitude that our overall views concerning the nature of reality are of little or no importance. All that counts in physical theory is supposed to be the development of mathematical equations that permit us to predict and control the behaviour of large statistical aggregates of particles. Such a goal is not regarded as merely for its pragmatic and technical utility: rather, it has become a presupposition of most work in modern physics that prediction and control of this kind is all that human knowledge is about. This sort of presupposition is indeed in accord with the general spirit of our age, but it is my main proposal in this book that we cannot thus simply dispense with an overall world view.[412]

And:

> The proposal for a new general form of insight is that all matter is of this nature: That is, there is a universal flux that cannot be defined explicitly but which can be known only implicitly, as indicated by the explicitly definable forms and shapes, some stable and some unstable, that can be abstracted from the universal flux. In this flow, mind and matter are not separate substances. Rather, they are different aspects of one whole and

unbroken movement. In this way, we are able to look on all aspects of existence as not divided from each other, and thus we can bring to an end the fragmentation implicit in the current attitude toward the atomic point of view, which leads us to divide everything from everything in a thoroughgoing way.[413]

This is not to say that any theory or proposal whatsoever must be accepted as being valid, but, rather, there are differing views or theories which may be mutually coherent and enhancing rather than absolutely antagonistic.

The Holomoving Alterverse & the Consciousness-Only Three Natures of Reality

Michael B. Menksy was a physicist and professor working at the Lebedev Physical Institute of the Academy of Science in Moscow. He worked in fields such as quantum field theory, quantum gravity, quantum theory of measurement and the foundations of quantum physics. His interest in the foundations of, and metaphysical implications of quantum theory, led to him having some spectacular insights into the relationship between quantum theory and the 'mystical' claims of some religions. In his articles and his book *Consciousness and Quantum Mechanics: Life in Parallel Worlds - Miracles of Consciousness from Quantum Reality* he provides a comprehensive spiritual psycho-metaphysics based on the discoveries of quantum physics and theory. As we shall see many of his insights overlapped with those of David Bohm.

Mensky called his quantum-potentiality and quantum-consciousness based account of the functioning of reality the Quantum Concept of Consciousness (QCC) and also the Extended Everett Concept (EEC). This latter name indicates that Mensky's account extends Hugh Everett's 'Many Worlds' quantum metaphysics to incorporate consciousness as a primary aspect, and therefore naturally gives rise to the necessity of the presence of life, intelligence and design. These arise from an internal mechanism of 'unfolding' due the the operation of an internal aspect of consciousness.

The usual version of the Many Worlds scenario involves the view that all potentialities within a global wavefunction are equally 'real'. Because of this viewpoint it is asserted that there are many sets of interconnected potentialities, each set making up a different 'world'. Thus a global wavefunction i.e. a wavefunction of a universe, consists of many potential 'worlds', each world being inhabited by multitudes of slightly different 'copies' of the 'same' persons who also inhabit all the 'other worlds' at the same time. This also means that at every moment in time each sentient being is 'splitting' into different copies of itself, perhaps an infinite number of copies, each one taking up residence in one of the potential worlds, before then instantaneously 'splitting' again. The number of 'copies' and 'worlds' in this science fiction scenario quickly passes into the absurd! Physicist Bryce DeWitt, who was actually a supporter of the Many Worlds perspective, referred to this idea as 'schizophrenia with a vengeance'.[414] In a 1970 paper DeWitt wrote:

I still recall vividly the shock I experienced on first encountering this multiworld concept. The idea of 10^{100+} slightly imperfect copies of oneself all constantly splitting into further copies, which ultimately become unrecognizable, is not easy to reconcile with common sense.[415]

Very rarely is it pointed out that such a viewpoint removes any coherence to the idea what a 'person' might be! Would, for example, a 'person' after 10,000 many-worlds split events still be the 'same' 'person'? Or would it take a million splittings to change person-identity to a different identity?

Of course, a central aspect of the notion of a 'person' involves a continuity of both physical and mental aspects, but neither Everett nor his admirers bothered to, or bother to, spend much attention, in any deep sense, on the issue of the nature of consciousness and its role and functioning in the Many Worlds scenario. However, among a significant section of the physics and philosophy academic community there is a very anti-later-Bohmian denigration of the idea that consciousness has any place in the elucidation of quantum functioning or a role to play in an explanation of the fundamental functioning of reality. Thus the philosopher Simon Saunders claims that:

> ...if the Everett interpretation recovers the elements of quantum chemistry, solid state physics and hydrodynamics, as the effective theories governing each branch, questions of mentality can be left to the biological sciences.[416]

But claims such as this are not based on evidence, they are based on a prior investment in the assumption that a quantum-materialist account of reality must be correct, whatever contrary evidence may come to light.

Another example is provided by physicist Sean Carroll, one of the most prominent proponents of the Many Worlds fantasy today, who writes in his book *The Big Picture*, with absolutely no evidence but lots of quantum-materialist-physicalist faith:

> ...even the most optimistic neuroscientist doesn't claim to have a complete and comprehensive theory of consciousness. Rather, what we have is an expectation that when we do achieve such an understanding, it will be one that is completely compatible with the basic tenets of the Core Theory - part of physical theory, not apart from it.[417]

It seems that mindless 'physical' molecules banging around in magical configurations can generate a lot of faith! After all, if we do not have a "complete and comprehensive theory of consciousness" how can we know that consciousness is a magical transformation of mindless matter?

Mensky asserts that consciousness is a fundamental aspect of quantum functioning. He asserts the reality of the *potential worlds*, but also asserts that any particular sentient being will, due to the endowment of consciousness, select a pathway through the possible worlds from the alternatives through the operation of the ability, generally unconsciously, but sometimes consciously, to make 'selections'. 'Unconscious' selections are made by an individuated structure of consciousness at levels not operative at the level of immediate awareness. A single individual, then, will not 'split', but, rather, will navigate a path through what Mensky calls the *Alterverse* of possible worlds. As we shall see, Mensky's Alterverse can be identified, within limits, with Bohm's holomovement.

For Mensky, then, the 'multiverse' of alternative quantum potentialities is termed the "Alterverse", the infinite scope of all *alternative possible worlds*. Any individuated mind associated with a sentient being will occupy only one of these worlds depending on their previous actions and mind-states and the choices they then make. This indicates the operation of what, as we have seen, in Buddhism is termed *karma-vipaka,* action and result. Intentional actions and cultivated mind-states in any particular lifetime will leave traces within a subtle quantum consciousness, and this subtle rebirth consciousness will influence which potentialities are activated within the Alterverse in future lifetimes. Mensky also gives a detailed account of a quantum mechanism which underlies the functioning of the process of reality as described by his quantum psycho-metaphysics. In a nutshell, in his EEC and QCC Mensky identifies a transcendent sphere of universal Mind as extending across all the worlds within Everett's many-worlds, or Bohm's holomovement, and he specifies a mechanism by which individual beings 'navigate' a pathway through the possibilities.

Mensky appreciated the fact that the quantum scenario is entirely consistent with the spiritual insight that, not only do the actions, perceptions and intentions of sentient beings, in particular human beings who have a much widened sphere of free will beyond that of animals, create future feedbacks in a current life, they also condition future potentialities for future lives. In this insight Mensky has seen that quantum physics indicates the reality and significance of two of the central doctrines for Buddhist practice: *karma* and *rebirth*. According to Mensky evidence for his 'Extended Everett Concept' (EEC):

> ...may be found in the spiritual sphere of knowledge (oriental philosophies, world religions and deep psychological practices). As a result, a much closer unification of the material and spiritual spheres of knowledge [can be] achieved.[418]

As we shall see, Mensky's quantum-spiritual psycho-metaphysics is entirely consistent with, coherent with, and supportive of Buddhism, not that Buddhism requires such support. And it also has some close affinities with some aspects of Bohm's worldview.

In this context it is useful to be aware that Bohm was sometimes a little ambiguous about some of his own concepts. We have seen in the previous chapter that Bohm endorsed free will in an imprecise fashion, whereas both Stapp and Mensky gave much more detailed and precise indications of their views. We can find a similar situation with Bohm's notion of the holomovement. In his book *Wholeness and the Implicate Order* for example Bohm tells us that: "Thus, *the holomovement is undefinable and immeasurable*." The italics are Bohm's. But, at the same time as telling us about the undefinable nature of the holomovement, he does a pretty good job of defining its general aspects, he likens it metaphorically to a radio carrier wave with its encoded 'enfolded' content:

> To indicate a new kind of description appropriate for giving primary relevance to implicate order, let us consider once again the key feature of the functioning of the hologram, i.e., in each region of space, the order of a whole illuminated structure is 'enfolded' and 'carried' in the movement of light. Something similar happens with a signal that modulates a radio wave .... In all cases, the content or meaning that is 'enfolded' and 'carried' is primarily an order and a measure, permitting the development of a structure. With the radio wave, this structure can be that of a verbal communication, a visual image etc., but with the hologram far more subtle structures can be involved in this way (notably three-dimensional structures, visible from many points of view). More generally, such order and measure can be 'enfolded' and 'carried' not only in electromagnetic waves but

also in other ways (by electron beams, sound, and in other countless forms of movement). To generalize so as to emphasize undivided wholeness, we shall say that what 'carries' an implicate order is the holomovement, which is an unbroken and undivided totality.[419]

This seems pretty clear, then, Bohm tells us that the holomovement carries the implicate order, or implicate orders, and this means that it carries all the potentialities for the possible explicate orders which can come into experiential 'existence'.

When we look into Mensky's idea of the 'Alterverse' we find a very close correspondence with Bohm's notion of the holomovement, which is hardly surprising as they both derive from a scientific investigation of the quantum level of reality. Mensky writes that:

> As a starting point for our analysis we need the concept of *quantum reality*. We shall formulate it here in a simple way as the *coexisting parallel classical realities* (classical alternatives), or, equivalently, as *parallel worlds*. Instead of the term "Universe", usually applied for our world ... we may apply the term "*Alterverse*" for the world in which quantum reality (presented as the superposition of the parallel classical worlds) rules.[420]

So here Mensky tells us that the fundamental reality of the process of reality is that it is comprised of a multitude of possible experiential worlds which 'exist' as a quantum superposition of 'parallel' or potential worlds. This does not correspond exactly to Bohm's notion of the holomovement because in Bohm's vision the holomovement is made up of a stack of increasingly more 'materialized' implicate orders which end up unfolding into the explicate order. There is, however, a very close similarity between the two ideas, wherein certain features may be extra or missing on one side or the other. The important point is that the correspondence is sufficient for there to be a general equivalence of functioning.

In the early part of his book *Consciousness and Quantum Mechanics: Life in Parallel Worlds*, Mensky tells us that:

> ...the phenomena of life and consciousness cannot be mechanistically reduced to the action of the laws of science as they are found in the course of exploring [inanimate] matter. The

> explanation of these phenomena on the basis of quantum mechanics requires [the] addition of a special independent element to the set of quantum concepts and laws. Such a new element of theory should directly connect quantum concepts with the concepts characteristic of life. The simplest way to find this element is to consider the phenomenon of consciousness and compare it with the description of observation (measurement) in quantum mechanics. Then it may be formulated as identification of consciousness with the "separation of the alternatives" - a concept relating to the "Many Worlds" interpretation of quantum mechanics. ... the addition of this element simplifies the conceptual structure of quantum mechanics instead of [rendering] it more complicated. If we consider not only the phenomenon of consciousness but [also the] more general phenomenon of life, this additional element may be called [the] *"life principle"*. It very naturally follows from the analysis of [the] theory of consciousness ... The life principle formulates [the] evolution of [a] living system in such a way that it is determined by ... goals as well as by causes. The main goal of living system[s] is survival so that their evolution provides their survival. However, for more sophisticated forms of life, the goals may include other criteria [such as] the quality of life.[421]

Mensky is emphatic that the phenomena of life and consciousness cannot be *reduced* to either quantum mechanics or "any other theory of [inanimate] matter." These aspects are, of course, involved in the processes and functions of living organisms, but:

> ...life and consciousness are not the direct consequence of these processes. Life is not the function of the body, and consciousness is not a function of the brain. Rather body is a realization of life, and brain is an instrument of consciousness.[422]

This is a beautifully formulated observation. It is the phenomenon of 'life', considered as a fundamental force within the process of reality, which evolves the bodies of sentient beings, and it is consciousness, again considered to be fundamental, that organizes the brain for the manifestation of individuated consciousness from a deeper layer of non-individuated primordial consciousness/awareness. Here Mensky's viewpoint clearly resonates with Bohm's worldview. Bohm, for example, says:

> ...life itself has to be regarded as belonging in some sense to a totality, ... It may indeed be said that life is enfolded in the totality and that, even when it is not manifest, it is somehow 'implicit' in what we generally call a situation in which there is no life. ... The above does not mean, however, that life can be reduced completely to nothing more than that which comes out of the activity of a basis governed by the laws of inanimate matter alone (though we do not deny that certain features of life may be understood in this way). Rather, we are proposing that as the notion of the holomovement was enriched by going from three dimensional to multidimensional implicate order and then to the vast 'sea' of energy in 'empty' space, so we may now enrich this notion further by saying that in its totality the holomovement includes the principle of life as well.[423]

This shows how close the perspectives of Bohm and Mensky are. As we can see in the above quotes, Mensky incorporates a 'principle of life' as an internal principle within the 'Alterverse', just as Bohm tells us that "the holomovement includes the principle of life".

According to Mensky, although life and consciousness are not *reducible* to quantum mechanics, it is nevertheless the case that they are "connected" in a deep and irreducible manner with "quantum reality." In fact, as Mensky's quantum psycho-metaphysical scenario unfolds it becomes apparent that he considers life and consciousness to be fundamental aspects of the universe that are internal to the quantum realm. Mensky, furthermore, considers the concepts of quantum reality, consciousness and the 'life principle' as being interconnected and mutually supporting. Quantum reality has an internal aspect of consciousness which manifests through the unfolding of life through the operation of the life-principle. The life-principle can be thought of as an internal 'pressure' within quantum reality that acts to 'unfold' the bodies of various organisms in order to manifest individuated consciousness.

This process occurs through a quantum evolutionary process, in order that individuated consciousness can be expressed through the brains which are "instruments of consciousness." Individuated consciousnesses are all fragments of the vast undifferentiated primordial consciousness-awareness which resides within quantum reality. The internal 'pressure' due to the life-principle operates in order to maximize the degree and qualitative nature of embodied consciousness.

The resonance with Bohm's perspective is remarkable. Here is Bohm's view of the significance of the evolutionary process:

> What we are saying is, then, that movement is basically such a creative inception of new content as projected from the multi-dimensional ground. In contrast, what is mechanical is a relatively autonomous sub-totality that can be abstracted from that which is basically a creative movement of unfoldment. How, then, are we to consider the evolution of life as this is generally formulated in biology? First, it has to be pointed out that the very word 'evolution' (whose literal meaning is 'unrolling') is too mechanistic in its connotation to serve properly in this context. Rather, as we have already pointed out above, we should say that various successive living forms unfold creatively. Later members are not completely derivable from what came earlier, through a process in which effect arises out of cause (though in some approximation such a causal process may explain certain limited aspects of the sequence). The law of this unfoldment cannot be properly understood without considering the immense multi-dimensional reality of which it is a projection … Our overall approach has thus brought together questions of the nature of the cosmos, of matter in general, of life, and of consciousness. All of these have been considered to be projections of a common ground. This we may call the ground of all that is, at least in so far as this may be sensed and known by us, in our present phase of unfoldment of consciousness.[424]

The overall general coherence of the perspectives of Mensky and Bohm are clear. For both, evolution is not random but, rather, internally directed towards the unfoldment of life. Bohm depicts the process as being one in which the later are "not completely derivable from what came earlier", and the process as being one involving "projections of a common ground", which clearly indicates a coordinated organized directed process, involving an aspect of creativity. As we have seen previously, Bohm also writes on this subject:

> …living beings are in a continual movement of growth and evolution of structure, which is highly organized (e.g., molecules work together to make cells, cells work together to make organs, organs to make the individual living being, individual living beings a society, etc) …. Here, it is important to emphasize the essentially dynamic nature of structation, in inanimate nature, in

living beings, in society, in human communication, ... The kinds of structures that can evolve, grow, or be built are evidently limited by their underlying order and measure. New order and measure make possible the consideration of new kinds of structure.[425]

Here again we find the implication that the dynamic, organized and dynamic structures underlying forms are implicit within the implicate order.

The implications of Mensky's EEC (Extended Everett Concept) are dramatic and spectacular, with applications to several foundational problems in the arenas of psychology, parapsychology, evolutionary-biology and spirituality. Within the arena of the Darwinian-evolution verses intelligent design debate, Mensky's perspective indicates that there is an intelligence *internal* to the process of reality, not an external guiding God. What appears to be a material evolution is in reality an unfolding of quantum potentialities driven by the internal quantum 'life-principle' which acts through the quantum realm of potentiality to produce the manifested apparently material world and sentient beings within it. Mensky's account naturally and coherently elucidates the origin of life as being due to the internal pressure of the 'life-principle' acting through the quantum level.

Within the arena of the so-called 'hard problem' of consciousness, the problem is resolved by Mensky naturally and coherently. Individuated consciousness is an embodied fragment of the universal non-differentiated unified consciousness-awareness which is an internal aspect of quantum reality. Furthermore, Mensky explicitly views consciousness as having a "mystical" depth. He speaks of the necessity for including within the realm of consciousness "deep mystical features". Within the arena of spirituality and mysticism, Mensky's EEC indicates that mystical states of awareness are experienced when individuated consciousness dissolves into the less differentiated state of the ground-quantum-consciousness. This aspect of the EEC corresponds precisely to the Buddhist *Yogācāra* (Yoga-practitioners-Consciousness-Only) distinction between *jnana,* which indicates the universal nondual consciousness-awareness, which is also called 'wisdom', and *vijnana,* or divided consciousness, dualistic consciousness. In *Yogācāra* psycho-metaphysics there are various levels of *vijnana,* the most fundamental being the deepest *alayavijnana* or the fundamental ground-consciousness.

According to Mensky consciousness is associated with the "separation of the alternatives" which reside within the 'many-worlds' of quantum reality:

> ...while consciousness cannot be understood in the context of chemistry, classical physics and physiology, it turns out that it (or at least its main features) can be understood in the context of quantum mechanics. More precisely, the essence of consciousness can be interpreted as a special type of perception of quantum reality by living beings.[426]

It is worth noting that Mensky, a physicist, asserts the exact opposite to the 'philosopher' Simon Saunders regarding whether purely 'physical' aspects of reality could account for consciousness, which shares no physical qualities.

Classical physics, of course, refers to the pre-quantum physics which operated from the assumption that the material world existed at some level as it appears, as solid bits of matter. This perspective viewed consciousness as some strange magical transformation of matter into a qualitative realm of experience. Such a view, however, produces the 'hard problem' of the arising of consciousness because matter itself is, in this 'classical' view, defined to be devoid of the qualities of consciousness. With the advent of quantum discoveries this 'hard' problem has dissolved because the illusion of the material world has been found to have dissolved into the deeper immaterial realm of quantum potentiality which has an internal qualitative aspect of consciousness. As the physicist Nick Herbert has pointed out:

> ...every quantum system has both an 'inside' and an 'outside', and ... consciousness both in humans as well as in other sentient beings is identical to the inner experience of some quantum system. A quantum system's outside behavior is described by quantum theory, it's inside experience is the subject matter of a new 'inner physics'....[427]

Individuated consciousness flows through the physical organisation of organic sentient beings from the deeper nondual undifferentiated quantum awareness-consciousness within quantum reality.

By the phrase "the essence of consciousness can be interpreted as a special type of perception of quantum reality by living beings" Mensky indicates that living beings perceive a tiny portion of the full sweep of all the possibilities within the quantum realm, as if putting on blinkers in order to occupy and experience just one of the possible worlds. And Mensky also tells us that:

> ... quantum mechanics ... attempts to represent the measurement process ... as completely objective, as absolutely independent of the observer who perceives the result of the measurement, [such attempts] have not met with success ... the description of quantum measurements ... must involve ... the observer or, to be precise, *the observer's consciousness*...[428]

Mensky indicates that the usual presentation of the Many Worlds Interpretation (MWI) has the implication that an observer's consciousness must split between the alternatives and he suggests that the term 'Many-Minds' is a better description of the situation. Furthermore:

> In Everett's interpretation there appears [to be] some [ambiguity]. All alternatives are realized, and the observer's consciousness splits between all alternatives. At the same time, the individual observers subjectively perceive what is going on in such a way as if there exists a single alternative, the one he exists in. In other words consciousness as a whole splits between the alternatives but the individual consciousness chooses (selects) one alternative. ... in [any one] of Everett's worlds, all observers see the same thing, their observations are consistent with each other...[429]

Here Mensky indicates a crucial feature of his Extended Everett Concept (EEC), which is the insight that consciousness not only arises as the separation of alternatives, *it also performs a selection between them.* This is an extremely clever insight which avoids the extreme views of the MWI and the Copenhagen Interpretations. Because there is a, mostly unconscious, 'choice' on the part of an individual mind of which alternative to experience, there will be an appearance of a 'collapse' of wavefunction into just one world, but in fact the quantum potentialities remain as they are, and the individual consciousness moves a course way through, experiencing just one of the coherent worlds, but the other 'worlds' still remain as quantum potentialities.

Because 'quantum reality', which is the fundamental reality of our universe, consists of a multitude of alternative quantum potential 'worlds', Mensky proposes the notion of a quantum "Alterverse" of 'parallel worlds', or 'alternative classical worlds'. Mensky then asks why observers are not aware of these vast numbers of 'parallel worlds'. The answer he gives is that "alternative classical realities are separated by consciousness". According to Mensky:

> ...as a result of such a separation of alternatives by consciousness, we have the illusion that only a single world exists. Such is our subjective impression, even if objectively many parallel worlds exist.[430]

And:

> ...subjectively an observer has an illusion that there is only one classical world around him. The reason [for] this illusion is that classical alternatives are separated in his consciousness so that they are perceived independently from each other. *This is the classical vision of the objectively quantum world.*[431]

Individuated consciousness 'selects' one of the alternatives. In everyday life ordinary people do not inhabit a multitude of 'classical' alternatives:

> The consciousness as a whole splits between alternatives, and a 'component' of consciousness lives within one classical alternative, perceives only this single alternative classical reality.[432]

An individuated consciousness "lives within one classical alternative", by selecting, mostly unconsciously, a pathway through the possible worlds.

Note the important point that Mensky suggests that an "observer has an illusion that there is only one classical world", and that it is a "classical vision of the objectively quantum world". Here Mensky is pointing out that there is, in actuality, no so-called 'collapse of the wavefunction', there is only an *appearance,* or an *illusion,* of the creation of a 'real', 'solid', 'external' world, which seems to involve a 'real' 'collapse'. The appearance of a collapse of the wavefunction occurs because of the, largely unconscious, 'choice' on the part of the 'observer' as to which of the alternative routes through the universal quantum Alterverse of potentialities to take. It is remarkable how close this Mensky perspective regarding the illusory nature of the apparent 'collapse of the wavefunction' is to the Bohmian perspective we have looked at previously. As we have seen, physicist Jean Bricmont, in a section of his *Making Sense of Quantum Mechanics* book titled 'What About the Collapse of the Quantum State?', declares that:

> The short answer ... is that there is never any collapse of the quantum state in the de Broglie-Bohm theory, but there is an effective collapse or a collapse "in practice", which coincides with the one in ordinary quantum mechanics ...[433]

In other words, in the Bohmian scenario there is an appearance of an 'effective collapse' which is not an ultimately 'existent' event. For Mensky in actuality an observing consciousness separates the alternatives and 'chooses' a route, almost completely unconsciously. Like Bohm, Mensky often uses the word 'consciousness' with an extended meaning, the term 'consciousness' refers to both the explicit and implicit, or unconscious, levels. So, on top of the mostly unconscious route-taking there is also a small degree of conscious free-will involved. This aspect has been covered in the last chapter.

A similar viewpoint has been proposed more recently by the physicist-philosopher Shantena Agusto Sabbadini in his book *Pilgrimages to Emptiness: Rethinking Reality through Quantum Physics.* To understand this perspective it is necessary to recall that John Wheeler proposed his thought experiment involving a backwards in time determination as to whether a light beam from a distant quasar arrived as particle or wave. This thought experiment involved Wheeler's universal eye diagram shown below. And Wheeler wrote of this:

> My diagram of a big U (for universe) attempts to illustrate this idea. The upper right end of the U represents the Big Bang, when it all started. Moving along down the thin right and up along the thick left leg of the U symbolically traces the evolution of the universe, from small to large - time for life and mind to develop. At the upper left of the U sits, finally, the eye of the observer. By looking back, by observing what happened in the earliest days of the universe, we give reality to those days.[434]

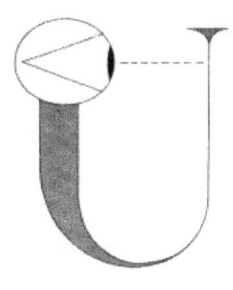

The Wheeler U

Such speculations gave rise to Wheeler' notion of a 'participatory universe'. Wheeler described his vision:

> The thing that causes people to argue about when and how the photon learns that the experimental apparatus is in a certain configuration and then changes from wave to particle to fit the demands of the experiment's configuration is the assumption that a photon had some physical form before the astronomers observed it. Either it was a wave or a particle; either it went both ways around the galaxy or only one way. Actually, quantum phenomena are neither waves nor particles but are intrinsically undefined until the moment they are measured. In a sense, the British philosopher Bishop Berkeley was right when he asserted two centuries ago 'to be is to be perceived'.[435]

And:

> We are participators in bringing into being not only the near and here but the far away and long ago. We are in this sense, participators in bringing about something of the universe in the distant past ...[436]

In the light of such claims Sabbadini suggests that a *persistence of information* approach is perhaps more appropriate. In this understanding, which is essentially the same as Mensky's proposal, there is not an actual 'collapse' taking place, rather, the consciousness of the observer selects a pathway within the alternative quantum possibilities, which are treated as being quantum information. Sabbadini writes that:

> The persistence of information approach fits well with the notion of a participatory universe, but this participation does not need to be understood in the sense that "we bring about something of the universe in the distant past". The apparent retroactive effect in time is merely a reflection of the inadequacy of our representations of matter, which are imbued with classical prejudice. ... Our choice of the setup in the cosmic delayed choice experiment therefore does not change anything in the distant past. What happened in the past is the creation of an entangled state ... What happens here now is that we can choose to perform on that state an observation conserving or destroying which path information ... But there is no retroactive change, ... Our choice of the experimental setup determines what we see today; it does not affect what happened millions of years ago ...[437]

Thus, according to both Sabbadini and Mensky, the quantum information about 'events', or possible 'events', in the distant past is available in the present and it is possible to 'read' this information in different ways by choosing different experimental arrangements. But the choice of the experimental setup does not change the physical situation in the past, and it does not destroy quantum information. It is the case, rather, that embodied minds experience the pre-existing quantum information in ways that are determined by the structure of the subjectivities involved and the associated environment.

This vision of the situation is clearly consistent to Mensky's understanding. Mensky states that individual subjectivities unfold experiential worlds from the quantum possible realities within the Alterverse, without changing the quantum pool of information within the Alterverse. As Mensky says:

> ...if parallel worlds coexist, why then do we see only a single world around us, but not a superposition of classically incompatible worlds. ... The answer ... is that the *alternative classical realities are separated by consciousness*. This means that an arbitrary observer perceiving the quantum world ... perceives different projections independently from each other; in the picture of one classical reality there is no place for the others (although they are objectively not less real than than the first one). It is evident that, as a result of such a separation of alternatives by consciousness, we have the illusion that only a single world exists.[438]

Here Mensky is implicitly referring to the separation of alternatives and the subsequent selection, through both unconscious and conscious 'choice', of which alternative to experience, thus creating the "illusion that only a single world exists". There is no 'collapse' of any wavefunctions, just a selection of which informational path to follow.

When we incorporate the *Yogacara* perspective, all 'decisions' regarding the quantum path travelled, which are determined by a combination of both unconscious and conscious factors, leave 'traces', or strengthened predispositions which will act as determining factors in future 'choices'. As Sabbadini says:

> ...all experience of an embodied observer is associated with persistence of information and that is the reason why the world appears objective to us. The world described by quantum physics

is not objective, is not realistic and local and is not made of things. But it appears to us as objective, it appears as a world of things, because all experience involves the formation of a trace. The appearance of an objective world is a consequence of an essential characteristic of the process of knowing; we experience the world from inside the world, we experience the world as embodied subjects.[439]

And, of course, we find echoes of this viewpoint in the writings of Bohm:

We may begin by considering the individual human being as a relatively independent sub-totality, with a sufficient recurrence and stability of his total process (e.g., physical, chemical, neurological, mental, etc.) to enable him to subsist over a certain period of time. In this process we know it to be a fact that the physical state can affect the content of consciousness in many ways. ... Vice versa, we know that the content of consciousness can affect the physical state ... This connection of the mind and body has commonly been called psychosomatic (from the Greek 'psyche', meaning 'mind' and 'soma', meaning 'body'). This word is generally used, however, in such a way as to imply that mind and body are separately existent but connected by some sort of interaction. Such a meaning is not compatible with the implicate order. In the implicate order we have to say that mind enfolds matter in general and therefore the body in particular. Similarly, the body enfolds not only the mind but also in some sense the entire material universe.[440]

A crucial point here is that mind and matter are not separate substances, they unfold as explicate aspects of the quantum implicate order which is closer to the nature of mind than matter, in fact, as we have seen, it is a kind of primordial mind-energy, which Dzogchen Buddhism simply refers to as Primordial Mind.

In an article, *Contiguity of Parallel Worlds: Buddhist and Everett's*, the Buddhist philosopher Andrey Terentyev refers to the "striking similarity of the views on reality in Buddhism and in the Extended Everett Concept by M. Mensky". In his conclusion Terentyev writes that:

I'd like to stress that we are not just considering analogies in different fields of human endeavour; in fact, both Buddhist thinkers and modern physicists, using very different methods, arrived basically at the same description of [the] reality we live in. This is the point where the parallel worlds of Buddhism and Physics unexpectedly touched each other, and the deeper meaning of this is yet to be appreciated by both parties.[441]

Certainly Mensky's perspective endorses the Buddhist understanding that the ultimate ground of the process of reality is a Primordial-Mind-Energy. Mensky tells us that:

> In psychology, only that which is subjectively perceived is termed the consciousness, i.e. only the 'classical component' of the consciousness, according to our terminology. Therefore, to identify the notion of 'consciousness' with some notion from ... quantum theory, we must broadly interpret consciousness as something capable of embracing the entire quantum world (alternative classical realities) rather than exclusively one [of] its classical projection[s].[442]

Here Mensky indicates the necessity for a notion of a global, universal consciousness, associated with all the possible worlds. This global consciousness is clearly equivalent to the Buddhist Primordial Mind, it underlies all manifested individuated consciousnesses. He then goes on to propound his 'identification hypothesis' that asserts the identity of "the ability of a human referred to as consciousness" and the "separation of the single quantum world into classical alternatives."[443] Thus he identifies individuated consciousness as arising when Primordial Mind 'separates'.

The above discussion is remarkably similar to the *Yogacara* doctrine of the three natures of the process of Consciousness-Only reality. The following description of the *Yogācāra* perspective is from the contemporary Buddhist scholar-practitioner Karl Brunnhölzl's translation and elucidation of the *Dharmadharmatavibhanga* which he has titled *Mining for Wisdom within Delusion*:

> ...the Yogacara system holds that what appears in a being's mind as the world is not a representation of an external world, but it is the world of this being as projected by the mind of this being. In the minds of beings of the same type (such as humans), similar but still individual images of the world arise due to similar latent tendencies, which are then misconceived as being an actual common world out there.[444]

According to the *Yogācāra* psycho-metaphysical account:

1) There is no external material world, each sentient being projects from their mind the illusion of an external world.

2) The details of the illusion that an individual sentient being projects as an external world are derived from latent tendencies which are themselves derived from previous projections in both the current life and previous lives.

3) All the sentient beings within a universe will project coordinated intersubjective projections, thus appearing to inhabit a shared environment. This is because they have shared coordinated intersubjectively created environments in previous lifetimes.

4) The source of these projections is a common fundamental ground-consciousness (*alayavijnana*). The ground-consciousness, or store-consciousness, receives the details of projections and actions of all sentient beings. These stored karmic 'seeds' determine future projections. The alayavijnana is described:

> The deepest, finest and subtlest layer of ... consciousness. It contains all the traces and impressions of past actions and all ... future potentialities.[445]

Thus what appears to be an external reality which contains a multitude of fellow sentient beings is actually a vast intersubjective illusion created by the thoughts, intentions and actions on the part of all sentient beings.

This is not to say that these projections are projected into nothingness. This analysis is part of the Yogacara Three Natures doctrine. The three intrinsic natures posited by Vasubandhu and the Mind-Only school are:

1. *Parikalpita*: the 'conceptually-constructed', 'imaginary', or 'imputational' nature;
2. *Paratantra*: the 'other-dependent' or 'other-powered' nature;
3. *Parinishpana*: the 'perfect', 'actual', 'consummate' or 'thoroughly-established' nature.

The three natures are interdependent modes of 'subjectivity' or 'mind', the 'perfect nature' being a kind of universal primordial 'subjectivity'. This view of the process of reality is metaphysically comprehensive and exhaustive, and, as we shall see, maps into Mensky's Alterverse quantum worldview.

In the Yogacara presentation the projections of the *Parikalpita* 'imputational nature' corresponds to the deep-seated belief that the dualistic worlds being experienced by sentient beings following various paths through Mensky's Alterverse really are comprised of 'real' 'external' things existing independently of mind. The reality of an external world is imputed onto the appearances which, in reality, are of the nature of mind. The 'fabricated' or 'imaginary nature', *Parikalpita,* is the powerful and yet false conviction that there really is an independent, external world "out there". This conviction is accompanied by the appearance of the mistaken division into a perceiver and perceived. Sentient beings under the sway of deep unconscious belief in the *Parikalpita*, which is pretty much all unenlightened beings, see the world as being really dualistic.

According to the Yogacara view the illusion of the subject-object dichotomy, which is termed 'grasper' *(grahya)* and 'grasped' *(grahaka),* emerges from an All-ground Consciousness *(alayavijnana)* which itself is an aspect of the non-dual ground of all phenomena *(dharmadhatu)*. It is noteworthy in this context that one of the important insights that Bohm focuses on is the quantum level illusory nature of the division between observer and observed:

> Thus, one can no longer maintain the division between the observer and observed (which is implicit in the atomistic view that regards each of these as separate aggregates of atoms). Rather, both observer and observed are merging and interpenetrating aspects of one whole reality, which is indivisible and unanalysable.[446]

Quantum discoveries have 'confirmed' psycho-metaphysical insights on the part of Buddhist philosophy.

The *Paratantra,* the 'other-dependent' or 'other-powered' nature is comprised of the vast array of karmic potentialities which derive from the karmic 'seeds' which arise from previous universes and previous actions, karmic 'seeds' are stored due to perceptive and intentional activities on the part of sentient beings over vast time periods in the current universe. This aspect comprises the potentialities in Mensky's Alterverse. These potentialities form the basis for the 'choices' made by sentient beings under the grip of the *Parikalpita*. Beings under the illusory sway of the *Parikalpita,* the deep seated grasping at the reality of a 'real' external reality, make Mensky-style 'choices' within the

Paratantra potentialities within the Alterverse.

The ultimate nature is the "perfected," or "thoroughly-established-nature" (Skt. *Parinispanna*). This is the ultimate luminous 'empty' field of potentiality which, in its own essence, is nondual awareness, self-luminous, unfabricated and unborn, i.e. eternal. This corresponds to *dharmata*, the ultimate nondual nature of phenomena, which is the ground out of which *dharmas*, the manifestations from *dharmata*, arise. As the Buddha said:

> There is, monks, an unborn, an unbecome, an unmade, unfabricated. If there were not that unborn, unbecome, unmade, unfabricated, there would not be the case that emancipation from the born, become, made, fabricated would be discerned. But precisely because there is an unborn, unbecome, unmade, unfabricated, emancipation from the born, become, made, fabricated is discerned.[447]

In quantum terms this is the ultimate fundamental quantum potentiality field which has an inner nature of intrinsic nondual awareness. In the objective terminology of quantum physics, which cannot directly detect the awareness aspect, this potentiality field corresponds to to the ground level quantum fields:

> Quantum field theory, the tool with which we study particles, is based upon eternal, omnipresent objects that can create and destroy those particles. These objects are the "fields" of quantum field theory. ... quantum fields are objects that permeate spacetime ... they create or absorb elementary particles ... particles can be produced or destroyed anywhere at any time.[448]

The 'thorough-established-nature', then, is the empty luminous nature devoid of any imprinted potentiality, it is the pure luminous mind-energy which can contain potentialities. The 'thoroughly-established' ground field of the process of reality carries, in a similar fashion to Bohm's holomovement, the potentialities of the *Paratantra*.

The *Ornament of Great Vehicle Sutra* (*Mahayanasutralamkara*) explains the 'three natures' as follows:

> The characteristics of reality are to be understood by means of one's own direct awareness. These characteristics are free from the four extremes of existence, nonexistence, sameness, and

difference. This, in turn, means that the ultimate reality does not exist in the way the characteristics of the imagined and dependent natures would make it seem. Nevertheless, in terms of the characteristics of the thoroughly established nature, the ultimate reality is not non-existent either. The term "all phenomena" does not refer to anything other than the phenomena of the dependent nature, which are profound, inner, dependent arisings. The concepts of these phenomena being real in the same way that they appear – as duality of perceiver and perceived – are the imagined nature. Although child-like, ordinary beings believe that such appearances exist in the way that they appear, the dualistic phenomena of apprehended and apprehender have never had any establishment. This is the thoroughly established nature, the ultimate reality, or the way things actually are. ... The dependent nature's lack of imagined duality of apprehended and apprehender is itself the thoroughly established nature. As there is no other "thoroughly established nature" aside from this, the latter is precisely the intrinsic nature of the phenomena that are the dependent nature.[449]

Here the 'thoroughly established nature' is identified with *dharmata*, the intrinsic nature of phenomena. *Dharmata*, the ultimate "thoroughly established" quantum "dream stuff", as the physicist Wojciech Zurek[450] refers to it, which underlies the process of reality is imprinted by the results of previous karmic actions and therefore generates further illusory images which are experienced by the multitude of mind-streams of sentient beings. The field of these dependent illusions is the "dependent-nature," so called because of the dependency on stored Alterverse potentialities and previous karmic activities. The imaginary-nature is the mode of perception which mistakenly treats these illusions to be 'real' and indicative of an actual 'external' world.

Mensky proposes a model involving two levels of consciousness. The first, which is identified with the quantum realm of parallel worlds in a state of quantum superposition, is a nondual, undifferentiated field of a 'universal', or collective primordial awareness-consciousness. Mensky also calls a significant aspect of this level of awareness, which resides within the quantum ground reality of 'parallel worlds', the 'super-consciousness'. In this realm of undifferentiated awareness, which manifests when individuated consciousness, i.e. waking everyday consciousness, is "turned off", it is possible for an intuitive "super-cognition" to operate. According to Mensky, such a supercognition, activated when the focused everyday consciousness is reduced, is able to have access to other parts of the Alterverse, even areas of potentiality lying in the past or the future. Mensky has written concerning this that such a faculty is able to "[supply] …information that is not available in the usual (conscious) state".

This feature of the relationship between consciousness and super-consciousness, the fact that when the focused awareness of individuated consciousness is reduced, consciousness (in its 'super-consciousness' mode) is then able to gain access to information which resides within parallel worlds - even some way back into the past or into the future - underlies the phenomenon of precognition. According to Mensky:

> …if consciousness is identified with the separation of the alternatives, then turning … consciousness off means [the] disappearance of the separation, i.e. [the] emergence of access to all alternative realities. The information from this enormous "database" makes feasible (in the state of unconsciousness) super-intuition, i.e. direct vision of the truth. Thus, extraordinary features of consciousness … should [be revealed] "*at the edge of consciousness*" when the consciousness (i.e. the separation of the alternatives) disappears or almost disappears. What appears then instead of consciousness (in the usual understanding of this word) may be called extended consciousness, or *super-consciousness*.[451]

Such a possibility of 'super-consciousness' insight into time, space and distant aspects of the universe is a reasonable assumption in a universe described by Bohm, wherein its inner nature is a deeper level of non-individuated consciousness, and:

> Ultimately, the entire universe (with all its 'particles', including those constituting human beings, their laboratories, observing instruments, etc.) has to be understood as a single undivided whole, in which analysis into separately and independently existent parts has no fundamental status.[452]

And:

> In terms of the implicate order one may say that everything is enfolded into everything. This contrasts with the explicate order now dominant in physics in which things are unfolded in the sense that each thing lies only in its own particular region of

space (and time) and outside the regions belonging to other things.[453]

In such an 'enfolded' universe, of course, one would expect that various para-psychological phenomena to be a possibility.

The currently prevailing paradigm in psychology, neuroscience and philosophy of mind still mistakenly considers that all aspects of human mind and consciousness are generated in some way by purely 'physical' processes occurring in the brain. However, the excellent book *Irreducible Mind: Towards a Psychology for the 21st Century*, published in 2007, demonstrates, with a wealth of empirical evidence, that the mainstream reductive materialism is false. The authors of this rigorously and extensively researched volume bring together detailed evidence from a large variety of psychological phenomena that are not only difficult, but in many cases impossible, to account for in the currently dominant physicalist-materialist worldview. In this context it is worth quoting an observation from the introduction to this excellent book:

> A useful principle that provides orientation ... was stated by Wind (1967): "It seems to be a lesson of history that the commonplace may be understood as a reduction of the exceptional, but the exceptional cannot be understood as an amplification of the commonplace." This lesson has not penetrated contemporary cognitive science, which deals almost exclusively with the commonplace and yet presumes – extrapolating vastly beyond what in reality are very limited successes – that we are progressing inexorably towards a comprehensive understanding of mind and brain based on classical physicalist principles. This serene confidence seems to us unwarranted. It is now evident, for example, that chess playing computer programs represent progress towards real intelligence in roughly the same sense that climbing a tree represents progress towards the moon.[454]

The kind of phenomena covered include such aspects as psychosomatic medicine, placebo effects, psychic healing, stigmata, multiple personalities, yoga, physiological effects in hypnosis, automatism, near death experiences, the nature of genius, and meditation. All these topics are presented with reference to thoroughly researched and validated, exhaustively referenced, phenomenon that are accepted by leading authorities in the various fields. One reviewer concludes that: "The topics that I have some knowledge of ... were covered with more than sufficient thoroughness and accuracy to support the point that consciousness, including subliminal consciousness, can produce physical effects and is not a mere epiphenomenon as proposed by philosophical materialists".[455]

Mensky addresses such issues from the basis of his Quantum Consciousness Concept (QCC) and Extended Everett Concept (EEC):

> Telepathy arises as an effect of quantum non-locality. The necessary condition for this is the purely quantum regime, i.e. quantum coherence, absence of decoherence. This condition is met not in consciousness, but in super-consciousness which is nothing else as the state of the quantum world as a whole...[456]

And:

> Another very important assumption accepted in [the] EEC is that consciousness has the ability to influence the alternative to be subjectively perceived. In a sense, this means that the ability exists to "control reality".[457]

And:

> If consciousness and separation of the alternatives are identified, then dimmed consciousness (in particular, in the state of sleep or trance) means an incomplete separation of alternatives, in which consciousness looks into 'other alternatives' and can single out the most favourable one among them.[458]

Mensky amplifies upon this by indicating that it is important to understand that quantum reality itself is not being altered, it is, rather, that the "subjectively perceived reality is controlled". In other words the subjective consciousness of a person is able to select, mostly unconsciously, more advantageous, or in some situations such as someone with a death wish less advantageous, pathways within the 'parallel worlds' of quantum reality.

This is in fact a kind of 'look-ahead' mechanism operating, again mostly unconsciously, through the quantum level where the focused 'higher' levels of conscious awareness are dampened or 'turned off', either completely or to some degree. Consciousness is thus able to "obtain information from the quantum world as a whole, i.e. to look into other alternatives, other realities". This is because in the Many Worlds perspective these alternative realities, 'parallel worlds' "exist objectively".[459]

And when "partitions between the alternatives vanish or become penetrable", in states "at the edge of consciousness" such as sleeping, trance or meditation for example, information from other worlds becomes available.

It is important to be aware of the various, flexible, uses of the term 'consciousness' in Mensky's exposition. Although he speaks of consciousness being "turned off" for example he also suggests that in this state, information from parallel worlds becomes available to 'consciousness'. This would seem, on the face of it, to be a contradiction. However, it is necessary to bear in mind that, although Mensky does speak of consciousness being 'on' or 'off', what we are dealing with is actually more like a continuum of states of awareness, or states of 'consciousness' on different levels. In this view, 'consciousness' should be understood as a field of energetic 'awareness' which can operate in both 'conscious' and 'unconscious' modes. Mensky sometimes uses the term 'consciousness' to denote only the 'focused' fully individuated mode, rather than the full field of consciousness.

When Mensky speaks of consciousness being 'turned off' he is referring to the 'high level' focused states of fully individuated consciousness. However, even in states generally considered to be 'unconscious' in Western terminology, the capacity for consciousness at 'lower levels' to function unconsciously is still present and, furthermore, any information which is accessed from 'parallel worlds' in such unconscious, or less conscious, states will be available to consciousness when it emerges in its more focused state. This is indicated by Mensky's use of terminology such as:

> ...when the explicit consciousness is disabled (in the [sphere] of [the] unconscious), the (implicit) consciousness witnesses, instead of the usual classical world, something quite different, including particularly all classical scenarios in all time moments.[460]

And, because the "implicit consciousness" has access to "all classical scenarios *in all time moments*", it has access to future quantum states, and subsequently it can orientate itself towards quantum states that are favourable to survival and enhancement of quality of life.

Mensky calls this important quantum mechanism *'postcorrection'*. When focused individuated consciousness is 'turned off' or reduced, then 'implicit consciousness', - the deeper layers of less individuated consciousness - has access to all parallel worlds, and some implicit knowledge from alternative worlds regarding future states can become available to both implicit and focused consciousness. Subsequently consciousness can make corrections in functioning, perhaps even functioning of the body, in the present moment in the light of information implicitly gained from the future. In this way an organism may make adjustments, or corrections, 'post', or after, the exploration of future quantum potentialities, to aim towards advantageous scenarios.

According to Mensky, consciousness can use this information gained through implicit consciousness from the future in order to steer a course through parallel worlds towards more favourable situations. And Mensky tells us that such a process underlies the evolutionary unfolding of quantum potentialities into embodied sentient life. It is important to bear in mind, during the remainder of this discussion, that, although Bohm does not explicitly refer to the mechanisms Mensky elucidates, nevertheless Mensky's ideas are fully coherent with Bohm's worldview.

Such insights led to Mensky's 'Quantum Concept of Life' (QCL), which suggests that: "a man (and a living being generally) can influence the subjective probabilities of the alternatives, increasing [the] probability to experience those alternatives that are favourable". Mensky also refers to this concept as the 'life principle' or 'principle of life', which he also identifies as a variety of the Anthropic Principle. "The Principle of Life", says Mensky, is "the statement that only favorable scenarios (i.e. those forming the life sphere) are realized for living beings".[461] Furthermore, this principle follows from the "properties of consciousness".[462] This means that Mensky's 'Quantum Concept of Life', extended into the 'life principle', tells us that, in the same way that a living being can use 'super-intuition' within the parallel worlds of 'super-consciousness' to navigate towards favourable future scenarios, so too Life, considered as a collective and global phenomenon, uses the same mechanism to drive the process of evolution: "the evolution of life may be expressed as the set of favourable scenarios instead of the wider set of all possible ... scenarios".[463] So, according to Mensky, the life force, which is internal to the quantum Alterverse, operates in order to unfold potentialities towards the "favourable scenarios" that enable sentient beings to flourish. This quantum metaphysical perspective leads us to the conclusion that there is a kind of primordial consciousness operating according to the quantum super-intuitive 'post-corrective' 'look-ahead' mechanism in order to unfold Life from the quantum potentialities within the 'Alterverse'.

Mensky, then, considers that Life and sentience are latent internal aspects within the quantum reality of the 'Alterverse'. One might say that there is an internal 'pressure', due to a primordial consciousness, to unfold life and sentience. A similar view has been proposed by the physicist Paul Davies, who proposes the possibility that the universe comes into existence as a *self-explaining universe*. Such a universe would necessarily contain organisms that embody the capacity for cogntion, which is to say consciousness, precisely because of the purpose of self-explanation, to use Davies' perspective, or self-cognition, is fundamental to the universe. According to Davies:

> …a good case can be made that life and mind *are* fundamental physical phenomena, and so must be incorporated into the overall cosmic scheme. One possible line of evidence for the central role of mind comes from the way in which an act of observation enters into quantum mechanics. It turns out that the observation process conceals a subtle form of teleology.[464]

And this is a similar perspective to Wheeler's 'self-perceiving' universe.

According to Mensky, Life and consciousness cannot be accounted for using the mechanistic methods appropriate for inanimate matter. This is because Life orientates itself towards goals such as survival and maximizing quality of life:

> Life is a phenomenon which is realized by living matter consisting of living organisms (living beings). Living matter differs from non-living matter in that its dynamics is determined not only by causes, but also by goals. First of all the goal of survival (prolongation of life) is important in this context. However in the context of sufficiently perfect forms of life more complicated goals are also actual. They can be formulated in terms of quality of life.[465]

These goals, within the context of evolution, can only be comprehended and elucidated within the context of a perspective like Mensky's EEC and QCL ('Extended Everett Concept' and 'Quantum Concept of Life') because, as we have seen, the EEC and QCL quite naturally provide a quantum mechanism for the 'look-ahead' capacity that is required for such natural goal-orientated behaviour. Materialist-mechanist accounts such as Neo-Darwinism are completely incapable of elucidating or accounting for the origin of life. The presence of the internal pressure for survival, and for the generation of higher levels of consciousness, however, explains evolution naturally within the EEC and QCL.

According to Mensky this fundamental 'postcorrection' look-ahead mechanism underlies the process of evolution:

> ...we shall introduce the mathematical formulism describing the principle feature of living matter (of its consciousness): the ability to correct its state making use of the information (about the efficient way of survival) obtained from the future. It will be assumed that the evolution of living matter includes a [look-ahead] correction [mechanism] providing survival at distant moments. This correction leaves in the sphere of life only those scenarios of evolution which are favorable for life. Unfavorable scenarios do not disappear from the quantum reality but are outside the sphere of life... this correction (selection of favorable scenarios) is represented by the special mathematical operator which is called *postcorrection*.[466]

The essential point is that living beings can, generally unconsciously (Mensky suggests this is one reason for the importance of sleep), feel out the favourable future scenarios and then amplify the quantum probabilities of the current situation in order to orientate organisms towards favourable conditions for survival and the enhancement of quality of life. This process of the quantum look-ahead mechanism operates on a global scale within the process of evolution.

Mensky's account of how the primordial 'Life-Operator' began its business of unfolding sentient organic life is illuminating:

> If the picture of the world as it appears in consciousness were far from classical, then, due to quantum non-locality, this would be a picture of a world with 'locally unpredictable' behaviour. The future of a restricted region in such a world would depend on events even in very distant regions. No strategy of surviving could be elaborated in such a world for a localised living being. Life (of the form we know) would be impossible. On the contrary, a (close to) classical state of the world is 'locally predictable'. The evolution of a restricted region of such a world essentially depends only on the events in this region or not too

far from it. Influence of distant regions is negligible. Strategy of surviving can be elaborated in such a world for a localised living being.[467]

Life unfolds from the primordial nonlocal quantum field of potentiality. The fact of quantum nonlocality means that all distant regions of the primordial quantum field are instantaneously interconnected. If Life were to remain nonlocal then events in very distant regions would instantaneously effect all other regions, and this would produce a locally unpredictable world where individuated consciousness could not predict local events in order for organisms to survive. Therefore, Life must produce predictable local, or 'classical' regions, wherein events have a degree of local predictability. Of course, in the Bohmian *Undivided Universe* scenario, nonlocality applies within the classical level reality. But, as we have seen, Bohm considers that the organisms of Life are relatively autonomous "sub-totalities" within the interconnected holo-movement, and such substructures can in essence break nonlocality and therefore function as if nonlocal. Thus we see, again, how Bohm's perspective is consistent with that of Mensky.

Mensky indicates that the level of consciousness at which the process begins is:

> ...the most primitive, or the most deep, level of consciousness, differing perceiving from not perceiving.[468]

This indicates that the first glimmers of the separations within the nonlocal quantum field take place through the operation of deep internal levels of the process of internal quantum perception. An important feature of this perspective is that originally the 'Alterverse' was entirely non-separated but contained all potentialities as 'possible worlds' awaiting unfoldment. In this state the field of potentiality is entirely 'nonlocal', which means that all points, irrespective of distance, are quantumly entangled. In this state, or a state close to this state of universal instantaneous interpenetration and interrelationship, an event anywhere in the field will have effects all over the field and, because of this, the world is "locally unpredictable". Classical level life could not function unless the nonlocal quantum 'Alterverse' starts to get 'local' through a separation of alternatives, which creates "locally predicable", or "locally stable", worlds of experience. Within such locally stable 'classical' regions: "the restricted region of such a world depends only on its state inside this region", rather than being determined by quantum fluctuations on the other side of the universe:

> It is only in [a] locally stable (therefore classical) world that the future can be predicted with relatively good reliability and ... consciousness separates the quantum world into its classical counterparts (alternatives) because (the only known for us) local form of life is feasible only in classical worlds.[469]

So, because of this situation consciousness must create a 'stable' 'classical' world which it can inhabit.

This process of unfoldment of potentialities begins when primordial consciousness begins a process of internal perception which starts the process of separation and thereby begins the process of the evolutionary unfoldment of organic sentient life of various forms and degrees. Mensky's account indicates how a deep internal 'pressure' due to primordial consciousness, Mensky's "Life-Principle", selects the structures conforming to a stable material world and the contained organic structures of sentience from the wealth of quantum possibility. This provides a view of evolution as an essentially quantum process which begins with the operation of the interior quantum cognition operating within the field of quantum potentialities.

The starting point, at the very base of the hierarchical cascade of what Bohm called quantum 'implicate orders' into material manifestation, is the glimmer of the division into perceiver and perceived. Subsequently this process produces increasingly 'materialized' organic structures embodying 'focused', 'separated' and individuated consciousness. This perspective is clearly suggested by Mensky's account of morphogenesis:

> There is one more unsolved problem in biology that also could obtain its explanation in EEC. This is the problem of morphogenesis. How an embryo is constructed starting from a single cell? Where is [the] plan of the process of constructing it, step by step, or how [is] constructing ... controlled and directed? ...consciousness (the primitive-level consciousness, or ability to somehow perceive, which is connected with a living being from the very beginning) periodically addresses to the quantum world as a whole, compare[s] various scenarios of constructing embryo (various 'building plans') and then, returning to the usual state, increase[s] probabilities of those scenarios that lead to the right construction. Of course, this is only a sketch of a possible explanation of the phenomenon, its main idea.[470]

This is a stunning insight into how the process of Life generates itself from quantum potentiality using the quantum 'look ahead' mechanism. The various possibilities for organic life 'exist' as quantum potentialities within the 'Alterverse'. The internal pressure of consciousness which is organizing the quantum potentialities into organic structures, structures capable of channeling the ground energy-awareness into the embodied individuated consciousness of sentient beings, is able to 'feel' its way ahead by addressing the "quantum world as a whole". The morphogenetic structures are already within the quantum ground as potentialities, they need to be actualised through being manifested into more 'explicate', 'solidified' or materialised versions.

Individuated consciousness, however, also maintains a fundamental connection to the universal background awareness of quantum potentialities for the future, the 'parallel worlds' of the 'Alterverse'. The brain acts as a kind of filter which sometimes allows deeper levels of consciousness to individuate into the 'separated' dualistic realm of experience:

> The brain is used by ... consciousness to control the body and obtain information about its state (and, through its perception, about the state of the environment). In other words, the brain (or rather some regions in it) is the part of the body which realizes its contact with ... consciousness, it is an interface between the consciousness and the body as a whole. In particular, when it is necessary the brain forms the queries that should be answered. Sometimes these queries are answered by the brain itself with the help of the processes of the type of calculations and logical operations. Other queries cannot be solved directly in the brain and are solved by the consciousness with the help of "direct sighting of truth."[471]

In other words sometimes the brain can figure out solutions with its own resources, so to speak, but on other occasions it needs to tap into deeper levels of awareness:

> ...unconscious [super-conscious] states of mind allows one to take information "from other alternatives" that reveals itself as unexpected insights, or direct vision of the truth.[472]

Such a mechanism underlies the phenomenon of precognitive dreams, for example.

Another dramatic consequence that Mensky derives from his analysis of quantum reality is the existence of 'probabilistic miracles':

> Probabilistic miracles essentially differ from "absolute" miracles that happen in fairy tales. The difference is that the event realized as a probabilistic miracle (i.e. "by the force of consciousness") may in principle happen in a quite natural way, although with a very small probability. This small but nonzero probability is very important. Particularly, because of the fundamental character of probabilistic predictions in quantum mechanics, it is in principle impossible to prove or disprove the unnatural (miraculous) character of the happening.[473]

In other words consciousness can in some circumstances amplify quantum potentialities which have small probabilities so that they, seemingly miraculously, become actualized. The important point that Mensky makes is that, although such events may appear to be 'supernatural', they are in fact entirely natural because they are the result of the quantum 'postcorrective' capacity of consciousness. An example of this is cases of individual or group prayer which may lead to a healing which appears miraculous. According to Mensky such events are not 'unnatural' because there is a quantum probability for the apparent miracle to occur, and this probability can be amplified by certain spiritual activities such as prayer.

In summary Mensky identifies the following aspects, or levels, of the functioning of the quantum postcorrection mechanism:

Life – the internal pressure of primordial consciousness acting through quantum potentialities which motivates the evolution of the apparently material world and the organic beings contained within it.

Survival – the natural unconscious operation of postcorrection in the survival processes of species and individual animals. The 'postcorrection' mechanism involves the ability of consciousness to employ a defocused 'super-consciousness' or 'sub-consciousness' to 'look-ahead' at future quantum potentialities and then alter current probabilities in order to steer towards favourable quantum scenarios.

Support of health – unconscious operation of postcorrection in determining the quality of health for an individual organism.

Free will – conscious operation of postcorrection in determining the quality of life and life environment for an individual organism. The use of intentionality in the decision making processes.

Control of appearing reality (probabilistic miracle) – postcorrection relating to objects external to the body. This is the ability of some people to seemingly cause events in their environments which have low, although non-zero, probabilities.

Super-intuitional insights – insights, foresights and "direct sighting of truth" from the "edge of consciousness," due to the operation of the 'super-consciousness' or 'sub-consciousness'.[474]

According to Mensky:
> In all these cases the operation of postcorrection does correct the present state, making it to be in accord [with] the criterion existing in the future. This results in the immediate choice of the correct solution of the problem, although its correctness can be confirmed only in the future. Consciousness, when [accessing] the regime of the unconscious, obtains the ability to look into the future, and makes use of the obtained information in the present.[475]

In his conclusion to his '*Postcorrection*' paper Mensky says that "the postulate of postcorrection broadens quantum mechanics, in the consideration [of] the law [of the] evolution of living matter".[476] In fact Mensky's new paradigm is capable of bringing dramatic insights into many crucial aspects of the process of reality. It accounts for the origin of life and the evolution of sentient beings.

According to Mensky the evolution of life involves goals as well as causes. For Mensky "the notion of goal (the basic goal is survival) is inherent in [the] living world".[477] And "for more sophisticated forms of life ... goals may include other criteria [such as] the quality of life".[478] Furthermore Mensky considers that spiritual goals are inherent within the process of life. As he points out:
> The phenomenon of consciousness demonstrates mystical features that are experienced by some people. All religions and spiritual schools that [have existed] for thousands of years include the mystical component as a necessary part of their message ...[479]

And Mensky considers such spiritual and mystical phenomena of consciousness to be actual, crucial and central. Because of this, the elucidation of the mystical features of consciousness within the domain of science is an important task. Furthermore, Mensky says that his paradigm, involving quantum consciousness, is a "bridge between the natural sciences and humanities, between matter and spirit".[480] His approach "unifies natural sciences with the sphere of spiritual knowledge including religion".[481] Within Mensky's Quantum Concept of Consciousness (QCC) "science, philosophy and religion meet together",[482] and the QCC explains "numerous strange phenomena, which are described in ... mystical teachings, including religion and oriental philosophies".[483] Furthermore the QCC shows that:

> QCC makes it possible to understand that there is no contradiction between science and mysticism. This makes it possible for [people] to believe in God, or in Truth, in ... Buddhism, and so on, and offers ... enormous possibilities, hidden in human beings, ... possibilities which make one truly free, without which he/she is only a slave of ... external circumstances.[484]

The QCC, then, elucidates and indicates the rational nature of the spiritual quest for enlightenment. Furthermore, Mensky concludes that "the picture of life after death given in Buddhism - the long series of earthly embodiments leading to ... enlightenment and ... nirvana" - is nearer the truth than that which Christianity gives".[485]

Mensky implicitly suggests a notion of a quantum 'soul'. We can interpret the concept of a 'soul', which is connected with the idea that there is some kind of life after death, with subtle quantum energy-potentialities which continue after the apparently 'material' body disintegrates. The explicit focused individuated consciousness which is associated with separation of the alternatives at the 'material' level disappears, but the quantum 'soul', which "in the period of death and immediately after ... is partially freed from the connections that she had with the life of the body",[486] inhabits, so to speak, the 'unconscious' realm of the parallel worlds of quantum reality. Furthermore, the way in which the parallel worlds are experienced depends upon the state of the soul, which also means it is dependent upon the activities carried out during the life just finished, (and the one before that, etc.). Mensky suggests that:

> We shall argue that this set of scenarios looks (for the soul of the dead man) as the paradise if the [dead] man was righteous, looks like hell for the sinner, and looks [like] purgatory in the

general case. ... Estimation is thus given to his life (or to his personality) and [his actions during his] stay in the world...[487]

The notion of 'purgatory' within the Catholic Church, or Russian Orthodox Church, is of an after-death sphere of purification. Mensky is suggesting that such a process can be understood as a process occurring naturally within the 'unconscious' quantum realms of 'parallel worlds' as they are experienced according to the activities carried on during embodied life.

Mensky suggests that this "turns out actually to be the judgement on the spent life."[488] In other words, the 'judgement' is not meted out by an external agency but is a result of the trace-potentialities within the quantum 'soul' naturally 'selecting' aspects of the 'parallel worlds'. In Buddhism this is karma and karmic consequences. In fact in Mensky's account, the way that afterlife is experienced is analogous, although on a more subtle level, to the way that apparent 'material' life functions. Mensky also tells us that:

> ...within the sphere of life [the] soul can select the niche in which she desires to exist. In order to make the selection, [the] soul investigates various scenarios. In this study, [the] soul can make use of her experience during the life of the body...[489]

The 'soul', then, which is a subtle quantum structure, navigates its way through the 'parallel worlds' of quantum reality, experiencing "bliss or sufferings",[490] through periods of embodiments and after-death states, towards more favourable states of existence:

> The soul tests various life criteria to find [the most favorable] set of them, which make her eternal existence comfortable. Testing any given set of life criteria is a stay in such [a] world, in which people are guided by precisely this set of criteria.[491]

And through this process the soul navigates towards the criteria of existence which produce the most 'blissful' state:

> Improving this criteria on the basis of this experience, the soul finally remains in that subset of the scenarios, which is determined by universal ... criteria. She understands after the experience of purgatory, what criteria led to the bliss, and she remains in the sphere, determined by these criteria. She finally settles into paradise and [experiences] eternal bliss.[492]

The quantum soul is in a quest for enlightenment and, as Mensky points out, this quantum perspective (EEC and QCC) leads to the psycho-metaphysical worldview of Buddhism:

> ...the soul experiences new earthly embodiment, reincarnation ... in which she is personified, its quality, depends on what *criteria of life quality* the possessor of this soul developed in the previous life ... This exactly corresponds to the Buddhist concept of *karma*. From what is the karma of the man in his past life, it depends, to what extent favorable will be the conditions for his next life. And from the fact whether he will improve his karma in the new life, his existence in the next embodiment will depend. Experiencing [a] long series of reincarnations, the man can be completely purged of sin, achieve enlightenment and taste nirvana i.e. infinite bliss. Then his soul will not experience the need for [a] new terrestrial embodiment and he will remain in "the other world" (in our terminology, he will be permanently existing in the quantum world, i.e. will always have an access to the entire set of parallel worlds).[493]

Thus we see that Mensky's EEC and QCC psycho-metaphysical account of the process of reality, is consistent with Bohm's worldview, and maps in dramatic and spectacular fashion into the spiritual psycho-metaphysics of the Buddhist worldview.

Quantum Samsara

A Bohmian Unity of Fragmentary Quantum Mirrors?

As we saw in an earlier chapter, Bohm wrote in *Wholeness and the Implicate Order* that:

> ...there is a universal flux that cannot be defined explicitly but which can be known only implicitly, as indicated by the explicitly definable forms and shapes, some stable and some unstable, that can be abstracted from the universal flux. In this flow, mind and matter are not separate substances. Rather, they are different aspects of one whole and unbroken movement. In this way, we are able to look on all aspects of existence as not divided from each other, and thus we can bring to an end the fragmentation implicit in the current attitude toward the atomic point of view, which leads us to divide everything from everything in a thoroughgoing way. Nevertheless, ... in spite of the undivided wholeness in flowing movement, the various patterns that can be abstracted from it have a certain relative autonomy and stability, which is indeed provided for by the universal law of the flowing movement.[494]

An important implication of this view is that we might expect that different proposals for 'interpretations' and views of quantum functioning may turn out to be differing "patterns that can be abstracted from" the "undivided wholeness in flowing movement" that constitutes Bohm's holomovement. If this view is correct then it will be possible to view various metaphysical proposals concerning the functioning and relationship between quantum and classical levels of the process of reality, including life and consciousness, as being interconnected coherent aspects of a higher unity. It is this vision we shall explore in this chapter by exploring commonalities within some differing perspectives.

This harmonizing task can be approached by drawing parallels, and showing connections, between David Bohm's notions of the holomovement and implicate order(s), which constitutes the context of unity, John Wheeler's vision of the 'Participatory Universe', Stephen Hawking's account of a 'Grand Design', Henry Stapp's account of the Heisenberg-von Neumann quantum 'ontology', Wojciech Zurek's 'Quantum Darwinism' proposal, Michael Mensky's Extended Everett Concept (EEC) and Quantum Concept of Consciousness (QCC) and the Buddhist Conscious-ness-Vehicle (*Vijnanavada*) concept of the *alayavijnana*, the ground or store consciousness. Of course, in this chapter it will only be possible to provide a sketch of the interconnections and implied harmony.

In the book *The Grand Design* Stephen Hawking and Leonard Mlodinow suggest such a kind of Bohmian Harmony (Bohm of course has precedence in this field of research) with their suggestion of the necessity for a "model-dependent reality":

> The naive view of reality ... is not compatible with modern physics. To deal with such paradoxes we shall adopt an approach which we call model-dependent realism. It is based on the idea that our brains interpret the input from our sensory organs by making a model of the world. When such a model is successful at explaining events, we tend to attribute to it, and to the elements and concepts that constitute it, the quality of reality or absolute truth. But there may be different ways in which one could model the same physical situation, with each employing different fundamental elements and concepts. If two such physical theories or models accurately predict the same events, one cannot be said to be more real than the other, we are free to use whichever model is more convenient.[495]

Such a view can easily be seen to be a variety of the ideas put forward by Bohm we have surveyed so far, with his notion that there can be 'fragmentary' views within an overall Wholeness which can validly elucidate different or similar aspects of the the Totality.

Bohm's concept of the quantum 'implicate order' can be conceived of as a universal field of nonlocal potential pre-experiential information which has been 'enfolded' into the field of potentiality by events and activities which have occurred previously within the operation of the field:

> The quantum field contains information about the whole environment and about the whole past, which regulates the present activity of the electron in much the same way that information about the whole past and our whole environment regulates our own activity as human beings, through consciousness.[496]

This field of potentiality, which is Bohm's generalization of the quantum wavefunction, is posited as being the common ground of the dualistic realm of subject-object experience, the experiential poles of mind and matter. As Bohm indicates, the information 'stored' within the 'implicate' field of potentiality depends upon activities which occurred in "the whole past". This view is clearly reminiscent of the Buddhist concept of 'karma', which is the view that actions of all kinds performed in the past (*karmas*) leave seeds (*bija*) or perfumings (*vasanas*) which shape future events (*vipaka*).

Bohm makes it quite clear that he conceives of the quantum implicate order as being closer to the nature of consciousness than that of the material world: "consciousness has to be understood in terms of an order that is closer to the implicate than it is to the explicate".[497] Thus, Bohm clearly leans towards the view that the ground of the dualistic world is of the nature of consciousness, although this must not be taken to imply that the type of consciousness which is the nature of the implicate order is on the same level as the consciousness exhibited by individuated sentient beings. It is, rather, a deeper level of unified awareness which has the potentiality of 'unfolding' into the 'explicate order' of the dualistic realm of individuated mind and experienced 'matter'.

The psychologist Karl Pribram collaborated with Bohm in the development of a 'holonomic brain theory', which is also referred to a the theory of the 'holographic brain'. According to this perspective the brain of each sentient being consists of an individual structured 'knot' of quantum field potentiality which is capable of storing information and traces of experience in a holographic manner. The experiences generated by the activities of sentient beings derive from the interaction between a brain's internal holographic structures and the information which resides in the external collective 'frequency domain'. On this view, learning is a result of the strengthening of informational experience-traces within the holographic field of brain structures, memories are generated through holographic resonance. Pribram referred to the information stored within holographic brain structures as being stored non-locally within the brain's quantum field. These brain-structures 'unfold' experience from the 'frequency domain'. Pribram and Bohm posited a model of cognitive functioning as being facilitated by an interference effect between individualized cognitive neurological quantum field structures, which are individuated aspects of the frequency domain, interacting with the global 'frequency domain' that is shared by all sentient beings.

In order to share a common apparently material world, each sentient being must unfold the same potentialities which underlie the appearance of the material world. Thus, there will be a completely individual unfoldments of a sentient being's psychological world, and there must be further 'common' or 'collective' unfoldments. Within the collective unfoldments there will be a species level and the basic appearance of the material world, to give two fundamental examples. Another example of collective unfoldments would be that of Jung's 'archetypes'. Bohm gave the example of the way in which the frequencies within a 'superposed' (frequencies laid on top of each other) assembly of radio waves can be tuned into, and thereby activated. This mechanism is a type of 'resonance'.

A similar approach is advanced by Henry Stapp's presentation of the Heisenberg-von Neumann perspective:

> The basic structure of orthodox (Heisenberg, von Neumann) quantum mechanics is very simple. The primary reality is a sequence of psychophysical events. Each such event has a psychological aspect and an associated physiological aspect. The connective support that links these events together is a field of potentialities that determines the objective tendencies (expressed in terms of probabilities) for specified psychophysical events to occur.[498]

In this depiction, Stapp's characterization of the basic field which gives rise to the psychophysical duality is the 'field of potentialities', which can be mathematically "expressed in terms of probabilities." This clearly denotes a global wavefunction which is somehow triggered to produce a sequence of moments of perception and experience. Furthermore, Stapp suggests that it is quite clear that this process involves a 'gap' which provides the necessity for conscious choices to be made that have a determining effect upon the nature of the immediate experience as well as the pool of potential experiences projected into the future:

> ...choices are not fixed by quantum laws; nonetheless each choice is intrinsically meaningful: each quantum choice injects meaning, in the form of enduring structure into the physical universe.[499]

The possibilities contained within the global 'frequency domain', 'implicate' wavefunction are determined by previous history, but an individual 'choice' has a degree of spontaneity and freedom.

Stapp's view clearly resonates harmoniously with the great twentieth century physicist John Wheeler's assertion that:

> Directly opposite to the concept of universe as machine built on law is the vision of *a world self-synthesized*. On this view, the notes struck out on a piano by the observer participants of all times and all places, bits though they are in and by themselves, constitute the great wide world of space and time and things.[500]

An observation which clearly indicates that Wheeler, in his later approach to the interpretation of the quantum evidence, considered that all the phenomena of the material world, space and time were the result of the congealed perceptions, so to speak, of all the sentient beings who have previously inhabited the Universe, presumably over vast time periods. In this view, the congealed perceptions would be stored as traces for future experience within the global wavefunction, which is equivalent to Pribram's 'frequency domain', or Bohm's implicate order.

As we have seen, the basic perspective which underlies these viewpoints corresponds closely with the following *Yogacara-Chittamatra* (Yogic-Practitioners/Mind Only) Buddhist perspective on the process of reality:

> …the mind is the principle creator of everything because sentient beings accumulate predisposing potencies through their actions, and these actions are directed by mental motivation. These potencies are what create not only their own lives but also the physical world around them. All environments are formed by karma, that is actions and the potencies they establish. The wind, sun, earth, trees, what is enjoyed, used, and suffered - all are produced from actions.[501]

Karma, a term often misunderstood in the West, simply means intentional activity of any kind on the part of sentient beings, or 'observer-participants', and such intentional activity will have future effects. Even minimalist perceptions of the seemingly material world are intentional activities which have karmic effects, in this case the effect being the production of a latency, or 'seed' for a similar perception to occur at a future time. Karma is the universal law of cause and effect which operates on all levels of reality, including the production of the structure of the seemingly independent material world. According to the Buddhist Mind-Only perspective the type of 'action' which is primarily responsible for the creation of apparent materiality is perception or cognition, which is also suggested by Wheeler's use of the term 'observer-participants.'

The Mind-Only mechanism which accounts for this remarkable mechanism, the transformation of Primordial Mind into the appearance of matter, as opposed to the impossible materialist alchemical dream of vivifying mindless matter into mind, can be called 'karmic resonance'. All actions and perceptions leave potencies within a deep level of collective mind called the *alayavijnana*, or ground-consciousness, a level of the process of reality which can be shown to correspond to the realm of quantum emptiness, or quantum potentiality. When these potencies are activated through being combined with potencies within the mind-streams of vast numbers of other sentient beings, an intersubjective creation of a shared material environment comes into being. This description of the process of reality, including the production of the intersubjective illusion of the material world, involves the mechanism of karmic cause and effect or quantum karmic resonance, the carrying forward and subsequent intersubjective activation of potencies within a deep collective mind-stream. When the subjective potencies resonate together in a reinforcing manner due their overall similarity, the collective experiential solidity of the apparently independent material world emerges. From this perspective, the 'objective' world of apparent material reality is an intersubjective creation on the part of all sentient beings who have ever existed and currently exist within the universe.

This Buddhist metaphysical perspective corresponds remarkably well with the work of Bohm and other quantum physicists. As we have seen, in his work *Wholeness and the Implicate Order* Bohm indicates that reality encompasses both the objective aspects and the subjective aspects of what is essentially an interconnected and undivided 'wholeness'. Bohm calls this totality the 'holomovement,' a notion which fits well with that of a ground consciousness:

> …what carries the implicate order is the holomovement, which is unbroken and undivided totality. In certain cases we can abstract particular aspects of the holomovement …, but more generally, all forms of the holomovement merge and are inseparable.

Bohm used the term 'implicate order' for the realm of potentiality from which the world of experience emerges. This is a nondual ground of potential experience which is activated to produce the dualistic realm of experience, which Bohm termed the 'explicate order'. The 'implicate order' is another label for what quantum physicists call the universal wavefunction, which is a quantum description of all the potentialities encompassing the manifestation of the experiential web of the entire universe. Such a wavefunction includes both the objective potentialities for experience and also the multitude of sentient experiencers moving through the global 'frequency domain' wavefunction, and thereby unfolding various experiences.

As mentioned previously Bohm also suggested a mechanism by which potentialities are unfolded as experienced actualities in the 'classical' dualistic world. An example that Bohm gave is that of the way in which a radio electromagnetic wave encodes the transmitted content within, or on top of, another frequency. The original content is 'unfolded' by tuning to the carrier frequency. We can think of various radio frequencies for instance as superposed into a global environmental composite waveform which becomes the overall radio wave environment, thus the radio environment contains multiple possibilities which can be tuned into and thus unfolded. Different tunings unfold different potentialities from the environmental frequency superposition (figure 1). In this picture, different experiences are unfolded out of the overall environmental frequency set through the mechanism of resonance, the actual resonance frequency depending on the tuning of the radio.

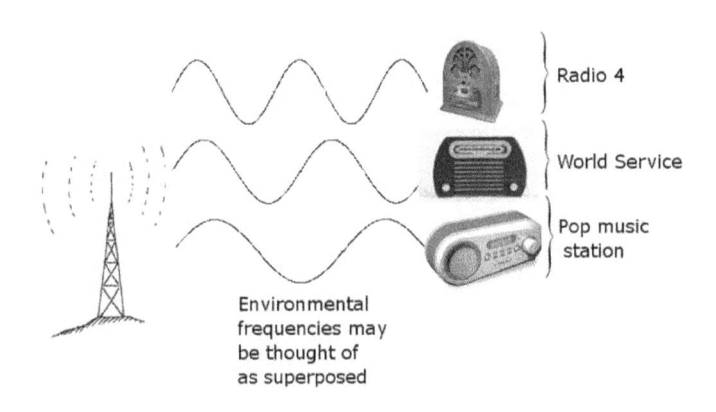

Figure 1

This viewpoint corresponds to physicist Max Tegmark's more recent presentation of the famous Everett's Many-Worlds scenario according to which it must be the case that there are a vast number of sentient beings trapped, so to speak, inside the global wavefunction. These can be thought of as quantum experiencing substructures, or 'sub-totalities' to use a term used by Bohm, which move through the overall wavefunction and in so doing unfold into experiential reality the potentialities within the wavefunction. It took the genius of Everett, Tegmark tells us:

> ... to realise that a single deterministically evolving wavefunction ... contains within it a vast number of ... perspectives where certain events appear to occur randomly.

Tegmark actually uses the term 'frog perspectives' because his original metaphor involved the notion of a frog being trapped inside the wavefunction, as opposed to the bird perspective adopted by theoretical 'external' observers, as shown in figure 2.

Bird represents outside objective observer

Universal wavefunction

Frog represents a 'subjective' sentient being moving inside the wavefunction

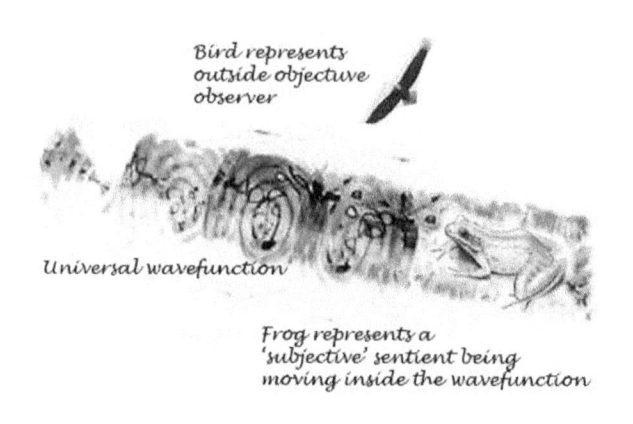

Figure 2.

But to be more precise, however, we must say that there are a vast number of frog perspectives, dog perspectives, cat perspectives, bird perspectives, fish perspectives, human perspectives, and so on, within the universal wavefunction. In fact, each sentient being is a 'subjective' quantum substructure, or 'sub-totality', moving, in fact cycling, through the universal wavefunction of quantum potentiality matrix of reality. Bohm describes this:

> We may begin by considering the individual human being as a relatively independent sub-totality, with a sufficient recurrence and stability of his total process (e.g., physical, chemical, neurological, mental, etc.) to enable him to subsist over a certain

period of time. ... In the implicate order we have to say that mind enfolds matter in general and therefore the body in particular. Similarly, the body enfolds not only the mind but also in some sense the entire material universe.[502]

Ultimately this process of reality is of the nature of consciousness. The fundamental field of awareness-consciousness, which creates the appearance of a material world, contains encoded, or enfolded, to use Bohm's terminology, tendencies, or potentialities, for matter-like experiences to occur. It is these potentialities, when manifested, which build up the illusion of an external material reality. This is precisely the view proposed in 1944 by the historical founder of quantum theory Max Planck, who said in a lecture that:

> All matter originates and exists only by virtue of a force... We must assume behind this force the existence of a conscious and intelligent Mind. This Mind is the matrix of all matter.[503]

This is an observation which is worth contemplating alongside the following assertion from the fourteenth century Tibetan Buddhist masterpiece *The Mountain Doctrine: Ocean of Definitive Meaning: Final Unique Quintessential Instructions* by Dolpopa Sherab Gyaltsen:

> I am called the matrix of attributes....
> I am called the pure matrix....
> The essence of ... of cyclic existence
> Is only I, self-arisen.
> Phenomena in which cyclic existence exists
> Do not exist-even particles-
> Because of being unreal ideation.[504]

Dolpopa's exposition is devoted to a lengthy and comprehensive elucidation of the nature of the 'matrix of phenomena', a fundamental Buddhist concept which is clearly analogous to the quantum wavefunction. When the above fragment is unraveled, and explicated, from within its own context, it turns out to be saying basically the same thing as Planck. 'Unreal ideation' is a Mind-Only term for the functioning of the fundamental mind-nature of reality. It is called 'unreal' because Dolpopa's style of Buddhist Mind-Only metaphysics attributes full reality to the 'nondual matrix' underlying the 'unreal ideation' of the world of duality. The term 'phenomena of cyclic existence' refers to the appearances of the apparent entities of the dualistic world, including the 'material' world. The 'matrix of attributes', or 'matrix of phenomena' corresponds to what quantum physicists call the 'wavefunction' of potentiality which underlies the manifestation of the appearance of the multitude of sentient-experiential worlds of cyclic existence, including the appearance of apparent materiality, which are experienced by the multitude of sentient beings.

In the following Dzogchen (the Buddhist 'Great Perfection' teachings) passage, the term 'intrinsic rigpa' refers to the fundamental awareness which resides as the ground quality of the universal wavefunction:

> The one intrinsic rigpa binds all experience: environments and lifeforms, infinite and unconfined, whether of samsara or nirvana, arise in spaciousness; spaciousness, therefore, embraces all experience at its origin.

In other words, the primordial awareness which is mathematically described by the universal wavefunction is the timeless source and origin of all the experiential continuums of lifeforms within their respective environments, which, of course, to a large extent overlap. *Samsara* indicates the experience of the process of reality from an unenlightened, or dualistic, perspective whilst *nirvana* is the experience from an enlightened, non-dual, perspective. The basis for both modes of experience is the quantum field of potentiality which has an internal nature of awareness-consciousness. One Buddhist term for this field of potentiality is *dharmata*:

> In many different places the Buddha said that all phenomena are empty, however, just saying that isn't sufficient. If one were to ask, "Does that word 'emptiness' indicate accurately and fully the nature of phenomena, the way in which phenomena abide?" No, it doesn't. "Does "luminous clarity" point out fully the way in which phenomena abide?" No, it doesn't. Does "wisdom" point out accurately the final nature of things?" No, it doesn't. There isn't a word that can properly describe dharmata. For that reason, it is said that dharmata or the true nature of things is inexpressible, meaning no matter what word one uses, one cannot express dharmata just as it is. If one attempts to think about dharmata, then it cannot be thought about accurately by the mind of an ordinary person. For that reason, it is said to "have passed beyond the sphere of the minds of ordinary persons."

According to the Wheeler-DeWitt quantum equation the universal wavefunction is timeless or 'frozen in time' and:

> From a God's-eye view, we can suppose, there is just a timeless universal state, which consists of a vast entangled superposition ... of states of subsystems of the universe. In these entangled superpositions, however, observables of certain subsystems of the universe are correlated with observables of other systems.[505]

It is these 'subsystems of the universe', or, again to use Bohm's term 'sub-totalities', which are sentient beings, the agents through which the universe perceives its own potentialities and creates the flow of time within the dualistic realms of existence. In other words, all sentient beings are agents through which the potentialities of the universal wavefunction of reality are unfolded. This is what Buddhist doctrine refers to as *samsara*, the endless cycle of conditioned and interdependent existence, which is driven by the causal karmic resonance mechanism of *karma-vipaka*, intentional actions or movements of consciousness and the subsequent effects (figure 3).

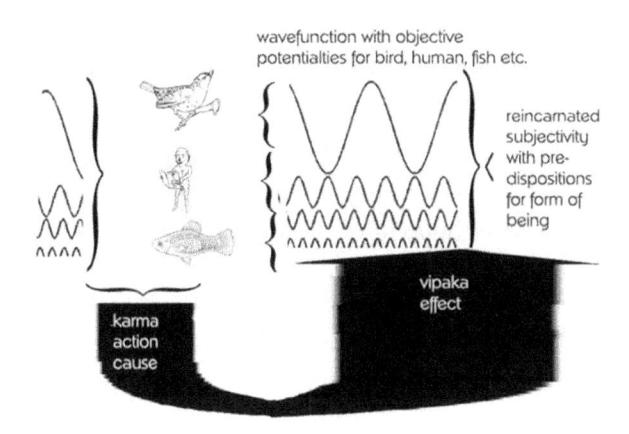

Figure 3

Each sentient being 'unfolds' a continuum of experience from out of the holomovement which takes place within the Mindnature of the quantum implicate order. For Bohm, the process of reality is the unfolding of an experienced world from the potentialities within the holomovement; this is the unfolding of the lived experience of the 'explicate order' from the 'implicate order'. This 'unfolding' from the 'implicate order' Bohm considered to occur through the operation of the same mechanism as a hologram is activated. In other words, our reality manifests as an interference pattern of wavefunctions interacting with each other in the same way as images are 'unfolded' from holograms. For Bohm, then, the nature of pre-experienced reality is considered to be an incredibly complex holographic wavefunction which encodes the potential experiences of a material world, and, of course, much more. When a subjective subsystem or 'sub-totality' of consciousness resonantly interacts at the quantum level with the objective possibilities, then a continuous stream of fleeting evanescent 'moments of experience' occur, and this process produces the illusion of a continuous experienced world.

Each moment of unfolded experience in turn leaves further traces in the ground-consciousness, which is the global wavefunction, and thereby the cycle of the 'self-excited' and 'self-synthesized' manifestation of the infinite multiplicity of the worlds of reality is kept cycling around a hub of emptiness-potentiality. As Donald D. Hoffman, Professor of Cognitive Science, University of California, says:

> The world of our daily experience – the world of tables, chairs, stars and people, with their attendant shapes, smells, feels and sounds – is a species-specific user interface between ourselves and a realm far more complex, whose essential character is conscious.[506]

The totality of all the different experiential continuums, which exist on many levels, species and individuals being the most obvious, are examples of levels of manifestation which make up the many-worlds of illusion generated from the functioning of the movement of individuated substructures or 'sub-totalities' of consciousness within the holomovement of the global wavefunction (figure 3).

Speaking in April 2003 to the American Physical Society, John Wheeler made the following remarkable, perhaps one might say 'mystical', sequence of remarks:

> The Question is what is the Question? Is it all a Magic Show? Is Reality an Illusion?
> What is the framework of the Machine? Darwin's Puzzle: Natural Selection? Where does Space-Time come from?

> Is there any answer except that it comes from consciousness?
> What is Out There?
> T'is Ourselves?
> Or, is IT all just a Magic Show?[507]

The Buddhist Mind-Only philosophers had come up with an affirmative answer to Wheeler's question as to the possibility that reality might be an illusory 'Magic Show' roughly one and a half thousand years ago:

> Phenomena as they appear and resound
> Are neither established or real in these ways,
>   Since they keep changing in all possible and various manners
>   Just like appearances in magical illusions.[508]

The *Yogacara-Vijnanavada* (Yoga-Practitioners / Consciousness-Way/ Mind-Only) philosophers' conceived of the epistemological-ontological process of reality as fundamentally consisting of a responsive energetic-experiential field within which all sentient beings are immersed and have their being. This field, the *alayavijnana* or 'ground consciousness' (also called the 'store-consciousness'), was conceived of as being of the fundamental nature of cognitive-awareness, and it is the 'stuff' from which all sentient beings, and the objects that they appear to experience, are constructed.

This fundamental ground of the process of reality is fundamentally experiential in nature and responds with great sensitivity to all intentional activity carried out by all sentient beings. Such actions can be performed by body, speech or mind but it is the intentionality behind any such action which is paramount in the mechanism by which all such activities leave traces within the fundamental cognitive-experiential field of reality, traces which will be activated at later moments when surrounding conditions within the field are resonantly conducive for the potentiality of the 'seed' or 'imprint' to emerge into full experiential reality. This psycho-metaphysical perspective anticipates some of the recent interpretations of the quantum evidence by nearly two thousand years. For instance, in a recent paper, 'Founding Quantum Theory on the Basis of Consciousness', Efstratios Manousakis writes:

> First, we conjecture that all human beings and the other living organisms have their own streams of consciousness. In order to gain an understanding of all of these related streams of consciousness together, and what precedes our human thoughts, and binds them together, we postulate the existence of the

Universal/Global stream of consciousness, as the primary reality that contains all of our individual streams, (which are sub-streams of the Universal conscious flow of events) and also conscious events that are not members of any human stream, but are like certain of our conscious events ... Note that the set of conscious events in consciousness must include all those that anyone has ever had, and for any personal stream of consciousness, all the events that have appeared in that person's stream of consciousness.[509]

This quantum perspective maps exactly on the *Yogacara-Vijnanavada* psycho-metaphysical worldview within which all sentient beings have their own continuum of consciousness cycling within the overall universal pool of energetic awareness-consciousness and potentiality. In some presentations of *Yogacara* the *alayavijnana* is identified with the individual ground-consciousness of sentient beings, each being having their own continuum which continues across lifetimes, whilst the all-encompassing, or 'Universal/Global' stream of energy-awareness-consciousness is called the *Alaya*.

This *Yogacara-Vijnanavada* psycho-metaphysics is also remarkably congruent to the excellent quantum psycho-metaphysics developed by the Russian physicist Michael Mensky. In the early part of his book *Consciousness and Quantum Mechanics: Life in Parallel Worlds*, Mensky tells us that:

...the phenomena of life and consciousness cannot be mechanistically reduced to the action of the laws of science as they are found in the course of exploring [inanimate] matter. The explanation of these phenomena on the basis of quantum mechanics requires [the] addition of a special independent element to the set of quantum concepts and laws. Such a new element of theory should directly connect quantum concepts with the concepts characteristic of life. The simplest way to find this element is to consider the phenomenon of consciousness and compare it with the description of observation (measurement) in quantum mechanics. Then it may be formulated as identification of consciousness with the separation of the alternatives" - a concept relating to the "Many Worlds" interpretation of quantum mechanics. ... the addition of this element simplifies the conceptual structure of quantum mechanics instead of [rendering] it more complicated. If we consider not only the phenomenon of

> consciousness but [also the] more general phenomenon of life, this additional element may be called [the] "life principle". It very naturally follows from the analysis of theory of consciousness ... The life principle formulates [the] evolution of [a] living system in such a way that it is determined by ... goals as well as by causes. The main goal of the living system[s] is survival so that their evolution provides their survival. However, for more sophisticated forms of life, the goals may include other criteria [such as] the quality of life.[510]

Mensky is emphatic that the phenomena of life and consciousness cannot be reduced to either quantum mechanics or "any other theory of [inanimate] matter". These aspects are, of course, involved in the processes and functions of living organisms, but:

> ...life and consciousness are not the direct consequence of these processes. Life is not the function of the body, and consciousness is not a function of the brain. Rather body is a realization of life, and brain is an instrument of consciousness.[511]

According to Mensky, although life and consciousness are not reducible to quantum mechanics, it is nevertheless the case that they are "connected" in a deep and irreducible manner with "quantum reality". As Mensky indicates, in order for the potentialities contained at the quantum level to come to life, so to speak, there must be an internal 'life-principle' which provides what we might call an 'unfolding force' which functions to unfold life and individuated consciousness.

As we have seen, this is a viewpoint also suggested by the recent notion of a "self-explaining universe" that the physicist Paul Davies has written about in his book *The Goldilocks Enigma*:

> ...a good case can be made that life and mind are fundamental physical phenomena, and so must be incorporated into the overall cosmic scheme. One possible line of evidence for the central role of mind comes from the way in which an act of observation enters into quantum mechanics. It turns out that the observation process conceals a subtle form of teleology.[512]

Such a universe would necessarily contain organisms that embody the capacity for cognition, which is to say consciousness, precisely because the purpose of 'self-explanation', to use Davies' terminology, or self-cognition, is fundamental to the universe. It is part of the "teleology" of the universe, its purpose is precisely to unfold life and consciousness.

In Mensky's psycho-metaphysics, which he calls the 'Extended Everett Concept '(EEC) with its associated 'Quantum Concept of Consciousness' (QCC), the global pool of awareness-consciousness-potentiality which Mensky calls the 'Alterverse', consists of the infinite quantum potentiality described by the Many-Worlds quantum perspective, also called the 'theory of the universal wavefunction', postulated by Hugh Everett III and extended by Bryce DeWitt. Within the overall sphere of the potentiality provided by the Many-Worlds contained within the universal wavefunction, the multitude of sentient beings cycle within the sphere of potentiality, in doing so they "separate the alternatives" and then choose an experiential path to follow within the overall infinite pool of possibilities.

Mensky's quantum psycho-metaphysical insight, which indicates that all sentient beings 'choose', mostly unconsciously, which path to navigate within the overall Many-Worlds quantum wavefunction of potentialities maps precisely onto the Buddhist Mind-Only/Dzogchen psycho-metaphysical worldview. In the following passage the term "Kun-gZhi" is the Tibetan for the *alayavijnana*:

> As the universal ground (Kun-gZhi) is the root of samsara, it is the foundation of all the traces, like a pond. As the Dharmakaya (ultimate body) is the root of nirvana, it is the freedom from all the traces, and it is the exhaustion of all contaminations... In the state of clear ocean-like Dharmakaya, which is dwelling at the basis, the boat-like universal ground filled with a mass of passengers – mind and consciousness and much cargo, karmas and traces – sets out on the path of enlightenment through the state of intrinsic awareness, Dharmakaya.[513]

The deepest level *alayavijnana*, the ground-consciousness, is pictured as a boat, filled with a mass of sentient beings, making its way through the clear "intrinsic awareness" of the ultimate *Dharmakaya*. The *Dharmakaya* can be thought of as the Many-Worlds container experienced from an enlightened perspective, the "clear ocean like" wisdom-awareness (*jnana*) of the uncontaminated enlightened state of being. As we have seen, Mensky was aware of such resonances between quantum psycho-metaphysics and Buddhism.

In a recent article, *Contiguity of Parallel Worlds: Buddhist and Everett's*, the Buddhist philosopher Andrey Terentyev refers to the "striking similarity of the views on reality in Buddhism and in the Extended Everett Concept by M. Mensky." In his conclusion Terentyev writes that:

> I'd like to stress that we are not just considering analogies in different fields of human endeavour; in fact, both Buddhist thinkers and modern physicists, using very different methods, arrived basically at the same description of [the] reality we live in. This is the point where the parallel worlds of Buddhism and Physics unexpectedly touched each other, and the deeper meaning of this is yet to be appreciated by both parties.[514]

As Terentyev says:

> The basic philosophical outcome of Everett based interpretations of the measurement problem consists in recognizing the fact that actually we live in a quantum world which is a superposition of macroscopically distinct states of different 'Everett Worlds' or 'classical alternatives' as Mensky would call it. ... Mensky identifies these classical alternatives with ... 'acts of consciousness' and this approach ... presupposes the existence of some kind of super-consciousness in the state of superposition while the classical consciousnesses of the observers in the 'classical alternatives' are illusory – as much as the 'classical' worlds themselves – because they mistakenly perceive their worlds as the 'whole' or 'real'.[515]

Terentyev then quotes Mensky and indicates deep parallels with Buddhist psycho-metaphysics:

> Everett's concept deals with two aspects of consciousness. The consciousness as a whole (we could compare this 'consciousness as a whole' or superconsciousness with Buddha's mind or jnana) splits between alternatives, and a component of consciousness (Buddhist vijnana) lives within one classical alternative.[516]

This distinction between the nondual, non-separated realm of *jnana*, which is nondual wisdom-awareness. and the divided and separated mode of individuated consciousness, designated in Buddhism as *vijnana* (vi-jnana = divided-jnana) is fundamental for *Yogacara* psycho-metaphysics.

From the *Yogacara* perspective the collective epistemological activity on the part of all sentient beings determines ontology, an insight which clearly prefigured the later quantum insights of John Wheeler and others. Furthermore, the *Yogacara* viewpoint requires that we assume that the 'ultimate' nature of all phenomena is cognitive in nature:

> Nothing, such as atoms and so on, exist externally, As anything other than cognition.[517]

It follows therefore that all phenomena are of the nature of consciousness:

> ...all these various appearances, do not exist as sensory objects which are other than consciousness. Their arising is like the experience of self-knowledge. All appearances, from indivisible particles to vast forms, are mind.[518]

The *Yogacara* cognitive 'stuff' which is conceived of as forming the fabric of reality bears an uncanny resemblance to the 'dream stuff' of quantum reality as portrayed by quantum physicist Wojciech Zurek:

> ...quantum states, by their very nature share an epistemological and ontological role – are simultaneously a description of the state, and the 'dream stuff is made of.' One might say that they are epiontic. These two aspects may seem contradictory, but at least in the quantum setting, there is a union of these two functions.

Here Zurek characterizes the quantum 'dream stuff' as being exactly the kind of cognitive medium capable of creating the appearance of a 'solidified' classical world through its own infinite web of internal acts of 'epiontic' interactions. Zurek refers to the inner process of quantum realm as consisting of the 'union' of the two functions of:

1) epistemology – which is the process of perception and knowing, and

2) ontology - the actuality of being.

The quantum 'epiontic principle', then, indicates that perception of quantum information, which is carried by quantum 'epiontic' 'dream-stuff', creates the illusion of an independent material reality. This does not imply, however, that the isolated conscious or unconscious (both being part of a structure of consciousness) perceptions of individual sentient beings create reality by lone whim or preference. As Zurek points out:

> ...while the ultimate evidence for the choice of one alternative resides in our illusive "consciousness," there is every indication that the choice occurs much before consciousness gets involved and that, once made, the choice is irrevocable.[519]

This indicates that consciousness is the ultimate source of selections of Many-Worlds alternatives. However, it is also important to point out that Zurek seems very reluctant to directly involve the notion that consciousness plays a primary role in his 'epiontic principle', despite the fact that some of his own insights suggest such a view.

The material world, as John Wheeler indicated, is a collective construction, but it is not the case that individual beings can beam consciousness rays in order to 'collapse' wavefunctions. But, as Wheeler also stated:

> Yes, oh universe, without you I would not have been able to come into being. Yet you, great system, are made of phenomena; and every phenomena rests on an act of observation. You could never even exist without elementary acts of registration such as mine.[520]

And:

> Beyond particles, beyond fields of force, beyond geometry, beyond space and time themselves, is the ultimate constituent the still more ethereal act of observer-participancy?[521]

As we have seen, Wheeler's later quantum metaphysical perspective is consistent with Bohm's later quantum metaphysics of a Participatory Undivided Totality.

Ultimately, as the significant physicist Roger Penrose has pointed out in his 1994 book *Shadows of the Mind*:

> At the large end of things, the place where 'the buck stops' is provided by our conscious perceptions.[522]

Like Zurek, however, Penrose discounts the notion that 'consciousness' actually does 'collapse' the wavefunction of possibilities into one actuality. In Penrose's case this is because he considers, mistakenly, that "consciousness is a rather rare phenomenon throughout the Universe". It may be the case that human and animal consciousness is "rare" within the Universe, but this does not entail that a deeper level of consciousness may not be operative. Such a mistaken view derives from the pervasive hold that a subtle materialism seems to have on the perspectives of most physicists.

However, there were, and are, intrepid physicists such as David Bohm, Henry Stapp, Michael Mensky and others who are advancing the increasingly inescapable conclusion that consciousness in some form, not necessarily fully manifested individual consciousness, is implicated in the appearance of the material realm. But, it still remains the case that the overall prejudice within both science and philosophy is towards a subtle materialism which seems to prefer to think of consciousness as arising from some kind of subtle material mechanism, rather than being a primary feature of the process of reality.

However, if Zurek's 'epiontic' paradigm is correct, and the evidence most certainly points in that direction, then the only reasonable conclusion must be that the ultimate ground of the appearance of the 'classical' realm of materiality must be, as Zurek says, some kind of 'epiontic' quantum 'dream stuff'. Furthermore, this quantum 'dream stuff', according to Zurek, has the fundamental feature that it is capable of preserving and proliferating cognitive activity, and preserves the quantum states that derive from such activity:

> ... the appearance of the classical reality can be viewed as the result of the emergence of the preferred states from within the quantum substrate through the Darwinian paradigm, once the survival of the fittest quantum states and selective proliferation of the information about them throughout the universe are properly taken into account.[523]

The insight that Zurek provides, which he probably dubs 'Darwinian' partly in order to downplay any 'mystical' connotations (Zurek was a student of Wheeler), is that "states that exist are the states that persist" and this is a persistence within a quantum realm which consists of, as Zurek puts it, "the dream stuff which reality is made of", and the mechanism that underlies this persistence is "an objective consequence of the relationship between the state of the observer and the rest of the universe".[524]

It might be thought that Zurek is a fully paid up 'consciousness is fundamental' quantum philosopher, but this would be incorrect. In fact he, like a few other quantum philosophers such as Wheeler, hedges his bets! Zurek is careful to distance himself from any New-Age type 'consciousness ceates reality' claims. Thus, in a 2011 book of interviews with quantum physicists on their views on various topics, *Elegance and Enigma: The Quantum Interviews*, Zurek indicated that:

> Naïve subjectivist accounts fail in one obvious way; the observer has to be outside of the quantum realm ...[525]

In his paper *Quantum Darwinism and Envariance,* Zurek explicitly rules out a role for conscious observers within the mechanism he is suggesting:

> Von Neumann has even considered the possibility that collapse [of the quantum wavefunction] may be precipitated by the conscious observers. The "anthropic" theme was later taken up by the others, including London and Bauer (1939) and Wigner (1963). The aim of this chapter is to investigate and - where possible – to settle open questions within the unitary quantum theory per se, without invoking any nonunitary or anthropic *deus ex machina*.[526]

Zurek also points out that:

> The natural sciences were built on a tacit assumption: Information about the universe can be acquired without changing its state. The ideal of "hard science" was to be objective and to provide a description of reality. Information was regarded as unphysical, ethereal; a mere record of the tangible....Quantum theory has put an end to this Laplacean dream about a mechanical universe. Observers of quantum phenomena can no longer be just passive spectators. Quantum laws make it impossible to gain information without changing the state of the measured object. The dividing line between what is and what is known to be has been blurred forever. While abolishing this boundary, quantum theory has simultaneously deprived the 'conscious observer' of a monopoly on acquiring and storing information. Any correlation is a registration; any quantum state is a record of some other quantum state. When correlations are robust enough, or the record is sufficiently indelible, familiar classical "objective reality" emerges from the quantum substrate.[527]

In other words, Zurek is blurring the line, or "boundary", as to what the process of 'observation' amounts to. By doing this Zurek is attempting to avoid what he calls "naïve subjectivist accounts" of quantum functioning. The tactic he employs, however, is dubious. Zurek makes his move by introducing a judicious, for his purpose, use of language, allowing the quantum 'epiontic' 'dream-stuff' to apparently 'observe' in a similar manner to sentient beings, although on a less conscious level of course. It is a sort of non-conscious, or perhaps 'semi-conscious', manner of 'observation'!

Zurek's underlying aim in this way of description seems to be to de-subjectivise, so to speak, the process of observation, hoping to remove it from the primary realm of consciousness. However, it is quite possible to take the complementary route and extend the realm of consciousness as being fundamental within the field of 'epiontic dream stuff', thus making the realm of quantum 'dream stuff' exactly the kind of 'stuff' required by Bohm's holomovement. And, as we have seen, such 'dream stuff' can be characterised as Primordial Consciousness.[528]

There are a couple of important points in the above passage from Zurek which need to be fully appreciated. In the last sentence there is the indication that the essential nature of 'Nature' is quantum, the 'substrate' from which "familiar classical 'objective reality' emerges" is quantum. Zurek begins by telling us that the notion of a 'mechanical universe' which is entirely independent of observation has to be abandoned. The fundamental revelation of quantum theory is that "it is impossible to gain information without changing the state of the measured object". But at this point Zurek wants us to conceive of this occurring without consciousness, even though, of course, "gaining information" is something which ultimately involves consciousness.

So Zurek also has to say elsewhere that 'the ultimate evidence for the choice of one alternative resides in our illusive "consciousness". For Zurek it seems that consciousness both is and isn't involved, it's quantum smoke and mirrors! It certainly *appears* that consciousness is implicated in the 'collapse' of the wavefunction, and because of this Zurek has no qualms in asserting that "the dividing line between what is and what is known to be has been blurred forever". Zurek, however, is another physicist who tries to distance himself from getting too enthusiastic about the role of consciousness in the process of reality, even though he knows he must make a brief nod of appreciation along the way.

Zurek describes his view of his Quantum Darwinism proposal as follows:

> The main idea of quantum Darwinism is that we almost never do any direct measurement on anything ... the environment acts as a witness, or as a communication channel. ... It is like a big advertising billboard, which floats multiple copies of the informtion about our universe all over the place.[529]

Here Zurek plays down the role of consciousness, as if suggesting it is only the quantum 'dream-stuff' of the environment doing any 'witnessing'. But, in this billboard advertising metaphor, the more appropriate image is that the more the observing punters buy into the advertisement and thereby make the product, which is in this case the solidified appearance of the apparently material world, more popular, the greater the number of billboards which spring up, and, as a consequence, the more punters are enticed to join in the product craze for a material reality.

When one sees through Zurek's subtle attempt to decapitate the quantum functional relevance of consciousness as being driven by a subtle quantum-materialist ideology, and also sees its invalidity, it becomes clear that Zurek's account is also consistent with the basic quantum metaphysical perspective contained within aspects presented by Bohm, Stapp, Wheeler and Mensky. In a consciousness-based Zurek-type quantum-epiontic-dream-stuff perspective, the process of the multitudinous perception-based creation of the material world 'billboard' takes off in a self-reinforcing, or self-resonating, manner. On this Zurek-based view, the creation of the classical, macroscopic, material world derives from a mechanism wherein the more often a perception of the appearance of materiality is made, either directly or indirectly, the more potent becomes the advertising billboard campaign, or the environmental template, or matrix, for that perception of material reality to occur again at some future point. Stapp made a similar point in the following way:

> Each subjective experience injects one bit of information into this objective store of information which then specifies ... the relative probabilities for various possible future subjective experiences to occur.[530]

And, of course, such a view is entirely consistent with John Wheeler's later quantum 'mystical' leanings. The physicist Anton Zeilinger refers to Wheeler's:

> ...realisation that the implications of quantum physics are so far-reaching that they require a completely novel approach in our view of reality and in the way we see our role in the

universe. This distinguishes him from many others who in one way or another tried to save pre-quantum viewpoints, particularly the obviously wrong notion of a reality independent of us.[531]

In the Guardian obituary for John Wheeler we read that:

> In 2002, he wrote: 'How come the universe? How come us? How come anything?' Although Einstein had once asked him whether, if no one looked at it, the moon continued to exist, Wheeler's answer to his 'how come?' questions was 'that's us'.[532]

So, Wheeler was well aware that acts of perception were the creative force behind the manifestation of the universe, this was clearly embodied in his self-perceiving universe graphic.

It only remained for the final step, the extraordinary knowledge known and realised by the great mystics of 'all times and all places', the fundamental nature of reality is Universal Self-perception. The phenomenon of the 'collapse of the wavefunction' is a subtle indication of the fundamental self-perceiving process of the universe. This notion that internal self-perception is fundamental within the process of the manifestation of life and the universe is also a central theme within the *The Grand Design* described by Hawking and Mlodinow:

> In this view, the universe appeared spontaneously, starting off in every possible way. Most of these correspond to other universes .... Some people make a great mystery of this idea, sometimes called the multiverse concept, but these are just different expressions of the Feynman sum over histories. ... The histories that contribute to the Feynman sum don't have an independent existence, but depend on what is being measured. We create history by our observations, rather than history creating us.[533]

The 'Feynman sum over histories' is a technique of quantum analysis in which all possible quantum paths are added together to determine probabilities, the details are not important here. The central significant claim, of course, is that "we create history by our observations".

Bohm and Hiley (B&H), however, are more restrained in their appraisal of the assertion that present observations determine the nature of the past. Essentially this is because they consider that the 'participation' referred to by Wheeler and others involves a mistaken absolute division and separation between an 'underlying reality' of the quantum wavefunction, and, on the other side, an independent 'observer' of some kind, either a measuring mechanism or a consciousness. For B&H this division is inappropriate because the process has to be located within a greater context of the universal holomovement, wherein both individuated consciousness and the appearance of material 'particles' 'unfold' from the deeper level of the implicate order. For B&H the assertion that the actuality of the past is fully determined in the present by observations of the past at the quantum level is not just an invalid over-extension of the available evidence, also it involves a metaphysical mistake.

In *The Undivided Universe* B&H present a quote from Wheeler:

> ...no phenomenon is a phenomenon until it is an observed phenomenon ... the universe does not 'exist out there' independently of all acts of observation. It is in some strange sense a participatory universe. The present choice of observation ... should influence what we say about the past ... The past is undefined and undefinable without this observation.

And they then respond to Wheeler's view as follows:

> We can agree with Wheeler that no phenomenon is a phenomenon until it is observed, because by definition, a phenomenon is what appears. Therefore it evidently cannot be a phenomenon unless it is the content of an observation. The key point about an ontological interpretation such as ours is to ask the question as to whether there is an underlying reality that exists independently of observation, but that can appear to an observer when he 'looks' ... We have proposed a model of such a reality in which we say, along with Wheeler, that the universe is essentially participatory in nature. However, unlike Wheeler, we have given an account of the participation, which we show ... to be rational and orderly and in agreement with all the actual predictions of quantum theory. In doing this we assume ... that the underlying reality is not just the wave function, but it also has to include the particles ... when we take this into account there is no need to say that the past is affected by our observation of the present.[534]

So here we see that B&H deny the validity of making an absolute division between an independent underlying wavefunction and observers, the interaction between which magically creates particles backwards in time. In B&H's model of the holomovement, particles are included in the model in a way that renders it unnecessary to make such an assertion.

As we have seen, in B&H's model of the holomovement of implicate orders and the derived explicate order, both individuated consciousness and the explicate functioning of the manifested realm of 'particles' arise from a deeper level of the functioning of the implicate orders. In such a situation, at the quantum level there may be an appearance of an interconnection between consciousness of the manifestation of the world of particles, but this does not justify the claim that particles are directly manifested backwards in time.

B&H write in the context of the delayed choice experiment:

> ...in our interpretation there is no need for this strange attribution of past properties. For the reality is that ... the wave is split to traverse both paths while the particle traverses one or other in each individual case. But ... the interaction with the measuring apparatus implies mutual participation in which the final state need not directly reveal the earlier properties of the system.[535]

So according to B&H the kind of 'participation' involved in the quantum world, and the arising of interaction with the explicate 'classical' world is mutual and not one-sided. But, as B&H also point out, consciousness is a significant participating ingredient in the interconnected functioning of the holomovement. As B&H write:

> The content of our consciousness is then some part of this overall process. It is thus implied that in some sense a rudimentary mind-like quality is present even at the level of particle physics, and that as we go to subtler levels, this mind-like quality becomes stronger and more developed. Each kind and level of mind may have a relative autonomy and stability. One may then describe the essential mode of relationship of all these as *participation*, recalling that this word has two basic meanings. To 'partake of' and to 'take part in'. Through enfoldment, each relatively autonomous level of mind partakes of the whole to one degree or another. Through this it partakes of all the others in its 'gathering' of information. And through the activity of this information, it similarly takes part in the whole and every part. It is in this sort of activity that the content of the more subtle and implicate levels is unfolded (e.g. as the movement of the particle

unfolds the meaning of the information of that is implicit in the quantum field ...)[536]

So, according to B&H, a mind-like quality is present at all levels of the holomovement but becomes stronger and more dominant at the deeper 'subtler' levels. And it this mind-like aspect that gathers information which is contained in the whole field and organises the manifestation and movement of particles. Thus it remains the case that in B&H's model, it is the mind-like aspect of the holomovement that manifests, organizes and co-ordinates particles, according to the information 'gathered' from the implicate orders of quantum fields.

However, it would seem that it would be incorrect to claim that the proposals made by other physicists should be considered to be *completely* false, but rather, they may be partial views which should not be taken as absolute and literal truths. For example, to claim that in the quantum context present observations create history need not amount to the claim that anything actually changes in the past, it can merely mean that our historical view is created in a certain way within the present time. This point is made by Shantena Agusto Sabbadini in his book *Pilgrimages to Emptiness: Rethinking Reality through Quantum Physics*, who, as we have seen, suggests that a *persistence of information* approach is more appropriate:

> The persistence of information approach fits well with the notion of a participatory universe, but this participation does not need to be understood in the sense that "we bring about something of the universe in the distant past". The apparent retroactive effect in time is merely a reflection of the inadequacy of our representations of matter, which are imbued with classical prejudice. ... Our choice of the setup in the cosmic delayed choice experiment therefore does not change anything in the distant past. What happened in the past is the creation of an entangled state ... What happens here now is that we can choose to perform on that state an observation conserving or destroying which path information ... But there is no retroactive change, ... Our choice of the experimental setup determines what we see today; it does not affect what happened millions of years ago ...[537]

As B&H write in the final pages of *The Undivided Universe:*

> In this whole process , it is clear that the observer along with his thoughts is contained in a inseparable way. For example, the

general theories of physics (including the very one we are proposing here) are an extension of this process. What such a theory produces is a kind of reflection of reality as a whole. ... this provides another aspect or appearance of the whole. We may usefully regard this appearance as a kind of mirror through which reality is perceived.[538]

Dharmakaya, Dharmadhatu & Buddhist Implicate Levels of Consciousness

The fundamental quantum ground of Bohm's holomovement is a nonlocally interconnected energetic field of potentiality which has an internal cognitive nature. This is clearly suggested by Bohm's several indications that individuated consciousness and the appearance of the material world must derive from a "common ground". Bohm's notion of an 'implicate order', which is his metaphysically extended description of the quantum wavefunction, can be conceived of as a universal field of nonlocal potential experiential information. This potential information has been 'enfolded' into the field of potentiality by events and activities which have occurred previously within the operation of the field:

> The quantum field contains information about the whole environment and about the whole past, which regulates the present activity of the electron in much the same way that information about the whole past and our whole environment regulates our own activity as human beings, through consciousness.[539]

Furthermore, this field of potentiality is posited as being the common ground of the dualistic realm of subject-object experience, the experiential poles of mind and matter. And Bohm also points out that: "consciousness has to be understood in terms of an order that is closer to the implicate than it is to the explicate".[540] Thus Bohm clearly leans towards the view that the ground of the dualistic world is *of the nature* of consciousness, although this must not be taken to imply that the type of consciousness which is the nature of the implicate order is on the same level as the consciousness exhibited by individuated sentient beings. It is a deeper level of mind-energy corresponding more closely to the Buddhist notion of nondual Mindnature, or Primordial Mind.

In the later development of his ideas Bohm added the notion of an all encompassing organising field which he called the 'super-implicate' order:

> ... the enfoldment is now seen on two levels, first an enfolded order of the vacuum with ripples on it that unfold; and second, a super information field of the whole universe, a super-implicate order which organises the first level into various structures ... The super-implicate order, which is the so-called higher field (the implicate order would be a wave function) ... would be a super-wave function.[541]

Bohm's description of the super-implicate order and its lower levels is redolent of the Buddhist notion of the *dharmadhatu / dharmakaya*, which is the ultimate space of phenomena from which all experiential phenomena arise. As Tulku Urgyen Rinpoche explains:

> The main image of dharmadhatu is that of space—the 'space of all things' within which all phenomena manifest, abide and dissolve back into.... Dharmadhatu is the basic environment of all phenomena, whether they belong to samsara or nirvana. It encompasses whatever appears and exists, including the worlds and all beings. ... The relationship between dharmadhatu, dharmakaya and the wisdom of dharmadhatu is like the relationship between a place, a person and the person's mind. If there is no place, there is no environment for the person to exist in; and there is no person unless that person also has a mind dwelling in the body. In the same way, the main field or realm called dharmadhatu has the nature of dharmakaya. Dharmakaya has the quality of the wisdom of dharmadhatu, which is like the mind aspect.[542]

Dharmadhatu, which has potentiality and mind aspects, is also identified as the purified mind in its natural state, free of obscurations and afflictions. It is the primordial nature of mind, the fundamental ground of consciousness. It would be remiss not to point out the significant similarity between Tulku Urgyen Rinpoche's description and Bohm's description, this passage has been presented in earlier chapters:

> The new form of insight can perhaps best be called Undivided Wholeness in Flowing Movement. This view implies that flow is, in some sense, prior to that of the 'things' that can be seen to form and dissolve in this flow. One can perhaps illustrate what is meant here by considering the 'stream of consciousness'. This flux of awareness is not precisely definable, and yet it is evidently prior to the definable forms of thoughts and ideas which can be seen to form and dissolve in the flux, like ripples, waves and vortices in a flowing stream. As happens with such patterns of movement in a stream some thoughts recur and persist in a more or less stable way, while others are evanescent.[543]

As we have seen in many passages, there is a deep correspondence between Bohm's vision of the interdependent functioning of a mind-like energy field of potentiality and those of Buddhist and Bon psycho-metaphysical perspectives.

In Bohm's discussions with Renée Weber the connection with the 'clear light' ultimate realm of the process of reality within Buddhist metaphysics is also indicated as significant:

> **Weber:** For the mystics there is always light. The primary clear light in the *Tibetan Book of the Dead* is the first thing the dying person is aware of. ...
>
> **Bohm:** Light enfolds the universe as well . ... Light in its generalised sense (not just ordinary light) is the means by which the entire universe unfolds into itself.
>
> **Weber:** Is this a metaphor for you or an actual state?
>
> **Bohm:** It's an actuality. At least as far as physics is concerned.
>
> **Weber:** Light is energy, of course.
>
> **Bohm:** It's energy and it's also information - content, form and structure. It's the potential of everything.[544]

The idea that the ultimate nature of the process of reality is of the nature of light is also an aspect of some Dzogchen teachings. The Dzogchen practitioner and translator Keith Dowman has given the main title to his book on *The Circle of Total Illumination Tantra* as *EVERYTHING IS LIGHT*, and he writes in the introduction:

> ...awareness is 'light' - the very cognizant light that is conscious experience, and constitutes inseparable knowing and the knowable. All phenomena in sensory perception are 'light'. ... Dzogchen is the vehicle of light - light is both source and manifestation. Such is the Dzogchen view ... With such insight, the definitive revelation of Dzogchen, a final holistic vision, becomes identical to the universal experience of nonduality.[545]

The original title of *The Tibetan Book of the Dead* is *Bardo Thodol*, which is translated to 'liberation by hearing on the after-death plane'. The purpose of the book is to help those who have died and whose subtle minds are subsequently in the intermediate state, between death and rebirth, to recognize and dissolve into primordial awareness and thus escape the cycle of death and rebirth. This is accomplished by reading aloud the text of the book, thereby assisting the dead individual navigate through the 'bardo' planes, which are the implicate levels of the unfolding of consciousness from the source primordial mind. The six bardos according to *Nyingma* and *Kagyu* schools of Tibetan Buddhism are:

Kyenay bardo: bardo of birth and life. This bardo commences from conception until the a person's mindstream withdraws from the body.

Milam bardo: bardo of the dream state. The milam bardo is a subdivision of the first bardo, because the dream state of sleep, of course, is included in the rhythm of the waking-dream and sleeping dream within a lifetime. Dream Yoga includes practices to integrate the dream state into Buddhist practice.

Samten bardo: bardo of meditation. This bardo is generally only experienced by meditators, though individuals may have spontaneous experience of it. Samten bardo is also an aspect of the kyenay bardo. These first three bardos relate to experiences within a lifetime.

Chikhai bardo: bardo of the moment of death. This is the transition from life to the after-life psychic state. The first experience of this transition is the 'clear light', at this point an experienced Buddhist practitioner who has prepared for this event may attain liberation. In many cases, however, the experience will be overwhelming, producing fear and confusion. Francesca Fremantle, in her book *Luminous Emptiness*, writes concerning this bardo:

> ...the luminosity of death appears to all sentient beings ... we are caught between the desire for continued existence and the fear of annihilation ... the confusion of most living beings is too great to face such an inconceivable dilemma, and they simply black out into unconsciousness.[546]

Chönyi bardo: bardo of the luminosity of the true primordial nature which commences after the final 'inner breath'. It is within this bardo that visions and auditory phenomena occur. Concomitant to these visions, there is a deep uprising of profound peace and pristine awareness. Sentient beings who have not practised during their embodied lives, who do not recognize the clear light at the moment of death, are often afflicted with 'delusion' throughout the fifth bardo of luminosity:

> A series of dream or trance-like visions and auditory sensations that each being experiences differently based on their remaining karmas, particularly their intense aversions or desires. An

experienced Buddhist practitioner, one able to recognize the clear light of the fourth bardo (even if liberation was not attained there), will be able to maintain an inner equanimity during this phase, and even experience transcendent realms of being, while others will be trapped in full delusion, as if immersed in a movie or dream.[547]

Sidpa bardo: bardo of karmic becoming or transmigration. This bardo precedes the next rebirth, a rebirth which is determined by the "karmic seeds" within the storehouse-ground-consciousness.[548]

It is illuminating to consider these 'planes' of consciousness, both embodied and those which are purely consciousness, from the point of view of Bohm's implicate-explicate holomovement worldview. Within this elucidation of the various modes of consciousness which manifest during various phases of the birth-life-death-rebirth cycle of a sentient being, who constitutes a relatively autonomous 'sub-totality' cycling within the overall holomovement, there are explicate and implicate phases. The grossest level of the first bardo of life, of course, is the explicate level of embodied existence. But even within the explicate level of human life consciousness has a background within the implicate level:

> …it is possible to comprehend both cosmos and consciousness as a single unbroken totality of movement.[549]

And:

> We begin by proposing that in some sense, consciousness (which we take to include thought, feeling, desire, will, etc.) is to be comprehended in terms of the implicate order, along with reality as a whole.[550]

So even in the explicate order of embodied everyday life, consciousness retains a connection to its deep roots in implicate orders.

In sleep the dreams that appear have their origin and location within implicate orders. In this state of implicate consciousness, as Mensky says, focused everyday consciousness is "turned off" and because of this:

> …dimmed consciousness (in particular, in the state of sleep or trance) means an incomplete separation of alternatives, in which consciousness looks into 'other alternatives' and can single out

the most favourable one among them.[551]

Here Mensky is referring to the 'alternatives' residing as potentialities within the quantum implicate order. Several scientific and mathematical discoveries were the result of dreams. For example Srinivasa Ramanujan was a self-taught Indian mathematician who was able to intuit complex numerical patterns and connections without producing proofs. Instead, the devout Hindu genius claimed that his findings came from a divine realm and were revealed to him in dreams by the goddess Namagiri.[552]

The structure of benzene was discovered by August Kekulé in the late 18th Century after he had a day dream about a snake biting its own tail, an ancient symbol called the 'ouroboros'. Kekulé was involved in research on this problem and quickly realised that the snake in his dream indicated the structure of benzene, a molecular structure unknown at that time. The psychologist Patrick McNamara has written in *Psychology Today*:

> For me the experimental and scientific case for the reality of precognitive dreams, dreams that contain images of the future, is now settled—they do occur. In fact, they appear to occur frequently not rarely. ... I have covered some of that experimental evidence and mentioned extensive meta-analytic reviews of these studies. In my opinion the discussion should now focus on why and how they occur. While skepticism concerning these dreams is always healthy, we need to avoid the kind of skepticism that devolves into a form of science denialism, obstructivism and dogmatism.[553]

The evidence is increasingly supporting the Bohm-Mensky quantum precognitive worldview.

The third bardo is the bardo of meditation, and here we meet a bardo which contains a range of implicate levels of consciousness because, as we shall see, there are a range of meditative states, each of which accesses subtly different qualitative levels of consciousness. In this context in it is worth noting that Mensky tells us that there are "at least two important features" of the Buddhist worldview that make it especially appropriate for his perspective:

> First, Buddhism does not require blind faith in the [teachings] it proclaims. Disciples are urged to believe only when they assure themselves in the course of the work on their own cons-

> ciousness that the doctrine is correct. Second, Buddhists consider their task to learn to perceive a special state …. which is impossible to exactly express by words and which may be characterized approximately as 'the root of consciousness', 'the origin of consciousness', or 'the preconsciousness'. This is an elusive state that precedes the emergence of consciousness. Learners are urged to work on their consciousness until they catch the sensation of 'being between consciousness and the absence of consciousness'. It is easily seen that the state of consciousness which is the goal of Buddhists bears much resemblance to the deepest or most primitive layer of consciousness (being "at the edge of consciousness") which is identified with the separation of alternatives in our Extended Everett's Concept.[554]

Here Mensky locates a crucial feature of the theory and practice of Buddhist meditation in the context of his quantum psycho-metaphysics. According to Mensky fully manifested focused consciousness is a result of the "separation of alternatives". On the other hand, the realm of the entire set of 'parallel worlds' of potentiality which makes up 'quantum reality' lies at the primordial level of nondual, i.e. non-separated, awareness, which embraces the entirety of quantum potentiality. Mensky explicitly identifies this with the non-separated realm. In the above quote he implicitly identifies "the most primitive layer of consciousness" with the origins of the "separation of alternatives in our Extended Everett's Concept". Furthermore, his analysis suggests that there must be a deeper nondual, non-separated, layer of primordial awareness prior to "the most primitive layer of consciousness" that arises as the alternatives separate.

Mensky's analysis fits perfectly with Buddhist doctrine. This structure of consciousness is precisely delineated within Buddhist traditions, in various formulations. Meditation techniques are described within the Buddhist traditions in order to directly experience the various 'states' described. In the Theravada tradition, which is derived from the earliest Buddhist teachings contained in the Pali Suttas, the more refined, more 'implicate', levels of consciousness, starting from the most gross towards the most subtle, are enumerated as the four lower *jhanas,* which are 'focused' meditative states producing a more expansive and still, i.e. non-disturbed, awareness, each succeeding one being more 'deep', subtle and accompanied with deeper blissful experience.

The first four *jhanas* are activated by concentrated focused attention on a meditation object. This focused attention is used to still the mind of thought and distraction, and at a certain point of stillness the states of *jhana* arise naturally. A description of the first *jhana* for example is:

> The experience is that the pleasant sensation grows in intensity until it explodes into an unmistakable state of ecstasy. This is Piti, which is primarily a physical experience. Physical pleasure this intense is accompanied by emotional pleasure, and this emotional pleasure is Sukha (joy) which is the fourth factor of the First Jhana. The last factor of the first Jhana is Ekaggata (one-pointedness of mind). Like Sukha, this factor arises without you doing anything, and as long as you remain totally focused on the physical and emotional pleasure, you will remain in the first Jhana.[555]

As the description indicates, the *jhanas* are very advantageous states of mind.

The four lower *jhanas* are followed by the four immaterial states of awareness, so-called because they correspond to immaterial planes of reality described in Buddhist metaphysics. These immaterial ''implicate' levels of consciousness are the 'base of boundless space', the 'base of boundless consciousness', the 'base of nothingness', and the 'base of neither-perception-nor-non-perception'. This final refined immaterial meditative state, the 'base of neither-perception-nor-non-perception', corresponds to that which Mensky means by his description "between consciousness and the absence of consciousness". The 'base of neither-perception-nor-non-perception' is an extremely rarefied and subtle state wherein just about all dualistic mental activity and factors are absent, however, there remains an expanse of nondual awareness. One description of the 'base of neither-perception-nor-non-perception' is:

> It is quite difficult to discuss because there is very little to discuss. Perception is a translation of the word 'Sanna' which refers to the categorizing, naming function of the mind. Hence in this state there is very little recognition of what's happening, yet one is also not totally unaware of what's happening. It is a very peaceful, restful state and has the ability to recharge a tired mind. It is entered from the Seventh Jhana by letting go of all the outward, infinite expanse and coming to rest in what seems to be a very natural calm quiet place. The mind seems to know a lot more about how to find this space than can be verbalized.[556]

Perhaps this is the point at which the first separation within the quantum ground occurs, which Mensky describes as:

> ...the most primitive, or the most deep, level of consciousness, differing perceiving from not perceiving.[557]

The next passage, which outlines a similar perspective from the point of view of an advanced Buddhist practitioner, is taken from the writings of the eighteenth century Tibetan Yogi Jigma Lingpa (1730-98). It describes how the *alayavijnana*, the ground-store-consciousness which underlies the arising of dualistic phenomena, itself arises from the deeper level of completely nondual, undifferentiated base-awareness, the *alaya*:

> When the alaya's own dynamic manifestation moves out from it, and awareness begins to enter its object, the alayavijnana rises up. It is as if the sensory elements of the alaya are awakening from a deep sleep. The objects that are grasped, the five sense objects, do not yet arise as substantial things, but a very subtle awareness that grasps them does rise up.[558]

The correspondence with Mensky's presentation is clear, here Jigma Lingpa describes the glimmering of the nondual *alaya* into its first level of separation - the *alayavijnana*.

A remarkable feature of such descriptions by accomplished Yogis is that they are descriptions based upon direct experience of these subtle levels of consciousness. The *alaya* is the completely non-differentiated ground of potentiality and awareness. The *alayavijnana*, the ground-store-consciousness, arises from the *alaya* as the separation towards individuated consciousness arises. The ground-store-consciousness hovers between nonduality and duality and carries the seeds of duality within it, it hovers on the edge of what Mensky calls "the separation of alternatives".

The following is a description of the meditative state of stilling the mind to the point where it rests within the base-ground or 'substrate consciousness', or 'all-base consciousness', given by the Third Karmapa, Rangjung Dorje:

> When the mind rests calmly within equipoise and meditates, when it relaxes and becomes stable, and when there arise barely any thoughts, then the consciousness rests within the mere clarity aspect of the all-base. Through relaxing the mind, mind consciousness and the all-base consciousness rest together in

mere continuous clarity aspect of mind. The clarity aspect of mind is an unfabricated cognition, a mere perception (sheer cognition) – the true nature of mind ... When the coarse thoughts of the mind consciousness are pacified, it comes to rest in the all-base, and both abide within the true nature of mind.[559]

Such direct experiences of the implicate levels of the holomovement are accessible through meditation practice.

As we have seen, the first three bardos are all related to an individual's embodied lifetime, so the second two are aspects of the first - the bardo of life. With the next bardo, the bardo of dying, we meet another of the three major bardos. The scholar and practitioner of *Nyingma* and *Kagyu* Tibetan Buddhism Dzogchen Ponlop describes this:

> In the bardo of this life, we may be very earnest in our contemplation of mind's ultimate nature, and try with great effort to gain some experience of it through meditation. At the time of death, however, this very experience arises effortlessly. When we finally reach the point of the dissolution of all dualistic appearances, we experience a moment of complete awareness, a moment of vivid clarity. It's like a shift in the weather, when the sky clears up; the dense covering of clouds is gone, and suddenly we see the vast sky. At this moment, mind arrives directly at its own ground. It's is just like coming home.[560]

The death process itself, according to Tibetan Buddhism, occurs through a dissolution through various levels: earth to water, water to fire, fire to wind, wind to consciousness. The terms 'earth','water', fire, 'wind', of course, do not indicate their literal denotations, they are labels given to energy fields becoming increasingly more subtle. In terms of the Bohmian worldview this is a description of how the death process is a dissolution through increasingly more subtle 'implicate' levels, moving towards the ground state of the process of reality.

According to Tibetan Buddhism the explicate 'physical' body is dependent upon, and is formed upon, a more subtle level of organisation, a level and type of organisation very similar to the notion of morphogenetic fields:

> From the perspective of Mahamudra, Dzogchen and the Vajrayana traditions, our ordinary physical body, composed of the five elements, is the "coarse body." The "inner essential

body," also known as the "subtle body," or the "vajra body," is not visible to the eye. The subtle body is composed of channels, or *nadis*, winds or energies, *pranas*, and essences of the physical body, or *bindus*. The channels are pathways through which the subtle energies, or winds, move. The winds carry the essences of the physical body.[561]

The death process involves a reverse pathway back through subtle levels. In the final stages of the death process the mind begins to dissolve, in the sense that coarser implicate levels of mind cease and subtler minds become manifest. Finally all that appears is a vacuity filled by blackness, during which the person eventually becomes unconscious. In time this is cleared away, leaving a totally clear emptiness, which is the mind of clear light. This is the final vision of death.

This description of the various internal visions described in the *Tibetan Book of the Dead* correlates to some extent with the literature on near-death experiences. People who have had a near-death experience often describe moving from darkness, for example a black tunnel, towards a brilliant, peaceful, loving light. A comprehensive study comparing death and near-death experiences of Tibetans and Euro-Americans has shown many similarities between the two. When the clear light vision ceases, the consciousness leaves the body and passes through the stages of dissolution in reverse order. As soon as this reverse process begins the person is reborn into an intermediate state between lives, with a subtle body that can go instantly wherever it likes, move through solid objects etc., in its journey to the next place of rebirth.[562]

So, a deceased individual, a Bohmian 'sub-totality' or substructure of the holomovement of reality, whose mind does not recognise and dissolve into the clear light expanse of the ultimate nondual nature of the process of reality, continues as an energy-potentiality for a future rebirth. Thus, a sentient organism which is a 'sub-totality' within the holomovement cycles through the implicate and explicate levels of the holomovement.

Bohm's view of the implicate and explicate orders of the holomovement, and also Mensky's EEC and QCC psycho-metaphysical account of the process of reality which we explored in the previous chapter, map in dramatic and spectacular fashion into the spiritual psycho-metaphysics of the Buddhist worldview. What Bohm called the holomovement, and what in Mensky's terminology is the 'quantum world', the "entire set of parallel worlds," is in Buddhist terminology the *Dharmadhatu*, the sphere of all phenomena, and the *Dharmakaya*, the absolute 'truth body' of the process of reality. In this worldview, the mind of an enlightened being becomes co-extensive with the nondual domain of universal consciousness-awareness.

This process of the dissolution of winds can be practised by practitioners of Highest Yoga Tantra, in which meditation techniques are employed to experience the death-process by someone who is not actually dying:

> It is therefore essential for the practitioner to know the stages of death and the mind-body relationship behind them. The description of this is based on a presentation of the winds, or currents of energy, that serve as foundations for various levels of consciousness, and the channels in which they flow. Upon the serial collapse of the ability of these winds to serve as bases of consciousness, the internal and external events of death unfold. Through the power of meditation, the yogi makes the coarse winds dissolve into the very subtle life-bearing wind at the heart. This yoga mirrors the process that occurs at death and involves concentration on the psychic channels and the channel-centres (*chakras*) inside the body.[563]

As one practitioner says:

> The thing about the dissolution of the winds is it is experientially verifiable and not just by enlightened beings - I've done it and who am I? Conceptual metaphysical subtleties can be ignored even if they are understood by people - they can dismiss the view as mere opinion. But an in-depth experience takes us from the reality of the everyday world to an experience of the non-dual luminous fundamental ground (albeit via a generic image until the last obstructions to omniscience are cleaned up) and back again - the whole universe (so to speak) dissolves (and our self with it) inwardly and then 'materialises' again, stage by stage. If you could make someone do this (implicate and explicate) they would know that Bohm was right ... They would know in moments. It proves the metaphysics.[564]

Such a view of the process or reality, and the experiential practices based upon it, are completely consistent with Bohm's view of the holomovement within which sentient beings are explicated into the everyday world through deeper layers of implicate mind-energy. The yogic practice of the 'dissolution of winds' is actually a practice of experientially becoming familiar with implicate levels of embodiment.

The following passage from Bohm clearly supports the Buddhist Yogic view that the human organism is manifested into explicate reality through layers of implicate subtle energy-potentiality structures:

> Rather, we are proposing that as the notion of the holomovement was enriched by going from three dimensional to multi-dimensional implicate order and then to the vast 'sea' of energy in 'empty' space, so we may now enrich this notion further by saying that in its totality the holomovement includes the principle of life as well. Inanimate matter is then to be regarded as a relatively autonomous sub-totality in which, at least as far as we now know, life does not significantly manifest. That is to say, inanimate matter is a secondary, derivative, and particular abstraction from the holomovement (as would also be the notion of a 'life force' entirely independent of matter). Indeed, the holomovement which is 'life implicit' is the ground both of 'life explicit' and of 'inanimate matter', and this ground is what is primary, self-existent and universal.[565]

In this holomovement of implicate and explicate orders, sentient beings are 'relatively autonomous sub-totalities' which cycle through the holomovement, enfolding into and unfolding from implicate levels over numerous lifetimes until they "dissolve in this flow." Such a description is reminiscent of the Buddhist *dharmadhatu / dharmakaya*.

When enlightenment occurs the particular being who becomes a buddha no longer has a stream of consciousness which takes rebirth, a buddha's mind-energy dissolves into infinite bliss of the empty nondual ground *dharmadhatu*. According to Mensky he or she will become co-extensive with the Alterverse. In Buddhist terminology the Alterverse in its enlightened form is called the *Dharmakaya*, the ultimate 'truth body' which is "the space-like intrinsic awareness unstained by samsara." Samsara is the cycle of dualistic conditioned existence, repeated death and rebirth. According to Buddhist psycho-metaphysics the holomovement, or the Alterverse, is the womb of buddhahood, or enlightenment *(Tathagatagarbha)*, and all sentient beings cycle thorough it for vast periods of time until they achieve enlightenment. In the following passage, which we have previously perused, the term "Kun-gZhi" is the Tibetan for the *alayavijnana*, the ground consciousness:

> As the universal ground (*Kun-gZhi*) is the root of *samsara*, it is the foundation of all the traces, like a pond. As the *Dharmakaya* (ultimate body) is the root of *nirvana*, it is the freedom from all

> the traces, and it is the exhaustion of all contaminations... In the state of clear ocean-like *Dharmakaya,* which is dwelling at the basis, the boat-like universal ground filled with a mass of passengers – mind and consciousness and much cargo, karmas and traces – sets out on the path of enlightenment through the state of intrinsic awareness, *Dharmakaya.*[566]

As we have previously noted the *dharmakaya* is the ultimate 'personal' aspect of the *dharmadhatu*, which is the ultimate space of potentiality for all phenomena, the ultimate level that encompasses Bohm's holomovement.

According to the Mahayana Buddhist *Trikāya* doctrine, the doctrine of the three 'bodies' (*kayas*) of a buddha, buddhas manifest to perform a teaching role on the material plane, from the *dharmakaya,* through an intermediate level of the *sambhogakaya*, the 'enjoyment or bliss body', and then appears as a buddha in the apparent material level as a *nirmāṇakaya*, the 'appearance body'. Thus, Tulku Thondrup describes the transmission of esoteric Tantric teachings and practices:

> The doctrine of the three bodies is important in all aspects of tantric teachings. The transmission comes from the ultimate body, the formless absolute, empty aspect of Buddhahood, the *dharmakaya*, the body of enjoyment, the *sambhogakaya*. The latter is the first of the two form-bodies. Its radiant, transcendent form ... can be perceived only by enlightened beings. The Buddhas of the *sambhogakaya* level dwell in inconceivably vast pure lands or Buddha-fields, whereas the other expression of the form-body, the *nirmanakaya*, enters samsara and manifests in various ways in order to free beings from suffering.[567]

In the introduction to the Dzogchen text *Everything is Light* Keith Dowman explains that:

> The three dimensions of trikaya may also be conceived as three levels of a smoothly evolving display, from the zero dimension through a subtle internal dimension to the level of unitary outside and inside appearances. These levels and all intervening levels are, therefore, congruous, all containing the product of what has gone before and the potential of what comes after. For this reason a symbol representing the outer dimension may enlighten the inner dimension and all stages of manifestation in between. Likewise, since the outer also contains the secret zero-

dimension, and, of course, the secret the outer dimensions, every symbol contains the totality. Such an understanding is predicated upon an existential understanding of the truth of non-duality as the one and only natural condition of being.[568]

And, according to Reginald Ray. Professor of Buddhist Studies, the *dharma,* the various forms of Buddhist doctrine, manifests from the *dharmakaya* through the 'implicate' levels of the form-bodies:

> …while the *dharma* ultimately derives from the *dharmakaya*, in order to be apprehended by human beings, it must be mediated by the form bodies. … the Buddha is understood to have taught in his two rupakayas, either the nirmanakaya, or sambhogakaya. The Vajrayana, in particular, was taught by the Buddha in his sambhogakaya, or nonphysical, body of light. For many modern people, the idea that a person could present spiritual teachings while in a nonmaterial form is implausible. … However, within the traditional Asian and particularly Buddhist context, it is taken for granted that such things can and do happen. It is routinely assumed that accomplished meditators can move about in an ethereal or subtle body and interact with other, similarly disembodied beings.[569]

If one truly takes the ideas of David Bohm concerning implicate orders of implicit-potential structures of consciousness-energy, which ontologically precede manifestation at the explicate level, seriously, then claims such as this are entirely plausible!

Visions of Totality & Unbounded Wholeness: West & East

In his book *Wholeness and the Implicate Order* Bohm laid out a plan of action for his work:

> ...it is shown that science itself is demanding a new, non-fragmentary world view, in the sense that the present approach of analysis of the world into independently existent parts does not work very well in modern physics. It is shown that both in relativity theory and quantum theory, notions implying the undivided wholeness of the universe would provide a much more orderly way of considering the general nature of reality. ... we go into the role of language in bringing about fragmentation of thought. It is pointed out that the subject-verb object structure of modern languages implies that all action arises in a separate subject, and acts either on a separate object, or else reflexively on itself. This pervasive structure leads in the whole of life to a function that divides the totality of existence into separate entities, which are considered to be essentially fixed and static in their nature. We then inquire whether it is possible to experiment with new language forms in which the basic role will be given to the verb rather than to the noun. Such forms will have as their content a series of actions that flow and merge into each other, without sharp separations or breaks. Thus, both in form and in content, the language will be in harmony with the unbroken flowing movement of existence as a whole.[570]

Here Bohm points out that a less fragmentary vision of the process of reality is implied and required by both relativity and quantum theory.

Remarkably, this conclusion was a natural result of a process of scientific investigation which began from a basis which assumed the contrary position, the assumed view of a fragmentary world, including, from the end of the nineteenth century, a fragmentary, atomic scientific-metaphysical account of the makeup of the process of reality. As Bohm points out:

> Consider, for example, the atomic theory, which was first proposed by Democritus more than 2,000 years ago. In essence, this theory leads us to look at the world as constituted of atoms, moving in the void. The ever-changing forms and characteristics of large-scale objects are now seen as the results of changing arrangements of the moving atoms. Evidently, this view was, in

certain ways, an important mode of realization of wholeness, for it enabled men to understand the enormous variety of the whole world in terms of the movements of one single set of basic constituents, through a single void that permeates the whole of existence. Nevertheless, as the atomic theory developed, it ultimately became a major support for a fragmentary approach to reality. For it ceased to be regarded as an insight, a way of looking, and men regarded instead as an absolute truth the notion that the whole of reality is actually constituted of nothing but 'atomic building blocks', all working together more or less mechanically.[571]

It is important to note here that, although it is clearly and crucially true that in his work Bohm enthusiastically advocated a holistic metaphysical vision of the process of reality, this did not mean that he also considered an analytical approach to understanding aspects of the functioning of reality to be as, a consequence, always inappropriate. As Bohm points out, the problem does not lie with the process of analysis in itself, it lies in the metaphysical over-extension of the assumptions or the results of the analysis. To put the issue a little more emphatically than Bohm, the problem lies in the dogmatic and determined attitude which thinks and asserts that science must be in some way atomic, fragmentary, deterministic, and at least subtly materialist in flavour in order to be science.

However, it is worthwhile keeping in mind that Bohm's 1952 approach in many ways suffered exactly the errors that Bohm's later approach sought to eradicate! This was the approach that Einstein, who disliked Bohr-style quantum theory, preached to Bohm and thus inspired the 1952 theory, and which Bohm later apparently rebelled against. Thus Bohm later wrote:

> ... physics has become almost totally committed to the notion that the order of the universe is basically mechanistic. The most common form of this notion is that the world is assumed to be constituted of separately existent, indivisible and unchangeable 'elementary particles'', which are the fundamental 'building blocks' of the universe. [572]

However, it would be wrong to assume, on the basis of Bohm's criticism of the the view of the ultimate 'particle' nature of reality, that the notion that the order of the universe has a particle aspect in some context, or at some level, has absolutely no validity. The crucial point is that such fragmentary approaches have limited validity. The over-estimation of validity derives from a scientific attitude wherein scientists and philosophers take the assumption that there is a foundational level of reality which is constituted by self-enclosed atomic, or sub-atomic, indivisible units as being an ultimate truth which rules out any other perspective. In the above quote Bohm indicates that the 'atomic' viewpoint is an 'insight' that is inappropriately given absolute status. But, although atomic theory is not appropriate as an absolute and definitive account of the process of reality, it is not entirely without scientific validity, whereas a mythical account of the Earth being carried on the back of a whale, for example, clearly lacks any scientific credence.[573]

In his discussion Bohm indicates that both Einstein's relativity theory and quantum theory have undermined the absolute role of the notion of ultimate and independent fundamental particles. Bohm tells us that these two theories undermine the 'atomic' vision in different ways because the two theories "differ radically in their detailed notions of order", each has its own notion of a "fragmentary mode of existence". Therefore:

> One thus sees that a new kind of theory is needed which drops these basic commitments and at most recovers some essential features of the older theories as abstract forms derived from a deeper reality in which what prevails is unbroken wholeness.[574]

Here Bohm suggests that Einstein's relativity theory and quantum theory should be dropped as absolute and ultimate accounts of reality, but, importantly, they can be reinstated as partial or relative truths which are coherently derived from a deeper interconnected level of 'unbroken wholeness'.

Thus, it is clear that Bohm, in his later post-1952 perspective, did not consider that the ideas of Bohr and Heisenberg were irredeemably wrong, even though he had in his 1952 paper attempted to provide an alternative perspective. As we have seen, this earlier attempt on Bohm's part actually led to a necessary development and acceptance of central aspects of the Bohr-Heisenberg perspective within a much more extended context of an interconnected nonlocal universe. Indeed, in their book *The Undivided Universe* Bohm and Hiley (B&H) say that:

> The context dependence of results of measurements is a further indication of how our interpretation does not imply a simple return to the basic principles of classical physics. It also embodies, in a certain sense, Bohr's notion of the indivisibility

of the combined system of observing apparatus and observed object. Indeed it may be said that our approach provides a kind of intuitive understanding of what Bohr was saying. He described a 'measurement' as a whole phenomenon not further analysable. ... it is not implied in Bohr's treatment that these values correspond to 'beables' that exist independently of the overall experimental context.[575]

B&H indicate that Bohr's viewpoint does not assert that a quantum measurement reveals a value of a pre-existing 'beable' that is independent of the experimental context. This means that the Bohmian view is not completely at odds with the Bohr 'Copenhagen' interpretation, but is, rather, an extension and elucidation of it. The Copenhagen perspective indicates that there is no actual 'beable' pre-existing, as well as no pre-existing value. The B&H view, however, indicates that there is a 'beable' pre-existing the measurement, but that values of properties of the 'beable' do not necessarily pre-exist. It is in this aspect, the claimed pre-existence of some kind 'element of reality', that the B&H view predominately differs from Bohr's perspective.

As we saw in an earlier chapter, the phenomenon of 'contextuality' is the feature of the Bohmian quantum worldview that requires that some aspects of reality have a very non-classical response to the act of 'measurement'. Simply put, contextuality is the phenomenon wherein a value of a 'beable' can be measured alongside some other feature and one specific result is obtained, but when measured alongside another different feature a different result for the same value of the beable is obtained. And in the Bohmian perspective this completely non-classical behavior is found in all aspects except the *position* of 'beables'. Physicist Jean Bricmont says of this:

> The statement that the measurement of an observable depends on the concrete experimental arrangement used to "measure" it is true, not only for spin and momentum, ... but for all quantum mechanical "observables", other than position. ... measurements do not in general measure anything pre-existing to them.[576]

And Florian J. Boge, in his book *Quantum Mechanics Between Ontology and Epistemology*, points out that:

> The notion of beable ... stems from Bell (1976), and Bell introduced the term to denote something that *is* and *is there*, in contrast to QMs observables. Essentially, Bohm and Hiley here

> deny ... the value definiteness rule ... Moreover, they seem to *embrace* contextuality ... and trace it back to the interaction of system and apparatus ...[577]

In other words, according to B&H, there are independent 'elements of reality', which John Bell gave the designation 'beables', but, for all properties other than position, what values of properties they give when measured depends upon the context of the measurement, these properties are not fully intrinsic to the 'beables'. This is a kind of having a 'beable' whilst at the same time not allowing the 'beable' to *be* very much in terms of having any properties which define its 'being'!

John Bell defined the term 'beable':

> The beables of the theory are those elements which might correspond to elements of reality, to things which exist. Their existence does not depend on "observation". Indeed observation and observers must be made out of beables.[578]

One might perhaps have expected that the term 'beable' should correspond to the 'classical' type term 'particle', but this is not the case. What Bell was objecting to was what he perceived as Bohr's reluctance to assert the existence of some independent aspect of reality, an 'element of reality', a phrase borrowed from Einstein, which was independent of observation. However, Bell in the following paragraph from his article 'Beables for Quantum Field Theory' states that:

> I will use the term "beable" rather than some more committed term like "being" or "beer" to recall the essentially tentative nature of any physical theory. Such a theory is at best a *candidate* for the description of nature. Terms like "being", "beer", "existent", etc., would seem to me lacking in humility. In fact "beable" is short for maybe-able" [579]

The point of the term 'beable', as coined by Bell, was to replace the term 'observable', which he seemed to detest as being characterized by "vagueness, subjectivity and indeterminism".[580] But it is difficult to see how a "maybe-able" really matches up to the full glory of an "element of reality". When one reads such discussions it is difficult not to recall Heisenberg's remark concerning his view that Bohm's 1952 pilot wave theory was little more than a re-branding of terminology for ideological purposes. Such 'ideological' readjustments of terminology seem to be a common feature of quantum disputes!

Bell wrote that:

> Bohm's 1952 papers on quantum mechanics were for me a revelation. The elimination of indeterminism was very striking. But more important, it seemed to me, was the elimination of any need for a vague division of the world into "system" on the one hand, and "apparatus" or "observer" on the other. I have always felt since that people who have not grasped the ideas of those papers ... and unfortunately they remain the majority ... are handicapped in any discussion of the meaning of quantum mechanics.[581]

So it seems that Bell may have been "humble" concerning his choice of words for what he thought must be ultimate "elements of reality", but he was far from humble in his evaluation of those who were dubious about Bohm's 1952 ideas.

Andrew Whitaker points out in his book *The New Quantum Age*, "Bell remained all his life a realist".[582] But one has to ask: how 'real' is a "maybe-able"? What Bell was actually asserting was his personal conviction that there *must be* some kind of bits and pieces existing, in some way, however quantumly strange, 'out there' in the process of reality. There *must be*, damn any contrary evidence, bits and pieces existing completely independent of consciousness. Although consciousness obviously would have an influence on our experience of them, consciousness could in no way have any role in creating the ultimate 'elements of reality'. This was, however, a personal conviction, not a scientifically established fact, and this conviction influenced the way that Bell described and interpreted physical processes.

In his early efforts Bohm was guided by similar concerns, the world surely could not be conjured into existence by sentient beings merely observing some kind of sea of quantum potentiality; there must be more 'realistic' aspects of the process of reality doing their own thing in a more deterministic fashion. His 1952 perspective resulted from Bohm's impressive mathematical ability to transform equations into new forms which conformed in restricted areas to a more 'realistic' and ' deterministic' form. But, as Heisenberg pointed out regarding the 1952 perspective, this attempt to produce a deterministic account provided an "ideological superstructure" for a realist account of limited value, until further development took place.

Science writer Jim Baggott, in his recent book *Quantum Reality: A Game of Theories,* apparently concurs with Heisenberg. Baggott compares Bohm's notion of pilot waves to Ptolemy's epicycles:

> Science though the ages has made a habit of stripping irrelevant or unnecessary metaphysical elements from theories that can be shown to work perfectly well without them. Examples include Ptolemy's epicycles, phogiston, the ether ...[583]

Claudius Ptolemy was a Roman philosopher who lived around 100 AD. At the time it was thought that the Earth was the centre of the universe. But, because in fact the Sun is the centre of the Solar System, Ptolemy had to produce a complicated model of how the planets moved which involved circles moving within circles. His Ptolemaic model accounted for the apparent motions of the planets by suggesting that each planet moved on a smaller circle, called an 'epicycle', that moved within a larger circle, called a 'deferent'. Later when the correct Copernican view was established with the sun as the center of the solar system it became apparent that the epicycles were actually an unnecessary addition due to viewing the situation through a skewed viewpoint.

However, it is useful to note that it would be incorrect to say that the Ptolemaic viewpoint was completely wrong, within its own context it produced very accurate predictions of planetary positions. From an ultimate quantum viewpoint Newtonian physics is incorrect, but on its own level it is very precise. As Bohm points out:

> ...one finds (e.g., as in the case of the Ptolemaic epicycles or of the failure of Newtonian concepts just before the advent of relativity and quantum theory) that older theories become more and more unclear when one tries to use them to obtain insight into new domains. Careful attention to how this happens is then generally the main clue toward new theories that constitute further new forms of insight. So, instead of supposing that older theories are falsified at a certain point in time, we merely say that man is continually developing new forms of insight, which are clear up to a point and then tend to become unclear.[584]

A developmental process like this occurred to Bohm's own 1952 attempt to produce a realist and deterministic account of quantum reality. The predicted paths of particles, which were originally conceived of 'real' independent particles, along with the guiding wave, being in a sense analogous to Ptolemy's epicycles. In the above quote we see that Bohm, in his later view, considered that the atomic and Newtonian view was perhaps analogous in some sense to Ptolemaic epicycles, so this would also have to apply to his 1952 'realist' theory. Baggott points out that the de-Broglie-Bohm theory could be so described because of its adherence to a quest for 'realism' and determinism which fits with classical intuitions, according to Baggott the de-Broglie-Bohm theory:

> ...is in all respects equivalent to quantum mechanics and yet it allows a profoundly different interpretation of events occurring at the quantum level, one which is more in tune with our more intuitive metaphysical preconceptions about the way reality ought to work.[585]

An observation which clearly endorses Heisenberg's "ideological superstructure" claim. But Baggott also points out:

> Those utilitarian physicists less concerned about causality and determinism, and less obsessed with interpretation and meaning, would have quickly dispensed with the theory's superfluous elements in the interest of more efficient calculation.[586]

An observation which perhaps does not reflect well on such physicists' concern for a full understanding of reality.

But, as Bohm saw the flaws in the claim of absolute realism, he developed his views much further, his view of such 'particles', which in *The Undivided Universe* are called 'beables', changed significantly. Subsequently, in *Wholeness* Bohm writes that:

> ...our ordinary notions of space and time, along with those of separately existent material particles, are abstracted as forms derived from the deeper order. These ordinary notions in fact appear in what is called the explicate or unfolded order,...[587]

And:

> Thus, the quantum theory shows that the attempt to describe and follow an atomic particle in precise detail has little meaning. ... The notion of an atomic path has only a limited domain of applicability. In a more detailed description the atom is, in many ways, seen to behave as much like a wave as a particle. It can perhaps best be regarded as a poorly defined cloud, dependent for its particular form on the whole environment, including the observing instrument.[588]

This shows a remarkable shift in perspective, especially when one considers Bell's enthusiastic endorsement of the claimed "elimination of any need for a vague division of the world into 'system' on the one hand, and 'apparatus' or 'observer' on the other" in Bohm's 1952 proposal. It is clear that Bohm must have felt the 1952 proposal had internal stresses which impelled him back into a more Bohrian perspective, a Bohr type perspective within a spectacular interconnected universal background of fundamental wholeness.

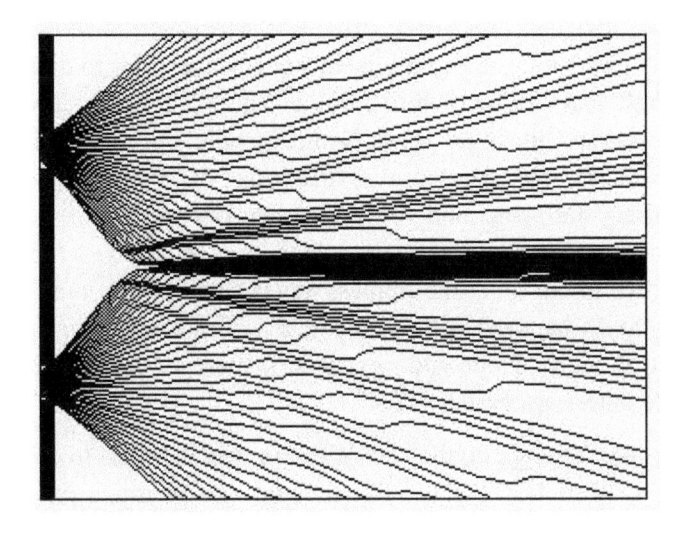

In order to understand this transition let us consider the experiment which verified the 'particle' paths in a two slit experiment as predicted by the Bohm equations as shown in the illustration. This diagram is the result of experimental researchers working at Birkbeck College in London in 1978 who ran a computer simulation of the Bohm equation. The non-Bohmian explanation for the manner in which the electrons behave is that the electrons pass through the slits as quantum mists of potentiality which interfere beyond the slits to produce stripes of electron density at the receiving apparatus. In the Bohmian worldview electrons are always 'particles', or 'beables', which are guided by a pilot wave, which is also called the 'quantum potential'. In the Bohm simulation the path taken depends upon the initial position of the electron. Jim Baggott tells us that "the resulting picture provoked gasps of astonishment".[589]

For someone who is emotionally invested, as John Bell evidently was, in a really 'real' reality, so to speak, then the 'particles' which seem to be taking part in the Bohmian dance of quantum paths must each be an 'element of reality' which actually exists as a self-enclosed bit of 'stuff', what in Buddhism is termed 'inherently-existent' bits of external reality. They certainly should not be merely appearance to the minds of observers. And it seems that Bohm in his 1952 Marxist-tainted mind actually thought something like this. But, as we have seen, by 1980, when *Wholeness and the Implicate Order* was published, he had changed his mind radically. In Bohm's later perspective 'things', or 'particles', emerge and dissolve, like thoughts, from a flowing background of potentiality:

> This view implies that flow is, in some sense, prior to that of the 'things' that can be seen to form and dissolve in this flow. One can perhaps illustrate what is meant here by considering the 'stream of consciousness'. This flux of awareness is not precisely definable, and yet it is evidently prior to the definable forms of thoughts and ideas which can be seen to form and dissolve in the flux, like ripples, waves and vortices in a flowing stream. As happens with such patterns of movement in a stream some thoughts recur and persist in a more or less stable way, while others are evanescent.[590]

In this move of thought on the part of Bohm, which began to conceive of 'particles' or 'beables' as more like thought-like emergent phenomenon appearing from within a deeper background flowing interconnected stream, Bohm began to approach his notion of 'active information'. In 1990 Bohm presented a paper, 'A new theory of the relation of mind and matter', published in the interdisciplinary journal *Philosophical Psychology*. In this vision Bohm conceived of an electron as as inseparable union of a both particle and a field. This view, that a particle, or 'beable', is both particle and field at the same time, is in contrast to the conventional view of 'quantum complementarity', established by Bohr, that it is either a wave or a particle, but not both at the same time.

The field aspect, in Bohm's new 'ontological' perspective, which is a dramatic reworking and revisioning of the 1952 pilot-wave approach wherein 'particles' were considered to be independent, rather than being simultaneously both wave and particle, is conceived of by Bohm as being a field of 'active information'. F. David Peat characterizes the role of information in Bohm's hidden variables pilot-wave perspective:

> The breakthrough in giving information a more "physical" role comes with Bohm's proposal that information plays an active

role in quantum systems. Bohm's 1952 Hidden Variable papers proposed an alternative approach to quantum theory in which the electron is a real particle guided by a new kind of force, the quantum potential. While at first sight Bohm's theory appears somewhat "classical" - electrons have real paths - the quantum potential is entirely novel. Unlike all other potentials in physics its effects do not depend upon the strength or "size" of the potential but only on its **form**. It is for this reason that distant objects can exert a strong influence on the motion of an electron.[591]

Thus the 'active information' contained within the guiding wave results from the non-local interconnected configuration of the entire universe.

On Bohm's later view, however, particles are not separate entities pushed around by the field but, rather, momentary pulses within the field. In quantum field theory particles are theoretically 'created' by applying the mathematical rules of the theory, but there is no further elucidation of any deeper causal process that indicates why such particles manifest from the underlying quantum field. Bohm suggests that there is a deeper level which he calls the 'super-quantum potential':

> ...the actuality will be the entire field over the entire universe ... we find that these equations are modified by what I called the super-quantum potential. This is related to the activity of the entire field as the original quantum potential was to that of the particles.[592]

In other words, in the early pilot-wave proposal the pilot-wave, which embodied the 'quantum potential', pushed the separately existing particles around the universe, but, in the new proposal, the new 'super-quantum potential' is able to 'inform' and thereby control the lower level quantum field, in an analogous way to which the pilot-wave controlled particles. In this new perspective, however, the high level 'informs', or gives form to, the lower level quantum field.

In conventional quantum field theory 'particles' are momentary appearances, or 'creations', produced by the interactions of wave aspects of quantum fields. As Bohm describes:

> ... solutions of the field equations represent waves that spread out and diffuse independently. ... there is no way to explain the origination of the waves that converge to a region where a

particle like manifestation is actually detected, nor is there any factor which could explain the stability and sustained existence of such a particle-like manifestation.[593]

Physicists manipulate their equations to gain solutions which represent waves which interact to produce 'particle-like' manifestations, but there is no explanation as to how and why such particle-like manifestations appear exactly in the way that they do in the real world. Bohm's 'super-quantum potential' is conceived of as solving this issue:

> ...this lack is just what is supplied by the super-quantum potential. Indeed, ... the non-local features of the latter will introduce the required tendency of waves to converge at appropriate places, the non-linearity will provide for the stability for the occurrence of the whole process. And thus we come to a theory in which not only the activity of the particle-like manifestations, but even their actualization, e,g. their creation, sustenance, and annihilation, is organized by the super-quantum potential.[594]

As Peat says:

> Later versions of Bohm's theory pictured the electron not so much as a real physical particle but as a process, a wave continually collapsing inward to a localized region and then expanding outward. This process is guided by a super-quantum potential. An activity of information is responsible for the existence of quantum particles and quantum events.[595]

Thus we see that the 'active information' contained within the 'super-quantum potential' 'informs' and organizes the the operation of the underlying fields. Bohm writes concerning this:

> So now we see quite generally that the whole universe not only determines and organizes its sub-wholes, but also that it gives form to what now has been called elementary particles out of which everything is supposed to be constituted. What we have here is a kind of universal process of constant creation and annihilation, determined through the super-quantum potential so as to give rise to a world of form and structure, in which all manifest features are only relatively constant, recurrent and stable aspects of the whole.[596]

So, the sub-wholes which are biological organisms are organized via the super-quantum potential acting through the lower level quantum fields.

This applies not only to the material aspects of their 'relatively constant' structure, but also the internal mental structures which also develop through the activity of the super-quantum potential which is a portal for active information pervading the universe. Bohm indicates that the two aspects, material and mental, develop in an interrelated manner:

> ... there is a kind of active information that is simultaneously physical and mental in nature. Active information can thus serve as a kind of link or "bridge" between these two sides of reality as a whole. These two sides are inseparable, in the sense that the information contained in thought, which we feel to be on the "mental" side, is at the same time a related neurophysiological, chemical, and physical activity...[597].

And, furthermore, Bohm goes beyond this assertion of a "two sides of reality" position. In fact, in the above quote Bohm is actually speaking of the explicate level wherein mind and brain operate. In *Wholeness* Bohm clearly indicates that "consciousness has to be understood in terms of an order that is closer to the implicate than it is to the explicate".[598] And Bohm in several places indicates that the implicate levels are more mind-like than matter-like:

> It is thus implied that in some sense a rudimentary mind-like quality is present even at the level of particle physics, and that as we go to subtler levels, this mind-like quality becomes stronger and more developed.[599]

Paavo Pylkkänen, in his book *Mind. Matter and the Implicate Order*, says of this:

> This reveals that Bohm accepts some kind of "panpsychism" or "panprotopsychism". Reality is deep down "psycho-physical". If we abstract a part from this reality, there will be some trace of this basic psycho-physical quality in it. If we have abstracted the "elementary particles" of physics, they are not completely physical, but demonstrate a rudimentary mind-like quality through their quantum field and the active information it contains.[600]

In fact, in Bohm's new 'ontological interpretation', mind and active information become primary aspects, residing at the deepest levels, of the process of reality. The term 'ontological' here indicates that the super-quantum potential and the quantum fields that it informs and configures through the active information it contains are the primary aspects of the process of reality; 'particles' are derived aspects. The entire process, however, is considered to be 'real'. Furthermore, mind is also a primary, ontological, aspect and quality of the inner implicate levels of quantum nature of the process of reality. As Bohm writes:

> One may ... describe the essential mode of relationship of all these as participation ... Through enfoldment, each relatively autonomous kind and level of mind to one degree or another partakes of the whole. Through this it partakes of all the others in its "gathering" of information, And through the activity of this information, it similarly takes part in the whole and in every part. It is in this sort of activity that the content of the more subtle and implicate levels is unfolded ... e.g., as the movement of the particle unfolds the meaning of the information that is implicit in the quantum field ...[601]

And this means that, as Bohm indicates:

> ...each human being similarly participates in an inseparable way in society and the planet as a whole. What may be suggested further is that such participation goes on to a greater collective mind, and perhaps ultimately to some yet more comprehensive mind in principle capable of going indefinitely beyond even the human species as a whole. (This may be compared with some of Jung's ... notions.)[602]

Jung's most significant notion in this context is that of the 'collective' unconscious'. Jung introduced the distinction between what Bohm would have termed the more 'explicate', 'personal', Freudian-type unconscious, and the deeper "collective" unconscious which comprises the common mental structures underlying the minds of all human beings, and even deeper than that, the collective mental structures for each species.

It is important to be aware that the term 'unconscious' in this context refers to 'implicate' levels of consciousness which are not readily or immediately available to the completely 'conscious' everyday mind. Even the 'conscious' everyday mind operates as a focus within the highest implicate level of the unconscious background; the conscious focus scans through contents which are at the surface of the individual implicate level of mind. But at deeper levels of implicate mind, as Jung wrote:

> ... in dreams, fantasies, and other exceptional states of mind ... mythological motifs and symbols can appear autochthonously at

any time, often, apparently, as the result of particular influences, traditions, and excitations working on the individual, but more often without any sign of them. These "primordial images" or "archetypes", as I have called them, belong to the basic stock of the unconscious psyche and cannot be explained as personal acquisitions. Together they make up that psychic stratum which has been called the collective unconscious.[603]

And:

> ... in addition to our immediate consciousness, which is of a thoroughly personal nature and which we believe to be the only empirical psyche ... there exists a second psychic system of a collective, universal, and impersonal nature which is identical in all individuals. This collective unconscious does not develop individually but is inherited. It consists of pre-existent forms, the archetypes, which can only become conscious secondarily and which give definite form to certain psychic contents.[604]

Here Jung indicates the fact of the common implicate mental background, which pre-exists as a set of common structures underlying individual minds. For the human species this provides the common mental structures for all human beings, structures which are given individual content by the life-experience of each being. There must also be common mental implicate mental structures underlying the minds of all other species, which accounts for instinctual behavior for example. And, of course, there are common physical organic structures which are potential at the quantum level which form the templates for the organic structure of all the species.

Jung had conducted a lifelong investigation of the symbolic and mythological material of the world's diverse cultures and as a result he was able to demonstrate that there are recurring themes and motifs which were exemplified in different specific features of culture and psychic and physical embodiment. In his work as a psychologist Jung was primarily concerned with working with archetypes, which are deep psychic structures of potentiality, which were relevant to the integration of the psychic functioning of his patients. But Jung also extended his interest to deeper religious and philosophical levels in his investigations into Alchemy with its emphasis on the interpenetration of psyche and the material world. This led to his significant concept of the *Unus Mundus*, the view of the physically, psychologically and spiritually deeply unified, interconnected and layered 'One World'. Jung also explored this perspective, in collaboration with the quantum physicist Wolfgang Pauli, in the context of the interpenetration of psyche and the physical world apparently revealed by quantum theory. Pauli wrote concerning his views to the Swiss physicist Markus Fierz in 1948:

> What now is the answer to the question as to the bridge between the perception of the senses and the concepts, which is now reduced to the question as to the bridge between the outer perceptions and those inner image-like representations. It seems to me one has to postulate a cosmic order of nature -- outside of our arbitrariness -- to which the outer material objects are subjected as are the inner images...The organizing and regulating has to be posited beyond the differentiation of physical and psychical... I am all for it to call this 'organizing and regulating' 'archetypes'. It would then be inadmissible to define these as psychic contents. Rather, the above-mentioned inner pictures (dominants of the collective unconscious, see Jung) are the psychic manifestations of the archetypes, but which would have to produce and condition all nature laws belonging to the world of matter. The nature laws of matter would then be the physical manifestation of the archetypes.[605]

Such a perspective clearly has a deep resonance with the later work of Bohm.

According to the most fundamental metaphysical insights of quantum discoveries all possible material organic structures exist as potentialities at the dawn of time. And this important insight has begun to gain a glimmer of attention in the realm of biology. As the biologist Adrian Woolfson in his book *Life Without Genes*:

> In the beginning there was mathematical possibility. At the very inception of the universe fifteen billion years ago, a deep infinite-dimensional sea emerged from nothingness. Its colourless waters, green and turquoise blue, glistened in the non-existent light of the non-existent sun ... A strange sea though, this information sea. Strange because it was devoid of location ...[606]

This is a poetic invocation of the infinite pool of quantum potentiality which resides at the deepest implicate order or level of the process of reality, this is the quantum repository for all 'archetypal' active information that will provide the basis for the unfolding of life over infinite time scales of enfolding and unfolding universes.

This new evolutionary viewpoint, which clearly undermines the crude Darwinian materialism which pervades the field of biology, has been validated by the Evolutionary-Developmental, or 'Evo-Devo', revolution in biology. As the Evo-Devo biologist Sean B. Carroll, in his book *Endless Forms Most Beautiful: The New Science of Evo Devo*, writes:

> The first shots in the Evo Devo revolution revealed that despite their great differences in appearance and physiology, all complex animals - flies and flycatchers, dinosaurs and trilobites, butterflies and zebras and humans - share a common "tool kit" of "master" genes that govern the formation and patterning of their bodies and body parts. ... The important point to appreciate from the outset is that this discovery shattered our previous notions of animal relationships and of what made animals different, and opened up a whole new way of looking at evolution.[607]

In other words, all animals, of whatever species whatsoever, share a fundamental genetic structure which derives from a hierarchical development of differentiation. This implies that there is a hierarchical set of universal virtual quantum possibility body-plans which underlie the manifestation all species. This is precisely what one would expect if there is a deep quantum-Platonic field of potentiality underlying the development of species, a view consistent with, and implied by, the Bohmian Implicate Order perspective. As Bohm says:

> So now we see quite generally that the whole universe not only determines and organizes its sub-wholes, but also that it gives form to what has until now been called the elementary particles out of which everything is supposed to be be constituted. What we have here is a kind of universal process of constant creation and annihilation, determined through the super-quantum potential so as to give rise to a world of form and structure, in which all manifest features are only relatively constant, recurrent and stable aspects of this whole.[608]

The universe has its own internal mechanism of enfoldment and unfoldment which means that "relatively constant, recurrent and stable" organic structures embodying fragments of universal consciousness, all sentient beings, come into explicate existence for limited periods of time and then are reabsorbed into the holomovement of the process of reality.

This Bohmian overarching view of sentient beings coming into limited 'existence' and then being reabsorbed back into the flow of the holomovement is very close to the Buddhist core vision of *samsara*, 'cyclic existence', with its emphasis on cyclic rebirth of a multitude of psycho-physical streams of manifestation. It is important to comprehend that in Buddhism that all moments of both the 'physical' (Pali - *rupa*, 'form') and 'mental' (Pali - *nama*, 'name') are momentary and each occasion of this pair is conditioned by the previous occurrence. On the basis of this 'enfolding' and 'unfolding' within the flow of a continuum of 'name and form' the appearance and experience of "relatively constant" sentient beings comes into being. Indeed the Buddhist paired categories of *nama-rupa* bears a close relationship with Bohm's notion of 'soma-significance':

> The notion of soma-significance implies that soma (or the physical) and its significance (which is mental) are not separate; rather they are two aspects of one overall reality.[609]

And:

> ...soma-significance means that the soma is significant to the higher or more subtle level. Signa-somatic means that significance acts somatically towards a more manifest level. ... in the unfoldment of matter there is a kind of soma-significance; that the soma may be significance and as a result something else unfolds.[610]

The unfoldment of the material substructures of the holomoving process of reality, substructures which embody the significance of fragments of universal consciousness, then, operates from more mind-like subtle implicate levels towards the explicate levels. Furthermore, manifestations at the explicate level condition further future manifestations in a karmic-like process.[611] Each sentient being 'unfolds' a continuum of experience from out of the holomovement-ground-consciousness which takes place within the mind-nature of the quantum implicate order. For Bohm the process of reality is the unfolding of an experienced world from the potentialities within the holomovement; the unfolding of lived experience from the implicate order.

The evidence for reincarnation (rebirth in a human or animal body, according to Buddhism rebirth can also be into immaterial planes of reality) today is so striking that only the disinterested and determined sceptics could avoid the implication of the evidence. For example the evidence presented by Ian Stevenson in his books such as *Where Reincarnation and Biology Intersect* is extraordinary. Especially remarkable is the dramatic evidence of recall of past lives by young children who have birthmarks that correspond to recalled violent deaths in the previous life.[612] In 2007 Jim B. Tucker published a paper entitled *'Children Who Claim to Remember Previous Lives: Past, Present and Future Research'*. In this paper he described the following case as typical:

> Kumkum Verma, a girl in India, is an example of the subjects that Stevenson studied (Stevenson, 1975). She was from a village, but when she was 3-1/2 years old, she began saying that she had lived in Darbhanga, a city of 200,000 people that was 25 miles away. She named the district of the city where she said she had lived, one of artisans and craftsmen, and her family did not know anyone from that district. Kumkum made numerous statements, and her aunt wrote down many of them. Some of her notes were lost, but Stevenson was able to get a copy of 18 of Kumkum's statements that her aunt had recorded. The detail in these statements included her son's name in the life she was describing and the fact that he worked with a hammer, her grandson's name, the town where her father had lived, and personal details, such as having an iron safe at home, a sword hanging near the cot where she slept, and a pet snake that she fed milk to. Kumkum's father talked to a friend who had an employee from the district in Darbhanga that Kumkum had mentioned. The employee went there to search for the deceased individual, the previous personality, that Kumkum was describing. He found that a woman had died five years before Kumkum was born whose life matched all of the details listed above. Of note is the fact that Kumkum's father, a landowner and homeopathic physician, visited the family in Darbhanga once but never allowed Kumkum to see them, apparently in part because he was not proud that his daughter seemed to remember the life of a blacksmith's wife.[613]

Stevenson began his research with an open mind and was impressed by the weight of evidence. A list of typical features of the cases collected by Stevenson include:

- Claims were made spontaneously at very young age, often starting at 2-3 years old and stopping by 6-7 years.
- Median length of time between death of previous personality and rebirth is 15 months.

- Claims involve ordinary lives, rather than grandiose fantasies. Sometimes the claim involves a previous life in humbler circumstances.
- The details which are given of previous lives are matched with evidence to a high degree.
- In about 70% of cases the mode of death involves unnatural, violent and sudden death.
- Many children show behaviours connected with the previous life.
- Many children showed emotional attachment to previous family members appropriate to the former relationship.
- In cases involving violent death over 35% of children displayed phobias related to the mode of death.
- Many of the children practised repetitive play linked to the previous life, acting out the occupation of the previous personality and occasionally re-enacting the death scene.

Cases such as these have been found wherever researchers have looked for them and they have been found on all continents except Antarctica, where no research has been done. Perhaps the most remarkable feature of Stevenson's research lies in his discovery of cases in which the child has birthmarks which correspond to the violent death of the previous person.

Within Buddhism the truth of rebirth is part of the Twelve Links of Dependent Origination (*Paticcasamuppada*). These links can clearly be seen as aspects of the enfoldment and unfoldment within the cyclic holomovement described by Bohm. These links have been described previously (pages 216-218). Within the Buddhist metaphysical worldview it is clear that the basic perspective involves an infinite mind-like pool of potentiality for experience which is triggered into manifestation by a deep internal 'craving' for dualistic experience within the 'explicate' world:

> Encircled with craving, people hop round & around like a rabbit caught in a snare. Tied with fetters & bonds they go on to suffering, again & again ... So a monk should dispel craving, should aspire to dispassion for himself.[614]

And here, with the Buddhist indication that the generation of a unsatisfactory mode of existence is caused by a mistaken mode of thought, we find ourselves projected back into some central insights of Bohm's thought. Although some details of the analysis differ, there are also significant similarities.

According to Bohm:

> Our fragmentary way of thinking, looking, and acting, evidently has implications in every aspect of human life. That is to say, by a rather interesting sort of irony, fragmentation seems to be the one thing in our way of life which is universal, which works through the whole without boundary or limit. This comes about because the roots of fragmentation are very deep and pervasive.[615]

In fact the the deepest root of the fragmentation clearly lies in the operation of the universe itself, for it seems to be the case that universe has its own internal mechanism of unfolding, but in this unfolding there is an unavoidable 'fragmentation'. As Bohm and Hiley indicate in *The Undivided Universe*:

> In a certain sense we could say that the overall quantum world measures and observes *itself*. For the classical 'sub-world' that contains the apparatus is inseparably contained within the subtle quantum world, especially through those nonlocal interactions that bring about the classical behaviour. In no sense is the 'observing instrument' really separate from what is observed. The relative autonomy of the classical level ... makes it possible for the total quantum world to manifest and reveal itself within itself in a measurement. Thus we should regard a quantum measurement as a manifesting process.[616]

In this extraordinary insight Bohm and Hiley endorse a version of John Wheeler's 'self-perceiving universe', a version which drops the claim that the creative self-perceiving activity operates backwards in time. B&H are indicating that the holomovement, the process of the universe, is driven at its deepest level by an internal mechanism of manifestation, driven by a natural mechanism of internal 'measurement', a mechanism of manifestation which operates in a way that naturally unfolds an explicate world of sentient beings and their environments.

Bohm indicates that this self-creative self-referring, or self-perceiving function is an aspect of 'Meaning', which Bohm conceives of as a creative, qualitative experiential energy that is intrinsic to the process of reality; it is a fundamental inner quality of the universe:

> Rather than ask what is the meaning of the universe, we would have to say that the universe *is* its meaning. ... And of course, we are referring not just to the meaning of the universe for us, but its meaning 'for itself', or the meaning of the whole for itself.[617]

And physicist Henry Stapp has made similar assertions:

> ...the quantum universe tends to create meaning: the quantum law of evolution continuously creates a vast ensemble of forms that can act as carriers of meaning; it generates a profusion of forms that have the capacity to sustain and refine themselves.[618]

Bohm presents meaning as the central stuff of reality. The three interrelated aspects of reality which Bohm isolates as fundamental for understanding experience are energy, matter, and meaning. Each of these three enfolds and implies the others. But meaning seems to be central. Without this inner quality of meaning being intrinsic to the universe from the start the universe could never mean anything, to itself or to anything within it. The function of meaning, then, can be looked at as the central source of the experiential polar aspects of mind and matter. Matter is an appearance of objective meaning to mind, and individuated consciousness, or awareness, is the ground of subjectively experienced meaning.

According to Bohm 'meaning' can be considered to be the most fundamental aspect of reality because it enfolds the other aspects, matter and mind, and it can also enfold itself, which is to say that it is possible to have multiple levels of meanings; higher level meanings can relate together meanings on levels beneath it, and so enfold them into a unity. So because the qualitative aspect of 'meaning' enfolds all three aspects, including itself, it is the fundamental aspect of experience:

> ... meaning refers to itself directly and this is in fact the basis of the possibility of that intelligence which can comprehend the whole, including itself. On the other hand, matter and energy obtain their self-reference only indirectly, first through meaning.[619]

Meaning makes all comprehension and understanding possible. And it also underlies the creation of the appearance of the material world. The quality of 'meaning', which results from the functioning of fundamental consciousness-awareness-energy, is able to refer to itself and thus can act as a basis for a self-referring or self-perceiving universe.

However, the fact that the universe manifests itself into an explicate order containing substructures which are perceiving, thinking beings; organic, conscious beings who, of necessity, must operate and survive in a fragmentary world, a world fragmented by the self-manifestation of the universe itself, means that these manifested sentient beings necessarily tend to think and operate in a fragmentary manner. As Bohm says:

> Of course, the prevailing tendency in science to think and perceive in terms of a fragmentary self-world view is part of a larger movement that has been developing over the ages and that pervades almost the whole of our society today: but, in turn, such a way of thinking and looking in scientific research tends very strongly to re-enforce the general fragmentary approach because it gives men a picture of the whole world as constituted of nothing but an aggregate of separately existent 'atomic building blocks', and provides experimental evidence from which is drawn the conclusion that this view is necessary and inevitable. In this way, people are led to feel that fragmentation is nothing but an expression of 'the way everything really is' and that anything else is impossible.[620]

One of the significant motifs in Bohm's later work is that most human thinking and behaviour is undermined and made false in a significant way because of the fragmentary nature of human research, thought, scientific enterprises, and institutions. However, it is also apparent that, according to Bohm's own analysis and presentation, such a fragmentary and analytic approach is often exactly appropriate in many circumstances. The problem arises when the fragmentary utilitarian approach to understanding the process of reality completely loses any awareness of its background of universal wholeness.

According to Bohm, modern science, philosophy, and society in general has lost contact with a necessary background of wholeness and this has led to serious problems for modern life. The extent to which Bohm's sociological analysis and prognostications can be defended is debatable, I for one consider that his sometimes doom-laden indications of societal and planetary breakdown were wildly overstated, and they have continued to be over-emphasised on the basis of various agendas. Bohm writes that :

> Rather, what should be said is that wholeness is what is real, and that fragmentation is the response of this whole to man's action, guided by illusory perception, which is shaped by fragmentary thought. In other words, it is just because reality is whole that man, with his fragmentary approach, will inevitably be answered with a correspondingly fragmentary response. So what is needed is for man to give attention to his habit of fragmentary thought, to be aware of it, and thus bring it to an end. Man's approach to reality may then be whole, and so the response will be whole.[621]

And:

> Men have been aware from time immemorial of this state of apparently autonomously existent fragmentation and have often projected myths of a yet earlier 'golden age', before the split between man and nature and between man and man had yet taken place. Indeed, man has always been seeking wholeness – mental, physical, social, individual.[622]

So, according to Bohm, it is necessary to find ways of loosening the fragmentary mode of engaging with the process of reality and find a method of reinstating awareness of wholeness in the lives of people. In the book *Thought as a System* Bohm stated that:

> Clear thinking implies that we are in some way awakened a little bit. Perhaps there is something beyond the reflex which is at work – in other words, something unconditioned. The question is really: is there the unconditioned? If everything is conditioned, then there's no way out. But the very fact that we are sometimes able to see new things would suggest that there *is* the unconditioned. Maybe the deeper material structure of the brain is unconditioned, or maybe beyond ... If there is the unconditioned, which could be the movement of intelligence, then there is some possibility of getting into this. We are saying that, perhaps unbeknownst to us, the unconditioned may have operated a little ...to be coherent we at least have to suppose that there may be the unconditioned. ... If we say that there cannot be the unconditioned, then it would be foolish for us to try to do anything with the conditioning.[623]

Bohm's suggestion here has a distinct Buddhist flavour. In the *Udana Sutta* the Buddha expressed the realm of the unconditioned, the nondual ground from within which the illusion of duality arises, in the following way:

> There is that dimension where there is neither earth, nor water, nor fire, nor wind; neither dimension of the infinite of space, nor dimension of the infinite of consciousness, nor dimension of nothingness, nor dimension of neither perception nor non-perception; neither this world, nor the next world, nor sun, nor moon. And there, I say, there is neither coming, nor going, nor staying; neither passing away nor arising: unestablished, unevolving, without support. This, just this, is the end of stress.[624]

And:

> There is, monks, an unborn, an unbecome, an unmade, unfabricated. If there were not that unborn, unbecome, unmade, unfabricated, there would not be the case that emancipation from the born, become, made, fabricated would be discerned. But precisely because there is an unborn, unbecome, unmade, unfabricated, emancipation from the born, become, made, fabricated is discerned.[625]

As we have seen in a previous chapter Buddhism has a very sophisticated set of meditation techniques which enables practitioners to directly experience deeper levels of mind. In this context, Bohm and F. David Peat, in their book *Science, Order and Creativity,* describe the Buddhist approach:

> ...each person is directed through reflection and meditation, to be aware, moment by moment, of the whole train of his or her thoughts. It is stated that in this process the fundamental "groundlessness" of the self can be seen. In this way a key piece of "misinformation" can be cleared up, i.e. the almost universal assumption that the self is the very ground of our being. This leads ultimately to Nirvana, in which there is a blissful unification within the totality.[626]

This is a description of a process of the dissolution of the constricting illusion that each individual's mind is entirely self-enclosed and separate from the deeper implicate levels of the universe. Such a process of the dissolving of the rigid strictures of fragmentary clinging to an illusionary 'self' moves towards becoming co-extensive with 'the totality'.

As we have seen, the view of the great significance of awareness of Totality is a central theme for Bohm. Thus, in *Wholeness and the Implicate Order* we read:

> ...the implicate order is particularly suitable for the understanding of such unbroken wholeness in flowing movement, for in the implicate order the totality of existence is enfolded within each region of space (and time). So, whatever part, element, or aspect we may abstract in thought, this still enfolds the whole and is therefore intrinsically related to the totality from which it has been abstracted.[627]

And from *The Essential David Bohm*:

> ...each entity is continually being formed from the infinite background and falls back into the background; to be generated again and again ... Thus each thing has its roots in the totality and falls back into the totality. Yet it still remains a thing having a certain degree of independent being. And this is possible because each thing contains in itself, its own special image of the totality (cosmos) out of which it formed itself and into which it is always dissolving (and reforming).[628]

This is the holomovement, apparently beginningless and endless, wherein vast numbers of sentient beings and their environments flicker in and out of explicate existence.

Once again we find Bohm's perspective resonates with Buddhist perspectives on the nature of Universal Totality. In the excellent book *The Buddhist Teaching of Totality*, by Garma C. C. Chang, the author portrays in outline the Buddhist vision of Totality:

> In the phenomenal world, the ever-flowing chains of events continuously interweave with one another, forming an immense "rimless net" rolling forward without cessation. But man, having only limited capacity and interest, cannot comprehend this vast intermeshing of events. With self-determination, he cuts off this "ever-flowing chain" and designates one point therein as the beginning and another as the end of a particular incident. .... A Martian, looking at our planet, does not see any sign of a beginning or an end; what he sees is a continuous, ever-

flowing chain of events. ... to say there is an absolute beginning of all events is meaningless.[629]

This is the vision of Totality, which is Bohm's holomovement, according to Hwa Yen Buddhism, which is derived from the 'Flower Garland' Sutra (*Avatamsaka Sutra*), and its scope is, to say the least, awesome:

> The Indescribable Indescribable
> Turning permeates what cannot be described
> It would take eternity to count
> All the Buddha's universes.
> In each dust-mote of these worlds
> Are countless worlds and Buddhas ...
> Indescribable are their wonders and names
> Indescribable are their glories and beauties
> Indescribable are the various Dharmas now being preached,
> Indescribable are the manners in which they ripen sentient beings ...
> Their unobstructed Minds are indescribable,
> There transformations are indescribable,
> The manners with which they observe, purify and educate
> Sentient beings are indescribable ...
> Each of them ripens sentient beings in indescribable manners, [630]

This indicates that it is the very purpose of the process of the universe, which is Bohm's holomovement, for the sentient beings within it to 'ripen' their minds, on the basis of Buddhist, or other, "Dharmas being preached" to them, so that they eventually become enlightened. And when they do become enlightened their fragmentary minds dissolve into, and become coextensive with, the Totality, the ultimate Mind-Energy of the universe.

According to Buddhism the ultimate purpose of the fragmentary embodied minds of human beings is to directly perceive and dissolve into the ultimate nature of reality. Thus the fragmentary meanings of the explicate world dissolve into the field of ultimate Meaning. Such an "ripened" mind is the mind of a buddha, a fully awakened being. The *Ornament of Sutras* says:

> Buddhahood is all phenomena,
> But it is no phenomenon whatsoever.[631]

And:

> With regard to the stainless expanse of dharmas,
> The explanation of the profound characteristics
> The state and the activity of the Buddhas
> Is nothing but sketching a colorful painting onto the sky.[632]

A beautiful poetic evocation of the merging of the fragmented mind with the Universal Mind-Energy.

However, it is important to be aware that, at least for most people, such a task is far from simple. Passages from many books discussing such possibilities sometimes give the impression that very simple practices can be guaranteed to achieve such a deep mystical union with the depths of the process of reality. But, for the majority of people this is not the case. Although, of course, very profound improvements in mental functioning can be achieved from quite limited, although committed, meditation practices, direct and immediate perception and union with the ultimate ground of the process of reality is a different issue, and most people would obviously require profound changes in daily life and attitudes to embark on a quest for 'ultimate' enlightenment. After all, great Tibetan Yogis such as Milarepa gave up all concerns with worldly life and withdrew to caves to meditate in order to achieve ultimate enlightenment.

However, within the fold of the Buddhist and pre-Buddhist Bon Dzogchen traditions there are some interesting debates concerning the nature and detail of both the analytic and meditation techniques employed. In particular, there is the significant and fascinating issue of how intellectual analysis and logic, which are aspects of the 'fragmented' world, point beyond themselves to the presence and vital significance of, and empower the experience of, the interconnected totality that encompasses them. Bohm expressed this seemingly paradoxical issue in *Wholeness*:

> So any notion of totality based on a fixed and permanent distinction between thought and reality must collapse when applied to the totality. The original form of the fixed distinction between thought and reality (i.e., non-thought) was:
>
> **T** is not **NT**
> All is either **T** or **NT**
>
> This form is characteristic of what is called Aristotelean logic ... This may be called the logic proper to things. Any particular

thought form that fits this logic can, of course, be applicable to a corresponding thing only under certain conditions which are required for that thing to be what it is. ... When we come to consider the 'totality of all that is', however, our primary concern is, as we have seen, not with conditioned things but with the unconditioned totality that is the ultimate ground of all. Here, the rules enunciated by Aristotle break down, in the sense that there is not even a limited domain or set of conditions under which they could apply: for, in addition to the Aristotelean rules, we have to assert the following:

> **T is NT**
> **NT is T**
>
> All is *both* **T** *and* **NT** (i.e., the two merge and flow into each other, in a single unbroken process, in which they are ultimately one).
> All is *neither* **T** *nor* **NT** (i.e., the ultimate ground is unknown, and therefore not specifiable, neither as T nor NT, nor in any other way).

If the above is combined with the original 'T is not NT' and 'All is either T or NT', and if we further suppose that 'T' and 'NT' are names of things, we will imply absolute self-contradiction.[633]

It is intriguing that Bohm's presentation of the logical configuration of the ultimate sphere of totality conforms to the Buddhist *Madhyamaka* assertion of the logical-existential configuration of the ultimate nature of reality, which is emptiness (*sunyata*). As the Madhyamika Bhavaviveka (1st-2nd century) indicated, the logical-existential configuration of the ultimate ground of reality according to Buddhist metaphysics is:

> Neither existent, nor nonexistent
> Nor both existent and nonexistent, nor neither.
> ...true reality ...is free from these four possibilities.[634]

Furthermore, as was demonstrated in the chapter **Bohmian Quantum Emptiness**, this configuration also corresponds to to the actual nature of quantum reality:

> ...a neutral K meson is typically not a K^0
> meson, not a $-K^0$ meson,
> not both and not neither.[635]

The quantum world hovers in an indeterminate state hovering between existence and nonexistence.

In his book *The Buddhist Teaching of Totality*, Garma C. C. Chang presents the following similar analysis which takes place through 3 phases. This analysis is: **Chi Tsang's Two Truths on Tree Levels**:[636]

Mundane / Conventional Truth	Ultimate Truth
1. Affirmation of being Δ	1. Denial of being $\sim\Delta$
2. Affirmation of either being or non-being Δ or $\sim\Delta$	2. Denial of either being or non-being $\sim(\Delta$ or $\sim\Delta)$
3. Either affirmation of either being or non-being or denial of either being or non-being $(\Delta$ or $\sim\Delta)$ or $\sim(\Delta$ or $\sim\Delta)$	3. Neither affirmation nor denial of either being or non-being $\sim[(\Delta$ or $\sim\Delta)$ or $\sim(\Delta$ or $\sim\Delta)]$

However, it is necessary to be careful drawing extreme implications from ultimate insights such as this. Such an ultimate perspective leads some New-Age style self-appointed pundits to be detrimental concerning the hard-edged Aristolelian type either-or logic, a logic on the basis of which science discovered the supposedly logic-threatening world of the quantum. For example the New-Age enthusiast Paul Levy seems to think that Buddhist four-valued 'paralogic' is the primary Buddhist logic, and that ordinary logic is somehow inferior:

> ...in Buddhism (also known as "paralogic") – we can hold seemingly contradictory statements as both being true simultaneously. This highest form of logic is characterised not by the two-valued logic of either/or, but by the four valued logic of both/and, where things can be true and false at the same time ...
> [637]

But this is inaccurate when extended into the conventional realm, which Levy does. Buddhist philosophy does not discard ordinary logic in favour of 'four-valued' 'dream' logic in philosophical analyses, such a claim is absurd.

The exemplary scholar in Buddhist philosophy (University of Oxford) Jan Westerhoff, has discussed this very issue in his translation of Nagarjuna's *Vaidalyaprakarana*, which he has translated as *Crushing the Categories*. This Buddhist philosophical text by the great Nagarjuna sets out to refute the notion that the logical-metaphysical categories of the *Nyaya* school of logic are substantially-established, rather than purely conceptual, categories. The question arises as to how extensive the refutation is? Does it apply at both the ultimate level and the ordinary everyday level of philosophical discourse? Westerhoff points out that:

> ... it seems unsatisfactory ... to rule out any employment of the categories at the ... conventional level, for in this case any talk of epistemic instruments or objects, inferences, examples, and so forth, would have to be given up, thereby effectively robbing oneself of the ability to put forward any structured philosophical arguments. When "refuting the Naiyayika's use of the sixteen categories" the aim is not to refute any use of the categories whatsoever, but to reject those kinds of usage that make unacceptable substantialist presuppositions... [638]

In other words, without the hardcore Aristotelian two-valued logic, it would not be possible to reason towards the ultimate sphere of Totality and Wholeness. Bohm could never have come to the conclusions he came to, or expressed them as he did, without Aristotelian style logic. As one Buddhist philosopher pointed out:

> Without reliance on conventions,
> The ultimate cannot be taught.[639]

It can be shown that even the great Buddhist philosopher Nagarjuna employed a precise and rigorous style of logic in his logical deconstructions to establish the nature of the ultimate. The four-valued logic applies to a particular, very restricted, mode of discourse, concerning the existential nature of ultimate reality. Using a dream-like, logically-fluid and indeterminate logic when constructing high-rise buildings or bridges clearly could have disastrous consequences! And it is equally essential to employ a precise logic in the large majority of normal conventional philosophical analysis.

However, there is a debate within Buddhism and Bon Dzogchen concerning the extent to which logic can lead to the experience of the ultimate nature, or capture in concepts the nature of the ultimate nature. This debate is the central theme of the wonderful book *Unbounded Wholeness: Dzogchen, Bon, and the Logic of the Nonconceptual* (which is a translation and introduction to the Bon Dzogchen text *Authenticity of Open Awareness*), by Anne Carolyn Klein and Geshe Tenzin Wangyal Rinpoche. This book examines the debate within the pre-Buddhist Bon Dzogchen tradition, wherein the ultimate nature of the holomoving process of reality is termed 'Mindnature':

> Dzogchen, or the Great Completion, is well known as the most revered system of thought and practice among the ancient Bon traditions in Tibet. In these traditions mindnature is at once the gaol of practice and its starting point. Being wholly uncontrived, mindnature nether improves on enlightenment nor becomes flawed in samsara. Always present in all beings, it is the abiding condition of every mind. Enlightenment is simply the full manifestation and experience of this abiding condition.[640]

This perspective is, clearly, consistent with Bohm's contention that the ultimate implicate level of the holomovement of the process of reality is of the nature of Mind and is experienced as ultimate Meaning. When experienced at the explicate level, which in Buddhist terminology is *samsara*, or cyclic existence, ultimate Mindnature is experienced as fragmented individuated consciousness, but this does not change its ultimate nature. It is worth pointing out here the close resonance with Bohm's analysis just presented on pages 360-361.

So how, in this tradition, is the ultimate nature of Mindnature introduced and accessed by a practitioner? According to *Unbounded Wholeness*:

> What is the student shown? Where is the mirror? The ultimate mirror is the natural, abiding condition itself, otherwise described as an unbounded wholeness. The principle of wholeness governs all of Authenticity's philosophical, soteriological, epistemological, and literary concerns. This is what the practitioner is shown and seeks to recognize. Wholeness defines liberation and determines its strategies, or lack thereof, that most facilitate it; wholeness also characterizes the awareness that recognizes wholeness as itself. Since wholeness does not, like logic, bifurcate the known universe into *is* and *is not* or any variation

thereof. *Authenticity* must deploy logic in a manner that somehow allows for this alogical perspective.[641]

And here is a central 'mystical' conundrum; the ultimate mindnature of the process of reality is, perhaps, beyond logic, but from a position of the fragmentary explicate point of view some form of logic or analysis may be necessary to elucidate and approach it.

The Geluk Tibetan Buddhist tradition asserts that the primary means of enlightenment is reasoning. As Daniel E. Perdue, in his book *Debate in Tibetan Buddhism*, indicates:

> ...Ge-luk-bas present a path to liberation which invariably involves reasoning. All Buddhists agree that in order to stop the cycle of birth, aging, sickness and death in which sentient beings suffer, one must defeat the foe of ignorance. Sentient beings have since beginningless time assented to normal appearances and conceptions and thereby been drawn into contaminated actions and afflictions which lead to suffering. Here, the Ge-luk-bas emphasize that the principle weapon to be used in the battle against ignorance is reasoning. ... the Ge-luk-bas issue a resounding call to analysis.[642]

According to this tradition it is possible for correct reasoning, combined with concerted and repeated meditation on the clearly perceived certainties generated by reasoning, to provide an internal path towards generating 'yogic direct perceivers', which are states of consciousness that directly perceive aspects of the ultimate nature of reality. As Perdue explains:

> The stated purpose for Buddhist reasoning is the development of yogic direct perceivers realizing subtle impermanence, the mind and body are selfless, etc. Only yogic direct perceivers can serve as the antidote to the ignorance that binds one in the suffering of cyclic existence. Such consciousnesses are produced in dependence on a very stable and insightful mind developed by the power of meditation...[643]

Each sense: ear, eye, nose, taste, touch has its own direct perceiver, these are sense consciousnesses that sense-monitor their objects directly and correctly and then subsequently induce mental consciousnesses ascertaining their objects as mental perceptions. 'Yogic direct perceivers', on the other hand, are internal mind direct perceivers that directly realize subtle, hidden phenomena. Such yogic direct perceivers only function when the person involved has developed the yogic perceiver through the necessary meditative development and stabili-zation. Such yogic direct perceivers enable practitioners to directly perceive aspects of implicate levels of the process of reality.

In order for this process to function it is necessary for a practitioner to achieve an advanced meditative state called 'calm abiding', a state of clear and alert thoughtless awareness which can be focused on a mental object such as a correct conviction regarding the result of a conceptual analysis. Through this method a conceptual-analytical certainty-mind-state regarding a hidden object, such as the emptiness of all phenomena, is transformed into direct non-conceptual immediate awareness of the aspect of reality in question. Thus, according to this perspective upon the development of enlightenment, a precise, rigorous and correct analysis is the precursor to a meditative transformation of the analytic phase into direct and immediate perception. Once the direct perception is stabilized the analysis is not longer needed, but can be used, of course, to instruct other people.

An important aspect of the analysis required by approaches such as the Geluk is the notion of *pramana*, or 'valid cognition'. The Buddhist epistemological view of the nature of 'valid cognition' derives from early Indian philosophers such as Nagarjuna, Vasubandhu, and became central at the beginning of the sixth century, when the master Dignaga wrote important works on this topic. In the seventh century another important Buddhist philosopher, Dharmakirti, also composed central treatises on valid cognition:

> The word pramana seems to be derived from the Sanskrit root ma, meaning to measure or ascertain; the prefix pra, meaning excellent or perfect; and the suffix ana, which indicates a method or instrument. So pramana is the study of methods for bringing about excellent knowledge. ... incorrect knowledge is the root cause for circling in samsara. In particular, this incorrect knowledge consists of mistaking what is impermanent to be permanent, mistaking what are causes of suffering to be causes of happiness, and mistaking what is not a self to be a self.[644]

On this view, then, precise analysis is a central feature of the 'path' to enlightenment.

However, the importance and sphere of effectiveness of the rigorous analytic *pramana* approach, in the context of its capacity to produce direct insight, is questioned by the Bon Dzogchen approach of *Unbounded Wholeness* (*Authenticity*):

> *Authenticity* vigorously puts forward well-known principles of authenticity (*pramana*) ... Words and concepts are a valid way of establishing one's view but cannot provide authentic realization of it. For ever locked into dualism of subject-object terminology, conceptual reasoning cannot realize the Dzogchen view. Yet, the texts emphasis on reasoning valorizes conceptuality in a way of coming to grips with issues raised by the category of unbounded wholeness. Though neither inference nor direct authentication is explicitly in the service of the other, they are tandem processes and, to a degree, complementary. The category of valid inference (*pramana*) so vital in, for example, Geluk discussions, does not exist here, for although it can establish the view, it cannot realize it.[645]

So, on this view, although conceptual analysis is capable of elucidating, clarifying, and conceptually describing to some degree, the nature of Totality and Wholeness, it cannot, contrary to the claims of the Geluk and other *pramana* traditions, play a direct role in the transformation of mind into direct awareness of the ultimate nature of the process of reality.

It is not the purpose of this work to resolve this debate regarding the extent of the appropriateness or effectiveness of conceptual analysis and comprehension in the task of moving one's mind in the direction of Totality and Wholeness. The aim is to highlight similar perspectives within, and similar concerns raised within, the contexts of Bohm's quantum philosophy of Wholeness and the 'mystical' spiritual traditions of Bon Dzogchen and Buddhism. In the above quote we see that for the Bon Dzogchen of *Authenticity* the role of language is limited because of its inherent subject-object structure. Here we find a direct correspondence with Bohm's later insight within the context of scientific philosophy, for in the quote from Bohm which opened this chapter we read about:

> ...the role of language in bringing about fragmentation of thought. It is pointed out that the subject-verb object structure of modern languages implies that all action arises in a separate

subject, and acts either on a separate object, or else reflexively on itself.

This situation, according to Bohm, renders language inappropriate to the sphere of Wholeness and Totality. As the *Lankavatara Sutra* proclaimed: "The vehicle of individual self-awareness is not the realm of logicians".[646]

Wholeness is, of course, undivided, there are no separated subjects and objects, actors and actors, observers and observed within the sphere of Totality. Again the *Lankavatara Sutra* is apposite:

> What appears to be external does not exist in reality, it is indeed Mind that is seen as multiplicity ; the body, property and abode - all these, I say, are nothing but Mind. All pairs of subject/object are manifestations of Mind ...[647]

We have already surveyed some other comments by Bohm which makes a similar point, but, to emphasize the central issue that this constitutes within Bohm's discourse, here is another of Bohm's explanations of this crucial issue:

> We can ask in a preliminary way whether there are any features of the commonly used language which tend to sustain and propagate this fragmentation, as well as, perhaps, to reflect it. A cursory examination shows that a very important feature of this kind is the subject-verb-object structure of sentences, which is common to the grammar and syntax of modern languages. This structure implies that all action arises in a separate entity, the subject, and that, in cases described by a transitive verb, this action crosses over the space between them to another separate entity, the object. ... This is a pervasive structure, leading in the whole of life to a function of thought tending to divide things into separate entities, such entities being conceived of as essentially fixed and static in their nature. ... The subject-verb-object structure of language, along with its world view, tends to impose itself very strongly in our speech, even in those cases in which some attention would reveal its evident inappropriateness.[648]

Bohm considered that such language and concept fragmentation issues are the hallmark of the Western scientific worldview and he indicated that Eastern philosophy and society still had a much more profound grasp of the relevance of a deeper perspective upon Wholeness and Totality. In the following quote it is worth noting Bohm's use of the term 'measure', which is used in precisely the same way as the Buddhist term *pramana*, which, as we have seen, is derived from the Sanskrit term, *ma*, which indicates the process of measurement. Bohm explicitly indicates his awareness of this resonance in his reference to the Sanskrit 'matra':

> Now, in the East the notion of measure has not played nearly so fundamental a role. Rather, in the prevailing philosophy in the Orient, the immeasurable (i.e. that which cannot be named, described, or understood through any form of reason) is regarded as the primary reality. Thus, in Sanskrit (which has an origin common to the Indo-European language group) there is a word 'matra' meaning 'measure'... But then there is another word 'maya' obtained from the same root, which means 'illusion'. This is an extraordinarily significant point. Whereas to Western society, as it derives from the Greeks, measure, with all that this word implies, is the very essence of reality, or at least the key to this essence, in the East measure has now come to be regarded commonly as being in some way false and deceitful. In this view the entire structure and order of forms, proportions, and 'ratios' that present themselves to ordinary perception and reason are regarded as a sort of veil, covering the true reality, which cannot be perceived by the senses and of which nothing can be said or thought.[649]

As we have seen, there are traditions within Bon Dzogchen and Buddhism, as well as elsewhere, which hold that the mind of a human being can break through the 'illusion' and become aware of, and activate, the ultimate 'implicate' nondual realm of the process of reality, and thereby ultimately at death dissolve into this realm, rather than being reborn.

Unless, on the other hand, such an enlightened being decides to return into a future life to help other sentient beings. Such a person is a bodhisattva, who, because they have generated *bodhicitta* (*bodhichitta*), which is a vast compassionate concern for the welfare of all sentient beings, are able to purposely retain certain subtle portions of their afflictive dispositions which prevent final dissolution, in order to postpone complete buddhahood and therefore continuously take rebirth in samsara for the benefit of sentient beings:

> The enhancing factor is not to relinquish the subtle afflictions that enable bodhisattvas to be willingly reborn in samsara for the sake of accomplishing the welfare of others. The activity of the knowledge of such bodhisattvas consists of liberating sentient beings continuously without manifesting their own buddhahood.[650]

Such beings of compassion continuously cycle through Bohm's holomovement in order to help all other beings who are trapped in the cycle of repeated samsaric explicate lives. It is important to recall that if Bohm's account of how each 'unfoldment' of a relatively stable organism is conditioned by preceding 'enfoldments' of previous events and activities, a process which as we have seen corresponds to the Buddhist concept *karma-vipaka*, cause and effect, wherein a continuity of changing, interconnected embodied moments of consciousness manifest in rapid sequence, then the Buddhist doctrine of rebirth, and the functioning of such bodhisattvas, is an entirely natural consequence.

As previously indicated, it is not the aim of this work to resolve any Buddhist metaphysical doctrinal issues, but, rather, to survey the remarkable interconnections of issues within Bohm's Western quantum philosophy of Wholeness and metaphysical visions of Wholeness within Bon Dzogchen and Buddhism. In the light of this, it is entirely appropriate to round off this chapter, and the book, by surveying some remarkable passages from the Bon Dzogchen text *Unbounded Wholeness,* and some others, and draw out their consistency with Bohm's quantum philosophy of Wholeness. In *Wholeness and the Implicate Order* Bohm wrote that:

> Men have been aware from time immemorial of this state of apparently autonomously existent fragmentation and have often projected myths of a yet earlier 'golden age', before the split between man and nature and between man and man had yet taken place. Indeed, man has always been seeking wholeness – mental, physical, social, individual.[651]

And:

> ... at each stage the proper order of operation of the mind requires an overall grasp of what is generally known not only in formal, logical, mathematical terms, but also intuitively, in images, feelings, poetic usage of language, etc. (Perhaps we could say that this is what is involved in harmony between the

> 'left brain' and the 'right brain'.) This kind of overall way of thinking is not only a fertile source of new theoretical ideas: it is needed for the human mind to function in a generally harmonious way, which could in turn help to make possible an orderly and stable society. ... this requires a continual flow and development of our general notions of reality.[652]

This is a prescription, requiring that the various fragmented modes of apprehension and engagement with the process of reality finds a location within an overarching and harmonising poetic and mythic view of Totality and Wholeness, which resonates powerfully with views expressed in *Unbounded Wholeness* (or *Authenticity*). For example:

> An important subtext of *Authenticity* is that reason and logic can, and apparently must, exist side by side with poetic, mythic and other voices. These do not cancel each other out; they are not even presented as contradictory. Like notes in a chord, enriched when they sound together, each also retains its unique resonance.[653]

And:

> Rather than taking shape around particular definitions, then, debates in Authenticity are usually initiated by setting two elements of an apparently binary pair against each other, for example, Buddhas and ordinary beings, or conditioned and unconditioned. The type of relation that could obtain between them is then questioned. Debate continues ...until the bifurcated issue suddenly breaks open into a kind of discursive wholeness, which is then often decorated by a poetic celebration ...[654]

And in a wonderfully apposite insight *Unbounded Wholeness* tells us that *Authenticity*, which refers to the view from the experiential totality of primordial awareness, indicates that, without division into relatively functional subunits and subsequent fragmentation, wholeness would never be comprehended, so we find that Wholeness is necessarily a Totality comprised of parts:

> Dzogchen explicitly sees itself as inquiring into the nature of an all-encompassing subjectivity; a playful and open plurality ...The fact of many diverse perspectives, *Authenticity* has argued, means that there must be wholeness in which they participate. Therefore, *Authenticity* will philosophically as well as mythically embrace philosophical incommensurability and the uncertainty that ensues. Thereby, its truth does not so much become subjectivized ... as

pluralised: "Since there are many diverse perspectives, it is impossible that there not be a whole, all-suffusing mind nature which is its basis."[655]

One can only marvel at the extraordinary resonance of this view echoing down to us through the ages from the Eastern lands of Tibet with Bohm's proposal that:

> ...when the whole field of measure is open to original and creative insight, without any fixed limits or barriers, then our overall world views will cease to be rigid, and the whole field of measure will come into harmony, as fragmentation within it comes to an end. But original and creative insight within the whole field of measure is the action of the immeasurable. For when such insight occurs, the source cannot be within ideas already contained in the field of measure but rather has to be in the immeasurable, which contains the essential formative cause of all that happens in the field of measure. The measurable and the immeasurable are then in harmony and indeed one sees that they are but different ways of considering the one and undivided whole. When such harmony prevails, man can then not only have insight into the meaning of wholeness but, what is much more significant, he can realize the truth of this insight in every phase and aspect of his life.[656]

Or, as one of the stanzas of the *Dwelling in the Very Heart of Space Tantra* observes:

> Essential heart of all that is,
> Mindnature, uncontrived and naturally pure,
> Exists from the first, without start or stop
> This is sure.
> Untouched by limits:
> Self-arisen open awareness, definitive pith
> Dwells as the heart of the sun
> This is sure.[657]

And as the Ch'an Master Hung Po proclaimed:

> This pure Mind, the source of everything, shines forever and on all with the brilliance of its own perfection. But the people of the world do not awake to it, regarding only that which sees, hears, feels and knows as mind. Blinded by their own sight, hearing, feeling and knowing, they do not perceive the spiritual brilliance of the source substance. If they would only eliminate all conceptual

thought in a flash, that source substance would manifest itself like the sun ascending through the void and illuminating the whole universe without hindrance or bounds.[658]

In this passage Hung Po indicates that the world as experienced only through the 'explicate' sense-faculties alone hides the "spiritual brilliance of the source substance", whereas there is a mode of perception which can perceive more deeply into what Bohm termed the 'implicate order'. This characterisation also clearly applies to Bohm's holomovement?

According to Hung Po the "source substance" is "pure Mind". The term 'pure' in this context means that the source Mind-energy is not disturbed by dualistic movements giving rise to the explicate experiential realm of subjects and objects. In *Wholeness and the Implicate Order* Bohm wrote:

> Intelligence and material process have thus a single origin, which is ultimately the unknown totality of the universal flux. In a certain sense, this implies that what have been commonly called mind and matter are abstractions from the universal flux, and that both are to be regarded as different and relatively autonomous orders within the one whole movement. ... It is thought responding to intelligent perception which is capable of bringing about an overall harmony or fitting between mind and matter.[659]

Furthermore, Bohm agrees with Hung Po that the nature of the "single origin" is close to the nature of mind, not matter. In discussing the basic metaphysical view of Descartes, Bohm says:

> By using the term 'thinking substance' in such sharp contrast to 'extended substance' he was clearly implying that the various distinct forms appearing in thought do not have their existence in such an order of extension and separation (i.e., some kind of space), but rather in a different order, in which extension and separations have no fundamental significance. The implicate order has just this latter quality, so in a certain sense Descartes was perhaps anticipating that consciousness has to be understood in terms of an order that is closer to the implicate than it is to the explicate.[660]

As we have seen in the investigations of the previous chapters, the most natural conclusion that can be reached with regard to Bohm's view of the nature of the implicate order is that it is most appropriately described as a field of Mind-energy which gives rise within the holomovement of fragmentary episodes of dualistic mind-embodiments wherein appearances of the material world manifest. As the Zen Patriarch Dogen described the situation, in paradoxical fashion:

> "All sentient beings," discussed now in the Buddha way, means all sentient beings possessing mind, for mind is itself sentient beings. Those beings not possessing mind should equally be sentient beings, because sentient beings are, as such, mind. Therefore, mind is invariably sentient beings; sentient beings are necessarily the Buddha-nature of existence. Grasses and trees, and countries and lands are mind. They are sentient beings in virtue of being mind, and are the Buddha-nature of existence on account of being sentient beings. The sun, the moon, and the stars - all are mind. They are sentient beings by reason of being mind, and are the Buddha-nature of existence because of being sentient beings.[661]

Bibliography

Aczel, Amir D. (2002) *Entanglement: The Greatest Mystery in Physics*. Basic Books.

Addiss, Stephen; Lombardo, Stanley; Roitman, Judith (2008). *Zen Source Book: Traditional Documents from China, Korea and Japan*. Hackett Publishing Company.

Aharanov, Yakir and Rohrlich, Daniel (2005). *Quantum Paradoxes: Quantum Theory for the Perplexed*. Wiley-VCH.

Allday, Jonathan (2009), *Quantum Reality: Theory and Philosophy*. CRC Press.

Ananthaswamy, Anil (2020), *Through Two Doors at Once: The Elegant Experiment that Captures the Enigma of our Quantum Reality*, Dutton (Penguin Random House).

Baggott, Jim (2004), *beyond measure*. Oxford University Press, Oxford.

Baggott, Jim (2020), *Quantum Reality: The Quest for the Real Meaning of Quantum Mechanics - a Game of Theories*, OUP Oxford.

Barbour, Julian (2001), *The End of Time: The Next Revolution in Physics*. Oxford University Press.

Barrett, Jeffrey A. (2001). *Quantum Mechanics of Minds and Worlds*. Oxford University Press.

Barrow, John D., Davies, Paul C. W., Harper, Charles L. (eds) (2004). *Science and Ultimate Reality*. Cambridge University Press.

Becker, Adam (2019), *What is Real?: The Unfinished Quest for the Meaning of Quantum Physics*. John Murray.

Bohm, David (1984), *Causality and Chance in Modern Physics*, Routledge; 2nd edition.

Bohm, David (1987), Hidden Variables and the Implicate Order, in - Hiley, Basil & F. David Peat (1991), *Quantum Implications: Essays in Honour of David Bohm*, Routledge.

Bohm, David (1990), A New Theory of the Relationship of Mind and Matter, *Philos. Psych.* **3**(2), 271-286.

Bohm, David (1994), *Thought as a System*, Routledge.

Bohm, David (2002), *Wholeness and the Implicate Order* (First published: Routledge & Kegan Paul, 1980; Routledge Classics).

Bohm, David, (2003), *The Essential David Bohm* ed. Nichol, Lee. (Routledge, London).

Bohm, David & Peat F. David (2000), *Science, Order and Creativity*, Routledge Classics. (First published 1987).

Bohm, David & Basil J. Hiley (1995), *The Undivided Universe: An Ontological Interpretation of Quantum Theory.* Routledge.

Boge, Florian J. (2019), *Quantum Mechanics Between Ontology and Epistemology*, Springer.

Bricmont, Jean (2016), *Making Sense of Quantum Mechanics*, Springer.

Bricmont, Jean (2017), *Quantum Sense and Nonsense*, Springer.

Brunnhölzl, Karl (2004), *Center of the Sunlit Sky: Madhyamaka in the Kagyu Tradition.* Ithaca: Snow Lion Publications.

Brunnhölzl, Karl (2007), *Straight from the Heart: Buddhist Pith Instructions.* Ithaca: Snow Lion Publications.

Brunnhölzl, Karl (2007), *In Praise of Dharmadhatu.* Ithaca: Snow Lion Publications.

Brunnhölzl, Karl. (2009), *Luminous Heart: The Third Karmapa on Consciousness*, Wisdom, and Buddha Nature. Ithaca: Snow Lion Publications.

Brunnhölzl, Karl (2010), *Gone Beyond Vol. 1.* Snow Lion Publications.

Brunnhölzl, Karl (2012), *Heart Attack Sutra: A New Commentary on the Heart Sutra.* Snow Lion Publications.

Brunnhölzl, Karl (2013), *Mining for Wisdom within Delusion: Maitreya's "Distinction between Phenomena and the Nature of Phenomena" and Its Indian and Tibetan Commentaries.* Snow Lion.

Capra, F (1975) *The Tao of Physics* Shambhala Publications.

Cabezón, José Ignacio (1992), *A Dose of Emptiness*, State University of New York Press.

Carroll, Sean B. (2006). *Endless Forms Most Beautiful.* Weidenfield & Nicolson

Carroll, Sean (2016), *The Big Picture: On the Origins of Life, Meaning and the Universe Itself*, One World.

Chandrakirti and Jamgon Mipham (2002), *Introduction to the Middle Way: Chandrakirti's* Madhyamakavatara *with Commentary by Jamgon Mipham.* Translated by the Padmakara Translation Group. Boston: Shambhala Publications.

Chang, Garma C.C. (1971), *The Buddhist Teaching of Totality: The Philosophy of Hwa Yen Buddhism*, Pennsylvania State University Press.

d'Espagnat, Bernard (2006). *On Physics and Philosophy*. Princeton University Press.

d'Espagnat, Bernard (Editor), Zwirn, Hervé (Editor) (2017), *The Quantum World: Philosophical Debates on Quantum Physics*, Springer (The Frontiers Collection).

Dalai Lama, H. H. (2008), *The Universe in a Single Atom: The Convergence of Science and Spirituality*. New York: Morgan Road, 2005. Abacus paperback: 2006, 2008.

Dalai Lama & Thubten Chodron (2019), *Samsara, Nirvana, and Buddha Nature*, Wisdom Publications.

Davies, Paul (2007), *The Goldilocks Enigma*. Penguin Books (First published 2006: Allen Lane).

Davies, Paul & Gregersen, Niels Henrik (eds) (2010), *Information and the Nature of Reality: From Physics to Metaphysics*. Cambridge University Press.

Devenish, R. P. (2012), *Principle Yogacara Texts*. Dharma Fellowship.

Dolling, L.M.; Gianelli, A. F. & Statile, G. N. (eds) (2003). *The Tests of Time: Readings in the Development of Physical Theory*. Princeton University Press.

Dowman, Keith (2017), *Everything Is Light: The Circle of Total Illumination*. CreateSpace Independent Publishing Platform.

Dzogchen Ponlop (2012), *Mind Beyond Death*, Snow Lion Publications.

Engle, Artemus B. (2009), *The Inner Science of Buddhist Practice: Vasubhandhu's Summary of the Five Heaps with Commentary by Sthiramati*. Snow Lion, New York.

Fremantle, Francesca (2003), *Luminous Emptiness: Understanding the "Tibetan Book of the Dead"*, Shambhala Publications Inc.

Freire, Olival (2020), *David Bohm: A Life Dedicated to Understanding the Quantum World*, Springer Biographies.

Garfield, Jay (1995), *The Fundamental Wisdom of the Middle Way (Nagarjuna's Mulamadhyamakakarika)*. Oxford University Press.

Geshe Kelsang Gyatso (2009), *Heart Jewel: The Essential Practices of Kadampa Buddhism*, Tharpa Publications: 2nd Revised edition.

Geshe Sonam Rinchen (2006) *How Karma Works* trans & ed. Ruth Sonam. Snow Lion Publications.

Ghirardi, Giancarlo, (2005), *Sneaking a Look at God's Cards: Unraveling the Mysteries of Quantum Mechanics, Revised Edition*, Princeton University Press.

Goswami, Amit (1995), *The Self Aware Universe: How consciousness creates the material world.* Tarcher/Penguin, (First published 1993)

Goswami, Amit (2008), *Creative Evolution.* Quest Books.

Greene, Brian (2004). *The Fabric of the Universe.* Allen Lane.

Gribben, John (1996). *Shrodinger's Kittens and the Search for Reality.* Phoenix.

Guenther Herbert V. (1984), *Matrix of Mystery: Scientific and Humanistic Aspects of rDzogs-chen Thought.* Shambhala Publications.

Guenther, Herbert V. (1995), *Ecstatic Spontaneity: Saraha's Three Cycles of Doha* - (Nanzan Studies in Asian Religions, Vol 4), Asian Humanities P.,U.S.

Guenther, Herbert V. (2001), *From Reductionism to Creativity: Rdzogs-Chen and the New Sciences of Mind*, Shambhala.

Guenther, Herbert V. (2012), *The Teachings of Padmasambhava*, Brill's Indological Library

Hawking, Stephen & Mlodinow, Leonard (2010), *The Grand Design: New Answers to the Ultimate Questions of Life.* Transworld Publishers – Bantum Press.

Hee-Jin Kim (2004), *Eihei Dogen: Mystical Realis.* Wisdom Publications.

Herbert, Nick (1985), *Quantum Reality: Beyond The New Physics.* Random House (Anchor Books), New York.

Hiley, Basil & F. David Peat (1991), *Quantum Implications: Essays in Honour of David Bohm*, Routledge.

Hopkins, Jeffrey (1996). *Meditation on Emptiness.* Wisdom Publications, U.S.A. (First published 1983).

Hopkins, Jeffrey (2006), *Mountain Doctrine: Tibet's Fundamental Treatise on Other-Emptiness and the Buddha Matrix by Dol-bo-ba Shay-rap-gyel-tsen.* Ithaca: Snow Lion Publications.

Hossenfelder, Sabine (2020), *Lost in Math: How Beauty Leads Physics Astray*, Hachette USA.

Hsing Yun, (Master) & Tom Graham (trans.) (2010), *Describing the Indescribable.* Wisdom Publications.

Kaku, Michio (2006), *Parallel Worlds: The Science of Alternative Universes and our Future in the Universe*. Penguin Books (First published by Doubleday 2005).

Kastrup, Bernardo (2020), *Decoding Schopenhauer's Metaphysics: The key to understanding how it solves the hard problem of consciousness and the paradoxes of quantum mechanics*, Iff Books.

Kelly E. E, Kelly E. W, Crabtree A., Gauld, A., Grosso M., Greyson B. (2007), *Irreducible Mind: Towards a Psychology for the 21st Century*, Rowman & Littlefield Publishers.

Kenchen Thrangu (2001), *Transcending Ego: Distinguishing Consciousness from Wisdom*. Namo Buddha Publication., Boulder, Colorado

Khenchen Thrangu (2012), *Pointing Out the Dharmakaya: Teachings on the Ninth Karmapa's Text*, Snow Lion.

Khendrup Norsang Gyatso (2004), *Ornament of Stainless Light: An Exposition of the Kalacakra Tantra*. Library of Tibetan Classics; Wisdom Publications; Boston.

Khenpo Tsultrum Gyamtso (2003), *The Sun of Wisdom* (Shambhala Publications)

Klein, Anne Carolyn & Tenzin Wangyal Rinpoche (2016), *Unbounded Wholeness: Dzogchen, Bon, and the Logic of the Nonconceptual*, Oxford University Press.

Kyabgon, Traleg (2010), *Influence of Yogacara on Mahamudra*, KTD Publications.

Levy, Paul (2018), *Quantum Revelation: A Radical Synthesis of Science and Spirituality*, SelectBooks.

Lingpa, Dudjom (2002) (Trans: Richard Barron & Susanne Fairclough), *Buddhahood Without Meditation*, Padma Publishing.

Lockwood, Michael (2005). *The Labyrinth of Time: Introducing the Universe*. Oxford University Press.

Longchen Rabjam (Author), Harold Talbott (Editor)(2014), *The Practice of Dzogchen: Longchen Rabjam's Writings on the Great Perfection*, Snow Lion Publications.

Longchenpa (trans. Lipmann, K. & Peterson, M.) (2000), *You Are The Eyes of the World*, Snow Lion.

Khenpo Shenga and Ju Mipham - Dharmachakra Translation Committee (2014), *Ornament of the Great Vehicle Sutras: Maitreya's Mahayanasutralamkara with Commentaries by Khenpo Shenga and Ju*

Mipham: 3 (Maitreya Texts), Snow Lion Publications; Translation edition.

Mensky, M. B. (2010), *Consciousness and Quantum Mechanics: Life in Parallel Worlds: Miracles of Consciousness from Quantum Reality.* World Scientific Publishing.

Mi-Pam-Gya-Tso and Jeffrey Hopkins (2006), *Fundamental Mind: The Nyingma View of the Great Completeness*, Snow Lion.

Norsen, Travis (2017), *Foundations of Quantum Mechanics: An Exploration of the Physical Meaning of Quantum Theory*, Springer.

Oerter, Robert (2006). *The Theory of Almost Everything*. Pi Press.

Oppenheimer, Robert (1954), *Science and the Common Understanding*. Oxford University Press.

Penrose, Roger (1995), *Shadows of the Mind*. Oxford University Press:1994, Random House-Vintage:1995

Penrose, Roger (1999), *Emperors New Mind*. Oxford University Press:1989, Oxford University Press paperback:1999

Penrose, Roger (2005), *The Road to Reality: A Complete Guide to the Laws of the Universe*. Vintage.

Perdue, Daniel E.(1992), *Debate in Tibetan Buddhism*, Snow Lion Publications.

Pylkkänen, Paavo T. I. (2006), *Mind, Matter and the Implicate Order: The Implicate Order Revisited*, Springer.

Rabjam, Longchen (trans. Tulku Thondup) (2002), *The Practice of Dzogchen*, Snow Lion.

Randall, L. (2012), *Higgs Discovery: The Power of Empty Space*, The Bodley Head Ltd.

Ray, Reginald A. (2002), *Secret of the Vajra World: The Tantric Buddhism of Tibet*, Shambhala Publications Inc.

Rongtonpa (Author), Christian Bernert (Translator) (2017), *Adorning Maitreya's Intent: Arriving at the View of Nonduality*, Snow Lion.

Rovelli, Carlo (2021), *Helgoland*, Allen Lane; 1st edition.

Sabbadini, Shantena Augusto (2017), *Pilgrimages to Emptiness: Rethinking Reality through Quantum Physics*, Pari Publishing

Saunders, Simon (Editor), Adrian Kent (Editor), David Wallace (Editor) (2012), *Many Worlds?: Everett, Quantum Theory, & Reality: Everett, Quantum Theory, & Reality*, Oxford University Press, USA.

Schlosshauer, Maximilian (Editor) (2011), *Elegance and Enigma: The Quantum Interviews (The Frontiers Collection)*, Springer.

Schwartz, Jeffrey M. & Sharon Begley (2003), *The Mind & The Brain: Neuroplasticity and the Power of Mental Force*. First Published: HarperCollins Publishers 2002; First Harper Perennial paperback edition 2003).

Shantarakshita (2005 - Padmakara Translation Group), *The Adornment of the Middle Way* (Madhyamakalamkara). Shambhala Publications.

Shar Khentrul Rinpoche Jamphel Lodro (2019), *The Great Middle Way: Clarifying the Jonang View of Other-Emptiness*, Dzokden Publications.

Sheehy, Michael R., Mathes, Klaus-Dieter (Editors) (2020), *The Other Emptiness, The: Rethinking the Zhentong Buddhist Discourse in Tibet*, State University of New York Press.

Sheldrake, Rupert (2009). *A New Science of Life* (Revised Edition). Icon Books.

Siderits, Mark & Katsura, Shoryu (2013), *Nagarjuna's Middle Way*, Wisdom Publications.

Smolin, Lee (2014), *Time Reborn: From the Crisis in Physics to the Future of the Universe*. Penguin.

Smolin, Lee (2019), *Einstein's Unfinished Revolution: The Search for What Lies Beyond the Quantum*, Allen Lane.

Stapp, Henry (2004), *Mind, Matter and Quantum Mechanics*. Springer-Verlag Berlin Heidelberg 1993, 2004.

Stapp, Henry (2007), *Mindful Universe*. Springer-Verlag Berlin Heidelberg.

Stapp, Henry (2017), *Quantum Theory and Free Will: How Mental Intentions Translate into Bodily Action*, Springer.

Thrangu Rinpoche, Kenchen (2001), *Transcending Ego: Distinguishing Consciousness from Wisdom*. Namo Buddha Publication., Boulder, Colorado.

Tulku Thondup Rinpoche (2005), *The Hidden Teachings of Tibet: An Explanation of the Terma Tradition: An Explanation of the Term Tradition*, Wisdom Publications,U.S.

Van Schaik, Sam, (2004), *Approaching the Great Perfection: Simultaneous and Gradual Methods of Dzogchen Practice in the Longchen Nyingtig*. Wisdom.

Vedral, Vlatko (2010). *Decoding Reality*. Dutton.

Waldron, William S. (2003). *The Buddhist Unconscious*. Routledge-Curzon.

Wallace, B. Alan (2011), *Minding Closely: The Four Applications of Mindfulness*. Snow Lion.

Walpola, Rahula (1974). *What the Buddha Taught*. Grove Press.

Wangyal, Tenzin Rinpoche (2000), *Wonders of the Natural Mind*, Snow Lion.

Wheeler, John Archibald & Ford, Kenneth (2000), *Geons, Black Holes, and Quantum Foam: A Life in Physics*, W. W. Norton & Company.

Wendt, Alexander (2015), *Quantum Mind and Social Science: Unifying Physical and Social Ontology*, Cambridge University Press.

Westerhoff, Jan & Nagarjuna (2019), *Crushing the Categories (Vaidalyaprakarana)*, Wisdom Publications.

Whitaker, Andrew (2016), *The New Quantum Age: From Bell's Theorem to Quantum Computation and Teleportation*, Oxford University Press.

Woolfson, Adrian (2000), *Life Without Genes*, HarperCollins.

References

1 Geshe Kelsang Gyatso (2009), 45

2 Geshe Kelsang Gyatso (2009), 49

3 Woit's blog - Not Even Wrong: www.math.columbia.edu/~woit/wordpress/?p=588

4 Wallace, B. Alan (2007) page ix

5 Quantum Emptiness -The Quantum Illusion-like Nature of Reality by Graham Smetham - Many Roads (bodhicharya.org)

6 membership.iop.org

7 https://www.infinitepotential.com

8 Rongtonpa (Author), Christian Bernert (Translator) (2017), 2

9 https://nitarthainstitute.org/2020/01/10/what-is-analytical-meditation/

10 https://www.lamayeshe.com/article/chapter/lam-rim-meditations-and-deity-practice

11 http://conscious.shift.over-blog.com/2017/04/dalai-lama-spirituality-without-quantum-physics-is-an-incomplete-picture-of-reality.html

12 Dalai Lama, H. H. (2008)

13 https://awaken.com/2019/07/dalai-lama-spirituality-without-quantum-physics-is-an-incomplete-picture-of-reality-2/

14 Khenchen Thrangu (2012), 5

15 Dowman, Keith (2017), 19

16 Dowman, Keith (2017), 22

17 Klein,Anne Carolyn & Tenzin Wangyal Rinpoche (2016), 3

18 Bohm, David (2002), 14

19 Klein,Anne Carolyn & Tenzin Wangyal Rinpoche (2016), 54

20 Bohm, David (2002), 191

21 Klein,Anne Carolyn & Tenzin Wangyal Rinpoche (2016), 56

22 Dowman, Keith (2017), 66-68

23 Hee-Jin Kim (2004), 99-100

24 Hee-Jin Kim (2004), 100

25 https://blogs.scientificamerican.com/cross-check/david-bohm-quantum-mechanics-and-enlightenment/

26 Becker, Adam (2019), 39 & https://arxiv.org/ftp/arxiv/papers/1603/1603.00353.pdf

27 http://philsci-archive.pitt.edu/1559/1/CosKraPL.pdf

28 http://philsci-archive.pitt.edu/1559/1/CosKraPL.pdf

29 http://philsci-archive.pitt.edu/1559/1/CosKraPL.pdf

30 N. Bohr, Speech on quantum theory at Celebrazionne del Secondo Centenario della Nascita di Luigi Galvani, Bologna, Italy, October 1937.

31 Bohm 1952 article - '*A Suggested Interpretation of the Quantum Theory in Terms of 'Hidden' Variables*'

32 Bohm, David & Basil J. Hiley (1995), 357

33 https://www.scienceandnonduality.com/article/david-bohm-implicate-order-and-holomovement

34 Bohm, David (2002), 243

35 Bohm, David (2002), 227

36 Bohm, David (2002)

37 Bohm, David & Peat, F. David (2000), 190

38 Pylkkänen, Paavo T. I. (2006), 20-21

39 Bohm, David (2002), 14

40 Penrose, Roger (1995), 237

41 Bohm, David & Peat, F. David (2000), 258-9

42 Bohm, David & Peat, F. David (2000), 259

43 Bohm, David (2002), 267

44 Kyabgon, Traleg (2010),109

45 Hsing Yun, Master & Tom Graham (trans)(2010), 113

46 Guenther, Herbert V. (1995)

47 Guenther, Herbert V. (2001), 190

48 Hawking, Stephen & Mlodinow, Leonard (2010),

49 Guenther, Herbert V. (2012), 13

50 Bohm, David (2002), 226

51 Klein,Anne Carolyn & Tenzin Wangyal Rinpoche (2016), 54-56

52 Bohm, David & Basil J. Hiley (1995), 381-382

53 Devenish, R. P. (2012), 2-3

54 Bohm, David (2002), 190-191

55 Bohm, David (2002), 188

56 Bohm, David (2002), 14

57 Bohm, David (2002), 250

58 N. D. Mermin. Hidden variables and the two theorems of John Bell. *Reviews of Modern Physics*, 65:803–815, 1993.

59 https://www.informationphilosopher.com/solutions/scientists/jordan/

60 Davies, Paul (2007), 280

61 Davies, Paul (2007), 281

62 Barrow, John D., Davies, Paul C. W., Harper, Charles L. (eds) (2004) 72 – Freeman J. Dyson: 'Thought-experiments in honor of John Archibald Wheeler.'

63 Wheeler, John Archibald & Ford, Kenneth (2000), 338

64 Wheeler, John Archibald & Ford, Kenneth (2000), 338

65 Wheeler, John Archibald & Ford, Kenneth (2000), 355

66 Rosenblum, Bruce and Kuttner, Fred (2006), 139

67 Rosenblum, Bruce and Kuttner, Fred (2006), 201

68 Rosenblum, Bruce and Kuttner, Fred (2006) website (quantumenigma.com)

69 Levy, Paul (2018), 109

70 Hossenfelder, Sabine (2020), 11

71 Kaku, Michio (2006), 148

72 Hossenfelder, Sabine (2020), 9

73 Levy, Paul (2018), 93

74 Goswami, Amit (2008), 22

75 Capra, Fritjov (1975), 152

76 Penrose, Roger (1995), 309

77 Penrose, Roger (1999), 293

78 Rosenblum, Bruce and Kuttner, Fred (2006), 179

79 Quoted in Stapp, Henry (2007), 161

80 Ghirardi, G. (2005), 403

81 Interview with Amit Goswami - Enlighten Next.

82 Hawking, Stephen & Mlodinow, Leonard (2010), 82-83

83 Hawking, Stephen & Mlodinow, Leonard (2010), 136

84 Hawking, Stephen & Mlodinow, Leonard (2010), 135

85 Hawking, Stephen & Mlodinow, Leonard (2010), 140

86 Hawking, Stephen & Mlodinow, Leonard (2010), 140

87 Bricmont, Jean (2017), 209

88 Bricmont, Jean (2017), 240

89 Bricmont, Jean (2017), 240

90 Bricmont, Jean (2017), 225

91 Levy, Paul (2018), 293

92 Zurek Wojciech H.(2002). ' Decoherence and the Transition from Quantum to Classical – *Revisited*' in *Los Alamos Science* Number 27 2002

93 Rosenblum, Bruce and Kuttner, Fred (2006), 75

94 D'Espagnat, Bernard, 'The Quantum Theory and Reality' *Scientific American*, Nov. 1979

95 Bohm, David & Basil J. Hiley (1995), 2

96 Penrose, Roger (1995), 313

97 Bricmont, Jean (2016), 16

98 Stapp, Henry (2004), 223

99 Stapp, Henry: 'Philosophy of Mind and the Problem of Free Will in the Light of Quantum Mechanics' p19

100 Stapp, Henry: 'Quantum Interactive Dualism', 18

101 Stapp, Henry: 'The Effect of Mind upon Brain', 12

102 Stapp, Henry (2004), 241

103 Stapp, Henry (2004), 241

104 Stapp, Henry (2004), 241

105 https://blogs.scientificamerican.com/cross-check/do-our-questions-create-the-world/

106 Bricmont, Jean (2016), 12

107 Lingpa, Dudjom (2002), 127-129

108 https://www.religion-online.org/article/the-implicate-order-a-new-order-for-physics/

109 Oppenheimer, Robert (1954), 8-9

110 Oppenheimer, Robert (1954), 8-9

111 Brunnhölzl, Karl (2004), 507

112 Barrett, Jeffrey A. (2001)

113 https://www.telegraph.co.uk/culture/books/bookreviews/9188438/Erwin-Schrodinger-and-the-Quantum-Revolution-by-John-Gribbin-review.html

114 Bricmont, Jean (2016), 12

115 Bricmont, Jean (2016), 12

116 Bricmont, Jean (2016), 13

117 Bricmont, Jean (2016), 16

118 https://journals.aps.org/pr/abstract/10.1103/PhysRev.85.166

119 Bricmont, Jean (2017), 138

120 Bricmont, Jean (2017), 140

121 Bricmont, Jean (2017), 140

122 Bricmont, Jean (2016), 19

123 Whitaker, Andrew (2016), 93

124 Bricmont, Jean (2016), 13

125 Smolin, Lee (2019), 110-111

126 Bohm, David & Basil J. Hiley (1995), 6

127 https://www.imperial.ac.uk/media/imperial-college/research-centres-and-groups/theoretical-physics/msc/dissertations/2009/Richard-Havery-Dissertation.pdf

128 Stapp, Henry (2004), 191

129 Stapp, Henry (2004), 191

130 Hiley, Basil & F. David Peat (1991), 15

131 Freire, Olival (2020), 127-128

132 Bohm, David (1984), ix

133 Bohm, David (1984), x

134 Bohm, David (1984), x

135 Lockwood, Michael (2005), 304

136 Penrose, Roger (1995), 309

137 Oerter, Robert (2006), 49

138 Aharonov, Yakir and Rohrlich, Daniel (2005), 1

139 Aharonov, Yakir and Rohrlich, Daniel (2005), 1

140 Allday, Jonathan (2009), 4

141 Geshe Sonam Rinchen (2006), 19

142 Max Planck: the reluctant revolutionary – Physics World

143 https://khaledbp.files.wordpress.com/2015/09/einstein_s-boxes.pdf

144 L. de Broglie, *The Current Interpretation of Wave Mechanics: A Critical Study* (Elsevier Publishing Company, 1964).

145 Greene, Brian (2004), 83

146 Greene, Brian (2004), 81

147 Einstein, Podolsky, Rosen 1935, 777 - Phys. Rev. 47, 777 (1935) - Can Quantum-Mechanical Description of Physical Reality Be Considered Complete? (aps.org)

148 Aczel, Amir D. (2002), 203

149 https://khaledbp.files.wordpress.com/2015/09/einstein_s-boxes.pdf

150 Penrose, Roger (1995), 300

151 Penrose, Roger (1999), 293

152 Penrose, Roger (2005), 591-593

153 Becker, Adam (2019), 97

154 Becker, Adam (2019), 98

155 Norsen, Travis (2017), 206-207 (quoted in)

156 Riggs P. J. - https://casinoqmc.net/local_papers/riggs_2008.pdf

157 Ananthaswamy, Anil (2020), 154

158 Smolin, Lee (2019), 207

159 d'Espagnat, Bernard (Editor), Zwirn, Hervé (Editor) (2017), 166

160 Norsen, Travis (2017), 188

161 Norsen, Travis (2017), 63

162 Norsen, Travis (2017), 206

163 Bricmont, Jean (2016), 148

164 Jeremy Butterfield: What is contextuality?

165 Bricmont, Jean (2016), 151

166 Kastrup, Bernardo (2020), 51-52

167 Boge, Florian J. (2019), 224

168 Boge, Florian J. (2019), 225

169 Baggott, Jim (2004), 217

170 Baggott, Jim (2004), 217

171 Bricmont, Jean (2016), 152

172 Boge, Florian J. (2019), 224-225

173 Bohm, David & Basil J. Hiley (1995), 6

174 Bohm, David & Basil J. Hiley (1995), 6

175 Hiley, Basil & F. David Peat (1991), 2

176 Wendt, Alexander (2015), 86

177 Bohm, David (1984), 33

178 Bohm, David (1984), 170

179 Hiley, Basil & F. David Peat (1991), 10

180 Bohm, David & Peat, F. David (2000), 91

181 Bohm, David & Peat, F. David (2000), 91-92

182 Bohm, David (2002), 191

183 Bohm, David & Basil J. Hiley (1995), 382

184 Avatamsaka Sutra

185 https://www.religion-online.org/article/the-implicate-order-a-new-order-for-physics/

186 Brunnhölzl, Karl (2012), 100

187 Brunnhölzl, Karl (2012), 101

188 Oerter, Robert (2006), 49

189 Herbert, Nick (1985), 16

190 Brunnhölzl, Karl (2012), 15

191 Allday, Jonathan (2009)

192 Vedral, Vlatko (2010), 200

193 d'Espagnat, Bernard (2006), 433

194 d'Espagnat, Bernard (2006), 440

195 Shar Khentrul Rinpoche Jamphel Lodro (2019), 9

196 Brunnhölzl, Karl (2004), 120

197 Shar Khentrul Rinpoche Jamphel Lodro (2019), 10

198 Rovelli, Carlo (2021), 131

199 http://www.nytimes.com/2012/03/25/books/review/a-universe-from-nothing-by-lawrence-m-krauss.html

200 Carroll, S. (2012), 35

201 Cabezón, José Ignacio (1992), 94

202 Siderits, Mark & Katsura, Shoryu (2013), 161

203 Shar Khentrul Rinpoche Jamphel Lodro (2019), 20

204 Kyabgon, Traleg (2010), 109

205 Bohm, David (2002), 267

206 Bohm, David (2002), 265

207 Garfield, Jay (1995), 2

208 Brunnhölzl, Karl (2004), 79

209 Brunnhölzl, Karl (2004)

210 *Abhidharmakosha* - Treasury of Abhidharma, VI, 4

211 Engle, Artemus B. (2009), 127

212 Bohm, David (2002), 233

213 Susuki, D. T, The Lankavatara Sutra: A Mahayana Text, 44

214 Shar Khentrul Rinpoche Jamphel Lodro (2019), 68

215 Bohm, David (1984), 164-165

216 Bohm, David (1984), 165

217 Sheehy, Michael R., Mathes, Klaus-Dieter (Editors) (2020), 1

218 Garfield, Jay (1995), 2

219 Bohm, David (2002), 152

220 Bohm, David (2002), 157-158

221 Smolin, Lee (2002)

222 Allday, Jonathan (2009), 408

223 https://www.religion-online.org/article/the-implicate-order-a-new-order-for-physics/

224 https://www.religion-online.org/article/the-implicate-order-a-new-order-for-physics/

225 Brunnhölzl, Karl (2004), 762

226 Shantarakshita (2005 - Padmakara Translation Group), 53

227 Shantarakshita (2005 - Padmakara Translation Group), 53

228 Shantarakshita (2005 - Padmakara Translation Group), 53

229 https://www.religion-online.org/article/the-implicate-order-a-new-order-for-physics/

230 Garfield, Jay (1995), 3

231 Chandrakirti and Jamgon Mipham (2002), 70

232 Khenpo Tsultrum Gyamtso (2003), 59

233 Brunnhölzl, Karl (2004), 84

234 Ghirardi, Giancarlo, (2005), 347

235 Ghirardi, Giancarlo, (2005), 348

236 Chown, Marcus (2007), 93

237 Kaku, Michio (2006), 148

238 Brunnhölzl, Karl (2004), 84

239 Geshe Sonam Rinchen (2006), 19

240 Kaku, Michio (2006), 148

241 Brunnhölzl, Karl (2004), 214

242 Bricmont, Jean (2017), 214

243 Bohm, David (2002), 236-237

244 Bohm, David (2002), 243

245 Bohm, David (2002), 243

246 Bohm, David (2002), 247-248

247 Bohm, David (2002), 225

248 Bohm, David (2002), xviii

249 Bohm, David (2002), 188

250 Bohm, David (2002), 190

251 Bohm, David & Peat, F. David (2000), 171

252 Bohm, David & Basil J. Hiley (1995), 354-357

253 Pylkkänen, Paavo T. I. (2006), 24

254 Pylkkänen, Paavo T. I. (2006), 26

255 Stapp, Henry (2004), 222-223

256 Hawking, Stephen & Mlodinow, Leonard (2010), 83

257 Hawking, Stephen & Mlodinow, Leonard (2010), 136

258 Bohm, David & Basil J. Hiley (1995), 314

259 Bohm, David & Basil J. Hiley (1995), 300

260 Stapp, Henry: 'Quantum Interactive Dualism', 18

261 Bohm, David & Basil J. Hiley (1995), 386

262 Dolling, L.M.; Gianelli, A. F. & Statile, G. N. (eds) (2003) p491 – John A. Wheeler (1978): 'The 'Past' and the 'Delayed Choice' Double-Slit Experiment.'

263 Bohm, David (2002), 244

264 Bohm, David (2002), 243

265 Pylkkänen, Paavo T. I. (2006), 44

266 www.gaianxaos.com/notes/morphogenetic_fields.htm

267 Sheldrake, Rupert (2009), 145

268 Smolin, Lee (2014), 146

269 Charles Sanders Pierce, 'A Guess At the Riddle' in *The Essential Pierce, Selected Philosophical Writings.*

270 https://www.sheldrake.org/files/pdfs/A_New_Science_of_Life_Appx_B.pdf

271 https://www.sheldrake.org/files/pdfs/A_New_Science_of_Life_Appx_B.

pdf

272 Susuki, D. T, The Lankavatara Sutra: A Mahayana Text p44 - elucidations added in square brackets,

273 https://www.researchgate.net/publication/2191061_Quantum_Darwinism_and_Envariance

274 Bohm, David (2002), 233

275 Bohm, David (2002), 249-250

276 Allday, Jonathan (2009), 493

277 Bohm, David (2002), 262

278 Pylkkänen, Paavo T. I. (2006), 139

279 Pylkkänen, Paavo T. I. (2006), 139

280 Bohm, David (2003) p146 (Extract from R. Weber: *Dialogues with Scientists and Sages: The Search for Unity* (1986) Routledge and Kegan Paul).

281 Bohm, David (2002), 263

282 https://plato.stanford.edu/entries/leibniz/

283 https://plato.stanford.edu/entries/whitehead/

284 Bohm, David (2002), 264

285 Rabjam, Longchen (trans. Tulku Thondup) (2002), 205-207

286 Brunnhölzl, Karl (2004), 84

287 http://www.ktgrinpoche.org/quote/self-arisen

288 Bohm, David (2002), 14

289 Rabjam, Longchen (trans. Tulku Thondup) (2002), 205-207

290 https://www.rigpawiki.org/index.php?title=Youthful_vase_body

291 Mi-Pam-Gya-Tso and Jeffrey Hopkins (2006)

292 Wallace, B. Alan (2011), 109

293 Wallace, B. Alan (2011), 237-238

294 Stapp, Henry (2004), 268

295 Kenchen Thrangu (2001), 34-35

296 Hopkins, Jeffrey (1996), 375

297 Barrow, John D., Davies, Paul C. W., Harper, Charles L. (eds)

(2004) p577 – Wheeler, J A (1999) 'Information, physics, quantum: the search for links.' In *Feynman and Computation: Exploring the Limits of Computers*, ed A. J. G. Hey, p309 (314). Cambridge, MA: Perseus Books.

298 Sarfatti, Jack 'Wheeler's World: It From Bit?' - Internet Science Education Project, San Francisco, CA..

299 Das Wesen der Materie" (The Nature of Matter), speech at Florence, Italy, 1944 (from Archiv zur Geschichte der Max-Planck-Gesellschaft, Abt. Va, Rep. 11 Planck, Nr. 1797)

300 Walpola, Rahula (1974)

301 Bohm, David (2002), 221

302 Bohm, David (2002), 264-265

303 Bohm, David (2002), 262

304 Oerter, Robert (2006), 59

305 Oerter, Robert (2006)

306 Hsing Yun, Master & Tom Graham (trans.)(2010), 113

307 Thrangu Rinpoche, Kenchen (2001), 106

308 Longchen Rabjam (Author), Harold Talbott (Editor)(2002), Tulku Thondup (Introduction)

309 Rongtonpa (Author), Christian Bernert (Translator) (2017), 7

310 Stapp H. Compatibility of Contemporary Physical Theory with Personality Survival, 13

311 Stapp H. Compatibility of Contemporary Physical Theory with Personality Survival, 14

312 Stapp H. Compatibility of Contemporary Physical Theory with Personality Survival, 15

313 Guenther, Herbert V. (1984), 24

314 Guenther, Herbert V. (1984), 38

315 Bohm, David (2003), 180

316 Guenther, Herbert V. (1984), 51

317 Bohm, David (2003), 180

318 Das Wesen der Materie" (The Nature of Matter), speech at Florence, Italy, 1944 (from Archiv zur Geschichte der Max-Planck-Gesellschaft, Abt. Va, Rep. 11 Planck, Nr. 1797)

319 Hopkins, Jeffrey (2006),138

320 Longchenpa (trans. Lipmann, K. & Peterson, M.) (2000), 38

321 Bohm, David (2002), 269

322 Bohm, David & Peat, F. David (2000), 226

323 Longchenpa (trans. Lipmann, K. & Peterson, M.) (2000), 37

324 Longchenpa (trans. Lipmann, K. & Peterson, M.) (2000), 36

325 Longchenpa (trans. Lipmann, K. & Peterson, M.) (2000), 39

326 Wheeler, J., A., 'Law Without Law'

327 Longchenpa (trans. Lipmann, K. & Peterson, M.) (2000),, 36

328 Guenther, Herbert V. (1984), 33

329 Hiley, Basil & F. David Peat (1991), 436

330 Hiley, Basil & F. David Peat (1991), 443

331 Hiley, Basil & F. David Peat (1991), 436

332 Hiley, Basil & F. David Peat (1991), 445

333 Hiley, Basil & F. David Peat (1991), 438

334 Bohm, David (2002)

335 Bohm, David (2003), 181

336 Hiley, Basil & F. David Peat (1991), 445

337 Brunnhölzl, Karl (2014), 128

338 Lingpa, Dudjom (2002), 25

339 Lingpa, Dudjom (2002), 41

340 Lingpa, Dudjom (2002), 95

341 Wangyal, Tenzin Rinpoche (2000),181

342 Bohm, David, (2003), 259

343 Bohm, David, (2003), 254

344 Bohm, David, (2003), 254

345 Bohm, David, (2003), 253

346 Pylkkänen, Paavo T. I. (2006), 126

347 Bohm, David, (2003), 259

348 Stapp, H. P. 'Minds and Values in the Quantum Universe' in

Davies, Paul & Gregersen, Niels Henrik (eds.) (2010), 117

349 Brunnhölzl, Karl (2009), 32

350 Waldron, William S. (2003),165

351 Thrangu Rinpoche, Kenchen (2001), 43

352 Nyanaponika Thera: 'Karma and its Fruit' in Samuel Bercholz (Editor), Sherab Chodzin Kohn (Editor): *The Buddha and His Teachings* – Shambhala (2002), 123

353 https://tricycle.org/magazine/cause-and-effect/

354 Nanamoli Thera & Bhikku Bodhi – http://www.accesstoinsight.org/tipitaka/mn/mn.009.ntbb.html

355 Thanissaro Bhikkhu – http://www.accesstoinsight.org/tipitaka/sn/sn12/sn12.002.than.html

356 Brunnhölzl, Karl (2007), 85

357 Waldron, William S. (2003), 168

358 Waldron, William S. (2003), 169

359 Stapp, Henry (2004), 223

360 Stapp, Henry (2004), 239

361 Stapp, Henry (2004), 197

362 Stapp, Henry (2004)

363 Stapp, Henry (2007), 22-23

364 Stapp, Henry (2007), 20

365 Stapp, Henry (2007), 10

366 Stapp, H. – 'Free Will' - http://www-physics.lbl.gov/~stapp/FW.pdf

367 Stapp, Henry: 'Quantum Interactive Dualism', 18

368 Stapp, Henry: 'The Effect of Mind upon Brain', 12

369 Stapp, Henry (2004), 241

370 Stapp, Henry: 'The Effect of Mind upon Brain', 2-3

371 Stapp, Henry: 'The Effect of Mind upon Brain', 2-3

372 Barrow, John D., Davies, Paul C. W., Harper, Charles L. (eds) (2004) p218 – Anton Zeilinger: 'Three challenges from John Archibald Wheeler.'

373 Stapp, Henry (2017), 40

374 Gribben, John (1996), 133

375 Gribben, John (1996), 135

376 https://www-physics.lbl.gov/~stapp/QID.pdf

377 Schwartz, Jeffrey M. & Sharon Begley (2003), 130

378 Schwartz, Jeffrey M. & Sharon Begley (2003), 163-164

379 Schwartz, Jeffrey M. & Sharon Begley (2003), 369

380 Waldron, William S. (2003), 24

381 Waldron, William S. (2003), 24

382 Stapp, Henry (2004), 191

383 Rahula, Walpola (1974) p32

384 Thrangu Rinpoche, Kenchen (2001), 28-29

385 Dalai Lama & Thubten Chodron (2019), 144

386 Khendrup Norsang Gyatso (2004), 78

387 Libet, B.- Do We Have Free Will – JCS **6** No 8-9,1999, p47

388 Stapp, Henry (2004), 167-168

389 Mensky, M. B. (2010), 12

390 Mensky, M. B. (2010), v

391 Mensky, M. B. (2010), 138

392 Mensky, M. B. (2010), 141

393 Mensky, M. B. (2010), 214-215

394 Mensky, M. B. (2010), 219

395 Mensky, M. B. (2010), 226

396 Mensky, Michael (2005): 'Concept of Consciousness in the Context of Quantum Mechanics'

397 Mensky, Michael: 'Reality in quantum mechanics, Extended Everett Concept, and Consciousness', 11

398 Sarfatti, Jack (1996). 'The Field of Qualia'

399 Goswami, Amit (1995), 140

400 https://link.springer.com/article/10.1007/s41470-019-00035-2

401 Stapp, Henry (2007), 33

402 Pylkkänen, Paavo - 'Henry Stapp Vs. David Bohm on Mind, Matter,

and Quantum Mechanics' - https://link.springer.com/content/pdf/10.1007/s41470-019-00035-2.pdf

403 Stapp, Henry (2007), 62

404 Pylkkänen, Paavo - 'Henry Stapp Vs. David Bohm on Mind, Matter, and Quantum Mechanics' - https://link.springer.com/content/pdf/10.1007/s41470-019-00035-2.pdf

405 Bohm, David (2002), 253

406 Bohm, David (2002), 259

407 Bohm, David (2002), 67-68

408 F. David Peat - 'Active Information, Meaning and Form' http://www.fdavidpeat.com/bibliography/essays/fzmean.htm

409 Stapp, Henry (2007), 62

410 Bohm, David (2002), 199

411 Stapp, Henry (2004), 207

412 Bohm, David (2002), xv

413 Bohm, David (2002), 14

414 Barbour, Julian (2001), 225

415 Barbour, Julian (2001), 225

416 Saunders, Simon (Editor), Adrian Kent (Editor), David Wallace (Editor) (2012), 9-10

417 Carroll,Sean (2016), 319-320

418 Mensky, M. B., 'Everett Interpretation and Quantum Consciousness' in *NeuroQuantology*, March 2013, Vol. 11, Issue 1, pages 85-96, 85

419 Bohm, David (2002), 190

420 Mensky, M. B. (2010), 76

421 Mensky, M. B. (2010), 12

422 Mensky, M. B. (2010), 12

423 Bohm, David (2002), 247

424 Bohm, David (2002), 269-270

425 Bohm, David (2002), 152

426 Mensky, M. B. (2010), 15

427 Herbert, Nick: 'Holistic Physics -or- Introduction to Quantum Tantra' – Internet document (www.southerncrossreview.org/16/herbert.essay.htm)

428 Mensky, M. B. (2010), 72

429 Mensky, M. B. (2010), 69

430 Mensky, M. B. (2010), 77

431 Mensky, M. B. (2010), 78

432 Mensky, M. B. (2010), 79

433 Bricmont, Jean (2016), 152

434 Wheeler, John Archibald & Ford, Kenneth (2000), 338

435 Scientific American, July 1992, p75

436 "The Anthropic Universe" Radio interview in 'Science Show' 18 Feb 2006

437 Sabbadini, Shantena Augusto (2017), 84-85

438 Mensky, M. B. (2010), 77

439 Sabbadini, Shantena Augusto (2017), 87

440 Bohm, David (2002), 265

441 Terentyev, Andrey, 'Contiguity of Parallel Worlds: Buddhist and Everett's' - http://www.neuroquantology.com/index.php/journal/article/view/640

442 Terentyev, Andrey, 'Contiguity of Parallel Worlds: Buddhist and Everett's' - http://www.neuroquantology.com/index.php/journal/article/view/640

443 Terentyev, Andrey, 'Contiguity of Parallel Worlds: Buddhist and Everett's' - http://www.neuroquantology.com/index.php/journal/article/view/640

444 Brunnhölzl, Karl (2013), 358

445 Rahula, Walpola - http://www.budsas.org/ebud/ebdha195.htm

446 Bohm, David (2002), 12

447 Udana Viii-3

448 Randall, L. (2006), 158

449 Khenpo Shenga and Ju Mipham - Dharmachakra Translation Committee (2014), 126

450 https://www.researchgate.net/publication/2191061_Quantum_Darwinism_and_Envariance

451 Mensky, M. B. (2010), 81

452 Bohm, David (2002), 221

453 Bohm, David (2002), 225

454 Kelly E. E, Kelly E. W, Crabtree A., Gauld, A., Grosso M., Greyson B. (2007), xxiv

455 Kennedy, J. E. Journal of Parapsychology, 2006, Volume 70, 373

456 Mensky, M. B. (2010)

457 Mensky, M. B. (2010), 81

458 Mensky, M. B. (2010), 133

459 Mensky, M. B. (2010), 82

460 Mensky, M. B. 'Post Correction and mathematical model of life in Extended Everett's Concept', 5

461 Mensky, M. B. (2010), 112

462 Mensky, M. B. (2010), 112

463 Mensky, M. B. (2010),167

464 Davies, Paul (2007), 275

465 Mensky, M. B. 'Post Correction and mathematical model of life in Extended Everett's Concept', 6

466 Mensky, M. B. 'Post Correction and mathematical model of life in Extended Everett's Concept', 3

467 Mensky, M.B.: 'Reality in quantum mechanics, Extended Everett Concept, and Consciousness', 6

468 Mensky, M.B.: 'Reality in quantum mechanics, Extended Everett Concept, and Consciousness', 6

469 Mensky, M. B. (2010), 83

470 Mensky, Michael: 'Reality in quantum mechanics, Extended Everett Concept, and Consciousness', 12

471 Mensky, M. B. 'Postcorrection and mathematical model of life in Extended Everett's Concept', 20

472 Mensky, M. B. - Logic of Quantum Mechanics and Phenomenon of

Consciousness.

473 Mensky, M. B. - Logic of Quantum Mechanics and Phenomenon of Consciousness.

474 Mensky, M. B. 'Postcorrection and mathematical model of life in Extended Everett's Concept'

475 Mensky, M. B. 'Postcorrection and mathematical model of life in Extended Everett's Concept', 18-19

476 Mensky, M. B. 'Postcorrection and mathematical model of life in Extended Everett's Concept', 23

477 Mensky, M. B. (2010), 122

478 Mensky, M. B. (2010), 12

479 Mensky, M. B. (2010), v

480 Mensky, M. B. (2010), 138

481 Mensky, M. B. (2010), 141

482 Mensky, M. B. (2010), 214-215

483 Mensky, M. B. (2010), 219

484 Mensky, M. B. (2010), 218

485 Mensky, M. B. (2010), 184

486 Mensky, M. B. (2010), 181

487 Mensky, M. B. (2010), 180-181

488 Mensky, M. B. (2010), 180-181

489 Mensky, M. B. (2010), 181

490 Mensky, M. B. (2010), 181

491 Mensky, M. B. (2010), 181

492 Mensky, M. B. (2010), 183

493 Mensky, M. B. (2010), 183

494 Bohm, David (2002), 14

495 Hawking, Stephen & Mlodinow, Leonard (2010), 7

496 Bohm, David (2003). *The Essential David Bohm* ed Nichol, Lee (Routledge, London) p146 (Extract from R. Weber: *Dialogues with Scientists and Sages: The Search for Unity* (1986) Routledge and Kegan Paul).

497 Bohm, David (2002), 249-250

498 Henry Stapp – The Mind is NOT What the Brain Does (2009), 6

499 Stapp, Henry (2004), 197

500 Barrow, John D., Davies, Paul C. W., Harper, Charles L. (eds) (2004) p577 – Wheeler, J A (1999) 'Information, physics, quantum: the search for links.' In *Feynman and Computation: Exploring the Limits of Computers*, ed A. J. G. Hey, p309 (314). Cambridge, MA: Perseus Books.

501 Hopkins, Jeffrey (1996), 375

502 Bohm, David (2002), 265

503 Das Wesen der Materie" (The Nature of Matter), speech at Florence, Italy, 1944 (from Archiv zur Geschichte der Max-Planck-Gesellschaft, Abt. Va, Rep. 11 Planck, Nr. 1797)

504 Hopkins, Jeffrey (2006),138

505 Lockwood, Michael (2005), 353

506 www.davidsmuse.co.uk/hoffman.html

507 Sarfatti , Jack 'Wheeler's World: It From Bit?' - Internet Science Education Project,

San Francisco, CA..

508 Brunnhölzl, Karl (2007), 25

509 Founding Quantum Theory on the Basis of Consciousness | SpringerLink

510 Mensky, M. B. (2010), 12

511 Mensky, M. B. (2010), 12

512 Davies, Paul (2007), 275

513 Longchen Rabjam (Author), Harold Talbott (Editor)(2002), Tulku Thondup (Introduction)

514 Terentyev, Andrey, 'Contiguity of Parallel Worlds: Buddhist and Everett's' -
http://www.neuroquantology.com/index.php/journal/article/view/640

515 Terentyev, Andrey, 'Contiguity of Parallel Worlds: Buddhist and Everett's' -
http://www.neuroquantology.com/index.php/journal/article/view/640

516 Ibid.

517 Thrangu Rinpoche, Kenchen (2001)

518 Thrangu Rinpoche, Kenchen (2001)

519 Zurek Wojciech H.(2002). ' Decoherence and the Transition from Quantum to Classical – *Revisited*' in *Los Alamos Science* Number 27 2002

520 Wheeler, J., A., 'Law Without Law', 199

521 Wheeler, J., A., 'Law Without Law', 199

522 Penrose, Roger (1995), 309

523 Barrow, John D., Davies, Paul C. W., Harper, Charles L. (eds) (2004) p121 – Wojciech H. Zurek: 'Quantum Darwinism and envariance.'

524 Barrow, John D., Davies, Paul C. W., Harper, Charles L. (eds) (2004) – Wojciech H. Zurek: 'Quantum Darwinism and envariance.'

525 Schlosshauer, Maximilian (Editor) (2011), 107

526 Barrow, John D., Davies, Paul C. W., Harper, Charles L. (eds) (2004) p125 Wojciech H. Zurek: 'Quantum Darwinism and envariance.'

527 Zurek Wojciech H.(2002). ' Decoherence and the Transition from Quantum to Classical – *Revisited*' in *Los Alamos Science* Number 27 2002, p21

528 An issue I discuss in detail in my book *Quantum Revelations of the Real and Unreal: Quantum Buddhist Metaphysics Rectifies New Age Propheteering & Subtle Quantum Materialist Madness.*

529 'The Evolution of Reality' – www.fqxi.org/community/articles/display/122 (The Foundational Questions Institute) November 10, 2009.

530 Stapp, Henry (2004), 268

531 Barrow, John D., Davies, Paul C. W., Harper, Charles L. (eds) (2004) p201 – Anton Zeilinger: 'Why the quantum? "It" from bit"? A participatory universe? Three far-reaching challenges from John Archibald Wheeler and their relation to experiment.'

532 Guardian obituary – Michael Carlson

533 Hawking, Stephen & Mlodinow, Leonard (2010), 136

534 Bohm, David & Basil J. Hiley (1995), 128

535 Bohm, David & Basil J. Hiley (1995), 130

536 Bohm, David & Basil J. Hiley (1995), 386

537 Sabbadini, Shantena Augusto (2017), 84-85

538 Bohm, David & Basil J. Hiley (1995), 389

539 Bohm, David (2003). *The Essential David Bohm* ed Nichol, Lee (Routledge, London) p146 (Extract from R. Weber: *Dialogues with Scientists and Sages: The Search for Unity* (1986) Routledge and Kegan Paul).

540 Bohm, David (2002), 249-250

541 Bohm, David, (2003), 140

542 Tulku Urgyen Rinpoche, *As It Is, Vol. I* (Boudhanath, Hong Kong & Esby: Rangjung Yeshe, 1999), pages 31-32.

543 Bohm, David (2002), 14

544 Bohm, David, (2003), 154

545 Dowman, Keith (2017), 17

546 Fremantle, Francesca (2003), 64

547 http://www.bellaonline.com/articles/art43350.asp

548 Bardo descriptions based on https://en.wikipedia.org/wiki/Bardo. See also - Fremantle, Francesca (2003)

549 Bohm, David (2002), 219

550 Bohm, David (2002), 249

551 Mensky, M. B. (2010), 133

552 https://www.science20.com/news_articles/divine_patterns_ramanujans_magical_mind_gets_math_formula-99186

553 https://www.psychologytoday.com/us/blog/dream-catcher/201911/the-reality-precognitive-dreams

554 Mensky, M. B. (2010), 96

555 http://www.katinkahesselink.net/tibet/jhana-2.html

556 http://www.katinkahesselink.net/tibet/jhana-2.html

557 Mensky, M. B. (2010), 96

558 Van Schaik, Sam, (2004), 140

559 Brunnhölzl, Karl. (2009), 123-4

560 Dzogchen Ponlop (2012), 119

561 Dzogchen Ponlop (2012), 126

562 https://www.hayagriva.org.au/wheel-of-life/death-process/

563 http://www.buddhanet.net/e-learning/history/deathtib2.htm

564 Personal communication

565 Bohm, David (2002), 247

566 Longchen Rabjam (Author), Harold Talbott (Editor), Tulku Thondup (Introduction) (2002)

567 Tulku Thondup Rinpoche (2005), 45-46

568 Dowman, Keith (2017), 33

569 Ray, Reginald A. (2002), 116

570 Bohm, David (2002), xiii-xiv

571 Bohm, David (2002), 10-11

572 Bohm, David (2002), 219

573 https://theislamissue.wordpress.com/2019/04/02/the-nun-whale-and-the-calamity-of-ibn-abbas/

574 Bohm, David (2002), xviiii

575 Bohm, David & Basil J. Hiley (1995), 122

576 Bricmont, Jean (2016), 151-152

577 Boge, Florian J. (2019), 221

578 https://www.informationphilosopher.com/solutions/scientists/bell/Beables_for_QFT.pdf

579 https://www.informationphilosopher.com/solutions/scientists/bell/Beables_for_QFT.pdf

580 Baggott, Jim (2020), 182 (quoted in)

581 https://www.informationphilosopher.com/solutions/scientists/bell/Beables_for_QFT.pdf

582 Whitaker, Andrew (2016), 91

583 Baggott, Jim (2020), 182

584 Bohm, David (2002), 6

585 Baggott, Jim (2020), 181

586 Baggott, Jim (2020), 182

587 Bohm, David (2002), xviii

588 Bohm, David (2002), 11-12

589 Baggott, Jim (2020), 177

590 Bohm, David (2002), 14

591 http://fdavidpeat.com/bibliography/essays/fzmean.htm

592 Bohm, David (1987), 42

593 Bohm, David (1987), 42

594 Bohm, David (1987), 42-43

595 http://fdavidpeat.com/bibliography/essays/fzmean.htm

596 Bohm, David (1987), 43

597 Bohm, David (1990), 282

598 Bohm, David (2002), 249-250

599 Bohm, David (1990), 190

600 Pylkkänen, Paavo T. I. (2006), 190

601 Bohm, David (1990), 283-284

602 Bohm, David (1990), 284

603 https://en.wikipedia.org/wiki/Collective_unconscious - Jung, Collected Works vol. 8 (1960), "The Significance of Constitution and Heredity in Psychology" (1929), ¶229–230 (p. 112).

604 https://en.wikipedia.org/wiki/Collective_unconscious - C. G. Jung, *The Archetypes and the Collective Unconscious* (London 1996) p. 43

605 https://paulijungunusmundus.eu/pauli_quotations_published.htm

606 Woolfson, Adrian (2000), 74

607 Carroll, Sean B. (2006), 9

608 Hiley, Basil & F. David Peat (1991), 43

609 Bohm, David, (2003), 160

610 Bohm, David, (2003), 172

611 https://www.metanexus.net/archive/conference2005/pdf/agocs.pdf

612 Stevenson, I., 'Birthmarks and Birth Defects Corresponding to Wounds on Deceased Persons' - http://www.scientificexploration.org

613 Tucker, Jim B., 'Children Who Claim to Remember Previous Lives: Past, Present, and Future Research' - https://www.scientificexploration.org

614 Dhammapada, 342

615 Bohm, David (2002), 20

616 Bohm, David & Basil J. Hiley (1995), 179

617 Bohm, David (2003), 181

618 Stapp, Henry (2004), 193

619 Bohm, David (2002)

620 Bohm, David (2002), 19

621 Bohm, David (2002), 9

622 Bohm, David (2002), 3

623 Bohm, David (1994), 72-73

624 Udana Viii-1

625 Udana Viii-3

626 Bohm, David & Peat, F. David (2000), 258

627 Bohm, David (2002), 218

628 Bohm, David (2003), 202

629 Chang, Garma C.C. (1971), xv

630 Chang, Garma C.C. (1971), 4-5

631 Brunnhölzl, Karl (2004), 332

632 Brunnhölzl, Karl (2004), 332

633 Bohm, David (2002), 76-77

634 Brunnhölzl, Karl (2004), 228

635 Barrett, Jeffrey A. (2001)

636 Chang, Garma C.C. (1971), 109

637 Levy, Paul (2018), 58

638 Westerhoff, Jan & Nagarjuna (2019), 15

639 Brunnhölzl, Karl (2004), 75

640 Klein, Anne Carolyn & Tenzin Wangyal Rinpoche (2016), 3

641 Klein, Anne Carolyn & Tenzin Wangyal Rinpoche (2016), 3-4

642 Perdue, Daniel E.(1992), 13

643 Perdue, Daniel E.(1992), 302

644 https://www.lionsroar.com/what-is-pramana/

645 Klein, Anne Carolyn & Tenzin Wangyal Rinpoche (2016), 6

646 Shar Khentrul Rinpoche Jamphel Lodro (2019), 38

647 Chang, Garma C.C. (1971), 173

648 Bohm, David (2002), 36-37

649 Bohm, David (2002), 29

650 Brunnhölzl, Karl (2010), 53

651 Bohm, David (2002), 3

652 Bohm, David (2002), xvi

653 Klein, Anne Carolyn & Tenzin Wangyal Rinpoche (2016), 21

654 Klein, Anne Carolyn & Tenzin Wangyal Rinpoche (2016), 20

655 Klein, Anne Carolyn & Tenzin Wangyal Rinpoche (2016), 63-64

656 Bohm, David (2002), 32-33

657 Klein, Anne Carolyn & Tenzin Wangyal Rinpoche (2016), 83

658 Addiss, Stephen; Lombardo, Stanley; Roitman, Judith (2008), 39

659 Bohm, David (2002), 67-68

660 Bohm, David (2002), 249-250

661 Hee-Jin Kim (2004), 129-130